Lecture Notes in Mathematics

Edited by A. Dold and B. Eckmann

T0234365

891

Logic Symposia Hakone 1979, 1980

Proceedings of Conferences Held in Hakone, Japan
March 21–24, 1979 and February 4–7, 1980

Edited by G. H. Müller, G. Takeuti, and T. Tugué

Springer-Verlag
Berlin Heidelberg New York 1981

Editors

Gert H. Müller
Mathematisches Institut, Universität Heidelberg
Im Neuenheimer Feld 294, 6900 Heidelberg, Federal Republic of Germany

Gaisi Takeuti
Department of Mathematics, University of Illinois
Urbana, IL 61801, USA

Tosiyuki Tugué
Department of Mathematics, College of General Education
Nagoya University, Chikusa-ku
Nagoya 464, Japan

AMS Subject Classifications (1980): 03Cxx, 03Dxx, 03Exx, 03Fxx, 03Hxx

ISBN 3-540-11161-1 Springer-Verlag Berlin Heidelberg New York
ISBN 0-387-11161-1 Springer-Verlag New York Heidelberg Berlin

This work is subject to copyright. All rights are reserved, whether the whole or part of the material is concerned, specifically those of translation, reprinting, re-use of illustrations, broadcasting, reproduction by photocopying machine or similar means, and storage in data banks. Under § 54 of the German Copyright Law where copies are made for other than private use, a fee is payable to "Verwertungsgesellschaft Wort", Munich.

© by Springer-Verlag Berlin Heidelberg 1981
Printed in Germany

Printing and binding: Beltz Offsetdruck, Hemsbach/Bergstr.
2141/3140-543210

Preface

Symposia on the Foundations of Mathematics, supported by a Grant-in-Aid for Co-operative Research, Proj. no. 234002 ('79) and 434007 ('80), were held at Gōra, Hakone in Japan, on March 21-24, 1979 and February 4-7, 1980.

These Proceedings record the numerous and extensive discussions that took place at the two meetings.

The Editors express their warmest thanks to the institutions and persons concerned. Their generous support enabled us twice to bring together almost all logicians of Japan for lively exchanges of ideas. We hope that everybody who was present was inspired to initiate new research.

The symposia were planned, convened and directed by the third Editor. The two first mentioned Editors - also in the name of all participants - express their heartfelt thanks for all the help and guidance they received from the third Editor!

The first Editor acknowledges the financial help of various universities in Japan, first of all Sophia University through Professor M. Yanase S. J. and Nagoya University through the third Editor. The Deutsche Forschungsgemeinschaft Bonn-Bad Godesberg provided for the travel expenses of the first Editor. He is most grateful for all the help and the overwhelming hospitality he received in Japan. He apologizes for not having been able - due to heavy administrative burdens - to include extended versions of the two papers

"Axiom of Choice Far Apart" and "Hierarchies and Closure Properties" which he gave at the Hakone-Symposium.

The second Editor visited Japan in 1979 as a visiting professor on a grant from the Gakujutsu Shinkokai (Japan Society for the Promotion of Science) through the third Editor and participated in symposia on the

Foundations of Mathematics held at Hakone '79 and at Kyoto University, Research Institute for Mathematical Sciences, May 30 - June 2, '79. Meeting old friends and younger logicians in Japan revived many happy memories for him. It is his happy duty to express his gratitude to the Gakujutsu Shinkokai, Nagoya University and his many friends in Japan.

We all hope that these Proceedings be stepping-stones for further research activities in mathematical logic in Japan.

G.H. Müller (Heidelberg)

G. Takeuti (Urbana, Ill.)

T. Tugué (Nagoya)

August 1981

まえがき

(PREFACE)

本報告集の Preface に，必ずしも「英文和訳」ではない日本語の，しかも肉筆の「序」を付する，という異例のことを行うこととなったのは，編者の一人である Heidelberg 大学教授 Gert H. Müller 博士の強い示唆に基づくものであって，われわれ（その一人はこれがそのままオフセット印刷となって世に出ることにいささかのしゅう恥を感じないわけではないが）は同教授の機知に富んだ発想と好意に満ちた配慮に従うこととしたのである．思うに，この報告集が Springer-Verlag 社の Lecture Notes の一つとして刊行される企てはそもそも Müller 教授の日本における数学基礎論グループに対する温かい好意から発したものであって，その実現は同教授の並々ならぬ熱意と御援助のたまものである．

1979年，文部省科学研究費補助金総合研究（A），課題番号 234002 のもとでの分担研究による集会として，3月21日～3月24日箱根町強羅において数学基礎論シンポジウムが開催された．この研究集会は編者の一人である竹内外史が，同年3月より4カ月間，日本学術振興会による昭和53年度招へい教授として，訪日した際もたれたものであって，3回にわたる招待講演のほか，集会の全日程にわたって出席して意見の交換及び助言を行った．次いで，1980年上智大学からの招へいによる Müller 教授の訪日があり，これにあわせて2月4日より2月7日までの間，同上総合研究（A）課題番号 434007 による研究の一環として，同じく箱根町強羅において数学基礎論シンポジウムが催された．同教授は心よく招待講演及び座長を引き受けられたのみならず，終始熱心に討論に参加，幾多の情報の提供，助言など多大な貢献をされた．

この報告集はこれら両年度の二つのシンポジウムにおける講演及び関連する研究の成果の中から集録したものである．これらの多くは上述の科学研究費課題番号 434007 による助成に負っている．また，両研究集会ではここに登載したもののほかに多くの講演者による貢献があった．その中にはすでに他の雑誌に掲載ないしは投稿されたものもある．

下記の編者の一人である竹内は，1979年に日本を訪れて，多くの数学基礎論の専門家と再会して親しく議論をする機会が得られ，また若い優秀な研究者の動向を知ることができて何よりもうれしく思ったのである．こうした機会を与えられた日本学術振興会会長茅誠司先生はじめ同会関係者各位，また受入研究者の所属大学として種々の便宜供与をいただいた名古屋大学をはじめお招きを受けた諸大学の関係者各位に対して，ここに改めて感謝の意を表するものである．

また，その翌年には Müller 教授の来日があって，日本の数学基礎論は大きな影響を受けた．同教授の訪日に多大な御援助を頂いた Deutche Forschungsgemeinschaft Bonn-Bad Godesberg，並びに上智大学の関係者各位，とりわけ上智大学理事長（当時，現学長）柳瀬睦男先生，同大学教授河田敬義先生に深じんなる謝意を表する．いま一人の編者である柘植が思うに，Müller，竹内両教授の引き続く来日という出来事がなければこの種の報告集が世に現れることは決してなかった．

ここに，われわれは二つのシンポジアムが日本の数学基礎論の発展に寄与したことの記念として，この報告集が出ることを心からの喜びとするものである．

1981年 8月

イリノイ大学　竹内外史
Takeuti, Gaisi

名古屋大学　柘植利之
Tugué, Tosiyuki

Table of Contents

Joint List of Participants in the Hakone Symposia '79 and '80

Norio ADACHI:	Dept. Math., Coll. Sci. & Eng., Waseda Univ., Nishi-Okubo, Shinjuku-ku, Tokyo 160
Katsuya EDA:	Inst. Math., Univ. of Tsukuba, Sakura, Niiharu-gun, Ibaraki 305
Masazumi HANAZAWA:	Dept. Math., Fac. Sci., Saitama Univ., Urawa 338
Susumu HAYASHI:	Inst. Math., Univ. of Tsukuba, Sakura, Niiharu-gun, Ibaraki 305
Shigeru HINATA:	Coll. Gen. Educ., Hosei Univ., Fujimi, Chiyoda-ku, Tokyo 102
Sachio HIROKAWA:	Fac. Eng., Shizuoka Univ., Johoku, Hamamatsu 432
Ken HIROSE:	Dept. Math., Coll. Sci. & Eng., Waseda Univ., Nishi-Okubo, Shinjuku-ku, Tokyo 160
Hiroshi HORIGUTI:	Dept. Math., Fac. Sci., Rikkyo Univ., Nishi-Ikebukuro, Tokyo 171
Tsutomu HOSOI:	Dept. Math., Coll. Sci. & Eng., Tokyo Sci. Univ., Yamazaki, Noda 278
Yoshinobu INOUE:	Dept. Math., Coll. Gen. Educ., Kobe Univ., Nada-ku, Kobe 657
Tsurane IWAMURA:	Dept. Math., Fac. Sci., Rikkyo Univ., Nishi-Ikebukuro, Tokyo 171
Yuzuru KAKUDA:	Dept. Math., Coll. Gen. Educ., Kobe Univ., Nada-ku, Kobe 657
Shigetoku KAWABATA:	Fac. Eng., Kyushu Univ., Hakozaki, Fukuoka 812
Kenji KAWADA:	Dept. Math., Aichi Univ., Machihata, Toyohashi 440
Toru KAWAI:	Dept. Math., Fac. Sci., Kagoshima Univ., Kagoshima 890
Moto-o KINOSHITA:	Dept. Math., Coll. Sci. & Eng., Waseda Univ., Nishi-Okubo, Shinjuku-ku, Tokyo 160
Yuichi KOMORI:	Dept. Math., Fac. Sci., Shizuoka Univ., Shizuoka 422
Reijiro KURATA:	Fac. Eng., Kyushu Univ., Hakozaki, Fukuoka 812
Shoji MAEHARA:	Tokyo Inst. of Tech., Oh-okayama, Tokyo 152
Kazuo MATSUMOTO:	Fac. Eng., Univ. of Osaka Pref., Mozuume, Sakai 591
Tohru MIYATAKE:	Inst. Math., Univ. of Tsukuba, Sakura, Niiharu-gun, Ibaraki 305
Chiharu MIZUTANI:	"
Nobuyoshi MOTOHASHI:	"

Akira NAKAMURA:	Fac. Eng., Hiroshima Univ., Higashi-Senda, Hiroshima 730
Kanji NAMBA:	Dept. Math., Coll. Gen. Educ., Univ. of Tokyo, Komaba, Meguro-ku, Tokyo 153
Toshio NISHIMURA:	Inst. Math., Univ. of Tsukuba, Sakura, Niiharu-gun, Ibaraki 305
Shigeo ŌHAMA:	Dept. Math., Toyota Tech. Coll., Toyota 471
Masao OHNISHI:	Dept. Math., Fac. Educ., Kobe Univ., Nada-ku, Kobe 657
Minolu OHTA:	Dept. Math., Aichi Univ. of Educ., Kariya 448
Takemasa OHYA:	Dept. Math., Fac. Sci., Tokai Univ., Hiratsuka 259-12
Hiroakira ONO:	Dept. Math., Integr. Arts & Sci., Hiroshima Univ., Hiroshima 730
Takeshi OSHIBA:	Nagoya Inst. of Tech., Showa-ku, Nagoya 466
Masahiko SAITO:	Dept. Math., Coll. Gen. Educ., Univ. of Tokyo, Komaba, Meguro-ku, Tokyo 153
Masahiko SATO:	Dept. Inf. Sci., Fac. Sci., Univ. of Tokyo, Hongo, Bunkyo-ku, Tokyo 113
Mamoru SHIMODA:	Fac. Eng., Kyushu Univ., Hakozaki, Fukuoka 812
Juichi SHINODA:	Dept. Math., Coll. Gen. Educ., Nagoya Univ., Chikusa-ku, Nagoya 464
Kokio SHIRAI:	Dept. Math., Fac. Sci., Shizuoka Univ., Shizuoka 422
Takakazu SIMAUTI:	Dept. Math., Fac. Sci., Rikkyo Univ., Nishi-Ikebukuro, Tokyo 171
Masayuki TAKADA:	Dept. Math., Tokyo Univ. of Agr. and Tech., Fuchu 183
Makoto TAKAHASHI:	Dept. Math., Coll. Sci. & Eng., Waseda Univ., Nishi-Okubo, Shinjuku-ku, Tokyo 160
Moto-o TAKAHASHI:	Inst. Math., Univ. of Tsukuba, Sakura, Niiharu-gun, Ibaraki 305
Saburo TAMURA:	Dept. Math., Fac. Educ., Kobe Univ., Nada-ku, Kobe 657
Hisao TANAKA:	Dept. Math., Fac. Eng., Hosei Univ., Kajino, Koganei 184
Nobutaka TSUKADA:	Inst. Math., Univ. of Tsukuba, Sakura, Niiharu-gun, Ibaraki 305
Tosiyuki TUGUÉ:	Dept Math., Coll. Gen. Educ., Nagoya Univ., Chikusa-ku, Nagoya 464
Tadahiro UESU:	Dept. Math., Fac. Sci., Kyushu Univ., Hakozaki, Fukuoka 812
Toshio UMEZAWA:	Dept. Math., Fac. Sci., Shizuoka Univ., Shizuoka 422

Mariko YASUGI:	Inst. Information Sci., Univ. of Tsukuba, Sakura, Niiharu-gun, Ibaraki 305
Masahiro YASUMOTO:	Dept. Math., Fac. Sci., Nagoya Univ., Chikusa-ku, Nagoya 464
Tsuyoshi YUKAMI:	Inst. Math., Univ. of Tsukuba, Sakura, Niiharu-gun, Ibaraki 305
Nobuo ZAMA:	Dept. Math., Coll. Gen. Educ., Rikkyo Univ., Nishi-Ikebukuro, Tokyo 171

Guests

1979	B.C. RAO:	Indian Statistical Inst.
	Gaisi TAKEUTI:	Dept. Math. Univ. of Illinois at Urbana
1980	Gert H. MÜLLER:	Math. Inst., Univ. of Heidelberg

VARIOUS KINDS OF ARONSZAJN TREE WITH NO SUBTREE OF A DIFFERENT KIND

Masazumi Hanazawa

In this paper, assuming V = L, we give various examples of Aronszajn ω_1-trees with certain very special properties. We describe the purpose of this paper more precisely near the end of this introduction after preliminary explanations.

Our study partly concerns general topology. The tree topology on a tree $\langle T, <\rangle$ is obtained by using the set of all open intervals in T as a basis for a topology on T. We assume always that a tree is a topological space by this topology. In this sense, we say that T is normal if the tree topology of T is normal. It should be noticed that the tree topology is different to the partial order topology which is used in the forcing. We write simply T instead of $\langle T,<\rangle$.

The historically first example of Aronszajn tree is a special Aronszajn tree, which was discovered by M. Aronszajn and F. B. Jones independently (see, M. E. Rudin [11]). It is also called a Jones space, which is a non-metrizable Moore space. Concerning the normal Moore space conjecture of himself, Jones asked whether it is normal. On the other hand, E. W. Miller [10] introduced the notion of Souslin trees and proved that the existence of such trees and the negation of Souslin hypothesis are equivalent. As is well-known, these problems turned out to be undecidable in ZFC after a long battle. These two kinds of Aronszajn trees are the most well-known examples of Aronszajn trees. Other kinds of Aronszajn trees which we deal with in this paper are given in Devlin and Shelah [4] and Baumgartner [1]. Devlin and Shelah introduced two kinds of Aronszajn trees by which they characterize (a) collectionwise Hausdorff trees and (b) (assuming V = L) normal trees. Those are almost Souslin trees and ω_1-trees with property γ. They

---

This research was partially supported by Grant-in-Aid for Scientific Research (No. 434007), Ministry of Education.

raised also the notion of ω_1-trees with no club antichain as a natural variant of the above two. In [1], there are two kinds of Aronszajn trees which relate to special Aronszajn trees. A special Aronszajn tree is characterized as a $\mathbb{Q}$-embeddable tree. So, we get the notion of $\mathbb{R}$-embeddable trees as its natural variant. Furthermore such a tree has the property that any uncountable subset contains an uncountable antichain. Following Baumgartner, we call an ω_1-tree with this property a non-Souslin tree; this naming is fairly reasonable because of the fact which we will point out in the last section. We deal with only these kinds of Aronszajn trees.

Under ZFC alone, various properties of Aronszajn trees are unclear; e.g. the normality of special Aronszajn trees is undecidable in ZFC (even the existence of normal Aronszajn trees can not be proved in ZFC). So, rather many facts about them have been given under ZFC + extra axioms. In this paper we assume V = L. As a matter of fact, $\diamondsuit^*$ (or even $\diamondsuit$ in many cases) suffices for our purpose.

We use the following abbreviations:

AT = the family of Aronszajn trees,

ST = the family of Souslin trees,

γST = the family of Aronszajn trees with property γ,

AST = the family of almost Souslin trees,

NCA = the family of Aronszajn trees with no club antichain,

SAT = the family of special Aronszajn trees,

RE = the family of $\mathbb{R}$-embeddable ω_1-trees,

NS = the family of non-Souslin trees.

The following are known results concerning our interest:

Fact 1. $\mathrm{ST} \cap \mathrm{NS} = \emptyset$ and $\mathrm{AST} \cap \mathrm{SAT} = \emptyset$,

Fact 2. $\emptyset \subseteq \mathrm{ST} \subseteq \gamma\mathrm{ST} \subseteq \mathrm{AST} \subseteq \mathrm{NCA} \subseteq \mathrm{AT}$,

Fact 3. $\emptyset \neq \mathrm{SAT} \subseteq \mathrm{RE} \subseteq \mathrm{NS} \subseteq \mathrm{AT}$,

Fact 4. $\mathrm{ST} \neq \emptyset \Longrightarrow \mathrm{ST} \subsetneq \gamma\mathrm{ST}$ (Devlin and Shelah [4]),

Fact 5. ($\diamondsuit$) $\mathrm{ST} \neq \emptyset$ (Jensen, see [3] & [4]),

Fact 6. ($\diamondsuit^*$) $\gamma\mathrm{ST} \subsetneq \mathrm{AST}$ (Devlin and Shelah [4]),

Fact 7. (◇) AST $\subsetneqq$ NCA $\subsetneqq$ AT (Devlin and Shelah [4]),

Fact 8. (◇) SAT $\subsetneqq$ RE $\subsetneqq$ NS $\subsetneqq$ AT (Baumgartner [1]).

In fact, Baumgartner announced Fact 8 under V = L. However ◇ suffices both for SAT $\subsetneqq$ RE ([2]) and for RE $\subsetneqq$ NS ([9]). From the above facts, we can classify AT as is figured in the following:

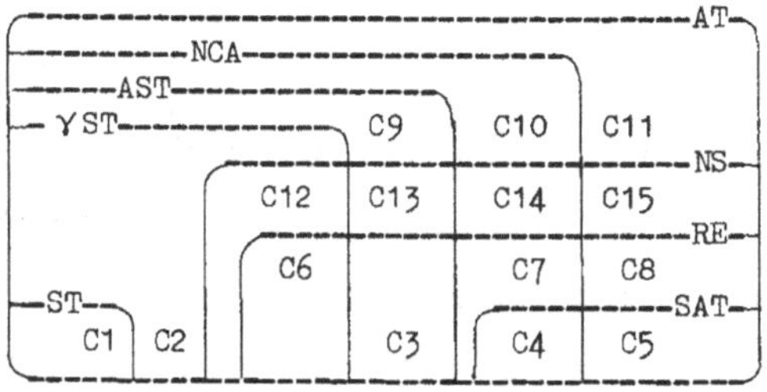

If $\diamondsuit^*$ holds, these classes C1, ..., C15 are all non-void (Hanazawa [8], [9]). Even ◇ suffices except five cases (C3, C6, C9, C12, C13). The fact that $\diamondsuit^*$ implies $\gamma\mathrm{ST} \cap \mathrm{RE} \neq \emptyset$ is used in this paper, so we sketch the proof in the last section to make this paper self-contained.

This work was motivated by Devlin and Shelah's proof of Fact 4. The tree $T \in \gamma\mathrm{ST} \setminus \mathrm{ST}$ which they constructed contains a subtree $\widetilde{T}$ such that $\widetilde{T} \in \mathrm{ST}$. This observation yields a question whether any tree $T \in \gamma\mathrm{ST} \setminus \mathrm{ST}$ has this property. In fact, it is refuted easily as will be shown later. But several more questions of this type arise; e.g. whether any tree $T \in \mathrm{AST} \setminus \gamma\mathrm{ST}$ contains a subtree $\widetilde{T}$ such that $\widetilde{T} \in \gamma\mathrm{ST}$. The purpose of this paper is just to answer these questions. Let us write $\widetilde{T} \lhd T$ to denote that $\widetilde{T}$ is an ω_1-tree and also a subtree of T. Then our results are described as follows:

<u>Theorem</u> ($\diamondsuit^*$). (1) $\exists T \in \gamma\mathrm{ST}\ \forall \widetilde{T} \lhd T\ (\widetilde{T} \notin \mathrm{ST})$.

(2) $\exists T \in \mathrm{AST}\ \forall \widetilde{T} \lhd T\ (\widetilde{T} \notin \gamma\mathrm{ST})$.

(3) $\exists T \in \mathrm{NCA}\ \forall \widetilde{T} \lhd T\ (\widetilde{T} \notin \mathrm{AST})$.

(4) $\exists T \in \mathrm{AT}\ \forall \widetilde{T} \lhd T\ (\widetilde{T} \notin \mathrm{NCA})$.

(5) $\exists T \in \mathrm{RE}\ \forall \widetilde{T} \lhd T\ (\widetilde{T} \notin \mathrm{SAT})$.

(6) $\exists T \in \mathrm{NS}\ \forall \widetilde{T} \lhd T\ (\widetilde{T} \notin \mathrm{RE})$.

(7) $\exists T \in \mathrm{AT}\ \forall \widetilde{T} \lhd T\ (\widetilde{T} \notin \mathrm{NS})$.

In fact, some of the above can be proved under the weaker assumption $\diamond$.

In §2, we point out that four of the above seven assertions are immediate consequences of some established results. The other cases are dealt with in §3. In the last section are supplementary remarks.

§1. Preliminaries.

We use the following constants: $\Omega = \{\alpha < \omega_1 : \lim(\alpha)\}$, $\mathbb{Q}$ = the set of rational numbers, $\mathbb{R}$ = the set of real numbers, $\mathbb{N} = \omega \smallsetminus \{0\}$ = the set of natural numbers. $\mathcal{T}$ = the family of stationary subsets of ω_1. $\mathcal{L}$ = the family of closed unbounded subsets of ω_1. $\widetilde{\mathcal{L}} = \{S: \exists A \in \mathcal{L}\ (A \subseteq S \subseteq \omega_1)\}$. We say that a set A is club when $A \in \mathcal{L}$. We say that a set is stationary if it belongs to $\mathcal{T}$. $|X|$ means the cardinality of X. We treat a function $f: X \to Y$ as a subset of $Y \times X$ as usual. $\{f(x): x \in X\}$ is denoted by $f''X$. A partially ordered set $\langle T, <_T\rangle$ is called a tree, if for every element x, the set $\hat{x} = \{y \in T: y <_T x\}$ is well ordered by $<_T$. We often write simply T instead of $\langle T, <_T\rangle$ and write $<$ instead of $<_T$. The order type of $\hat{x}$ is the height of x in T, ht(x). When $X \subseteq T$, $\hat{X} = \{y \in T: \exists x \in X\ (y \leq x)\}$. T_α is the set $\{x \in T: \mathrm{ht}(x) = \alpha\}$. We set $T \upharpoonright A = \bigcup_{\alpha \in A} T_\alpha$. A branch of T is a maximal linearly ordered (by $<_T$) subset of T. An antichain of T is a pairwise incomparable subset of T. An ω_1-tree is a tree T such that:

(1) $x \in T \Rightarrow \mathrm{ht}(x) < \omega_1$,

(2) $\alpha < \omega_1 \Rightarrow 0 < |T_\alpha| \leq \aleph_0$,

(3) $x \in T\ \&\ \mathrm{ht}(x) < \alpha < \omega_1 \Rightarrow \exists y, z \in T_\alpha\ (y \neq z\ \&\ x < y, z)$,

(4) $x, y \in T_\alpha\ \&\ \alpha \in \Omega\ \&\ \hat{x} = \hat{y} \Rightarrow x = y$.

If a subset T' of T is an ω_1-tree by $<_T$ and satisfies $\widehat{T'} = T'$, we say that T' is an ω_1-subtree of T and write $T' \lhd T$. A subset X of T is called unbounded (or stationary, club) if $\{\mathrm{ht}(x) : x \in X\}$ is unbounded (resp. stationary, club) in $\{\mathrm{ht}(x) : x \in T\}$. Tree topology of T is defined by taking as an open basis all sets of the form $(a, b] = \{x \in T: a < x \leq b\}$ for $a < b \in T$ and all sets of the form $\{x \in T: x \leq b\}$ for $b \in T$. We write also $[a, b]$ for $\{x \in T: a \leq x \leq b\}$, but use $[0, 1)$ to denote the interval of the real line $\{x \in \mathbb{R}: 0 \leq x < 1\}$. An

Aronszajn tree is an ω_1-tree with no unbounded branch. A Souslin tree is an ω_1-tree with no uncountable antichain. An ω_1-tree T has property γ iff, whenever X is an antichain of T, there is a club subset C of ω_1 such that $T \setminus T\restriction C$ contains a closed nbd (neighborhood) of X. An almost Souslin tree is an Aronszajn tree with no stationary antichain. A tree T is called $\mathbb{R}$-embeddable if there is a function $f: T \to \mathbb{R}$ such that $x <_T y \Rightarrow f(x) < f(y)$. $\mathbb{Q}$-embeddability is defined similarly. A special Aronszajn tree is an ω_1-tree which is the union of a countable collection of its antichains or equivalently is a $\mathbb{Q}$-embeddable ω_1-tree. An ω_1-tree T is called non-Souslin if every uncountable subset of T contains an uncountable antichain. ST, γST, AST, NCA, SAT, RE, NS and AT are defined as in Introduction. We denote by $\mathbb{T}$ the tree $\langle \bigcup_{\alpha<\omega_1} \omega^\alpha, \subset \rangle$, where $\omega^\alpha = \{f: f:\alpha \to \omega\}$. We fix the tree $\mathbb{T}$ in the rest of this paper. We write $x <_{\mathbb{T}} y$ or simply $x < y$ to denote $x, y \in \mathbb{T}$ & $x \subset y$. $\mathbb{T}$ is not an ω_1-tree since $|\mathbb{T}_\alpha| \not\leq \aleph_0$ if $\alpha \geqslant \omega$. Note that $0 \ (= \emptyset)$ is the least element of $\mathbb{T}$.

The following two lemmas supply our main tools:

<u>Lemma</u> 3. If $\diamondsuit$ holds, there exist two sequences $\langle Z_\alpha : \alpha < \omega_1 \rangle$ and $\langle e_\alpha : \alpha < \omega_1 \rangle$ such that (1) $Z_\alpha \subseteq \mathbb{T}\restriction\alpha$, (2) $e_\alpha \subseteq \mathbb{R} \times \mathbb{T}\restriction\alpha$ and (3) if T is an ω_1-subtree of $\mathbb{T}$, then for any subset $X \subseteq T$ and for any function $e: \mathrm{dom}(e) \to \mathbb{R}$, $\{\alpha < \omega_1 : Z_\alpha = X \cap T\restriction\alpha \ \& \ e_\alpha = e \cap (\mathbb{R} \times T\restriction\alpha)\} \in \mathcal{T}$).

<u>Proof</u>. By $\diamondsuit$, we have a sequence $\langle S_\alpha : \alpha < \omega_1 \rangle$ such that for every subset X of ω_1, $\{\alpha < \omega_1 : X \cap \alpha = S_\alpha\} \in \mathcal{T}$. $|\mathbb{T}| = |\mathbb{R} \times \mathbb{T}| = \aleph_1$ as $\diamondsuit$ implies CH. Take a bijective map $f: \omega_1 \to \mathbb{T} \cup \mathbb{R} \times \mathbb{T}$. Put $Z_\alpha = \mathbb{T}\restriction\alpha \cap f''S_\alpha$ and $e_\alpha = \mathbb{R} \times \mathbb{T}\restriction\alpha \cap f''S_\alpha$. We shall show that the sequences $\langle Z_\alpha : \alpha < \omega_1 \rangle$ and $\langle e_\alpha : \alpha < \omega_1 \rangle$ have the required property. Suppose that $X \subset T \subsetneq \mathbb{T}$ and e is a function $\subset \mathbb{R} \times T$. Put $A = f^{-1}{}''X \cup f^{-1}{}''e$. Define continuous functions g, h: $\omega_1 \to \omega_1$ by $g(\alpha) = \sup f^{-1}{}''T\restriction\alpha$ and $h(\alpha) = \sup f^{-1}{}''(e \cap \mathbb{R} \times T\restriction\alpha)$. Put $C = \{\alpha \in \Omega : g(\alpha) = h(\alpha) = \alpha\} \in \mathcal{L}$ and $E = \{\alpha \in \omega_1 : A \cap \alpha = S_\alpha\} \in \mathcal{T}$. By easy calculation, we obtain that

$$\{\alpha : Z_\alpha = X \cap T\restriction\alpha, \ e_\alpha = e \cap \mathbb{R} \times T\restriction\alpha\} \supseteq E \cap C \in \mathcal{T}.$$

Lemma 4. If $\diamondsuit^*$ holds, there exists a sequence $\langle W_\alpha : \alpha < \omega_1 \rangle$ such that (1) W_α is countable, (2) $X \in W_\alpha \Rightarrow X \subseteq \mathbb{T}\lceil\alpha$ and (3) if T is an ω_1-subtree of $\mathbb{T}$, then $\forall X \subseteq T \; \{\alpha < \omega_1 : X \cap T\lceil\alpha \in W_\alpha\} \in \widetilde{\mathcal{I}}$.

Proof. By an argument similar to the above.

§2. Easy cases.

Some of the seven assertions of Theorem described in Introduction follow immediately from some established results. We show such ones in this section.

First observe the following facts:

1. $T \in \gamma ST \;\&\; \widetilde{T} \lessdot T \Rightarrow \widetilde{T} \in \gamma ST$,
2. $T \in RE \;\&\; \widetilde{T} \lessdot T \Rightarrow \widetilde{T} \in RE$,
3. $T \in NCA \;\&\; \widetilde{T} \lessdot T \Rightarrow \widetilde{T} \in NCA$,
4. $T \in SAT \;\&\; \widetilde{T} \lessdot T \Rightarrow \widetilde{T} \in SAT$,
5. $T \in ST \;\&\; \widetilde{T} \lessdot T \Rightarrow \widetilde{T} \in ST$.

(AST and NS also have this kind of property obviously.)

Hence we have the following:

6. $T \in RE \cap \gamma ST \Rightarrow \forall \widetilde{T} \lessdot T \, (\widetilde{T} \in RE \cap \gamma ST)$,
7. $T \in NCA \cap SAT \Rightarrow \forall \widetilde{T} \lessdot T \, (\widetilde{T} \in NCA \cap SAT)$,
8. $T \in ST \Rightarrow \forall \widetilde{T} \lessdot T \, (\widetilde{T} \in ST)$.

As is described in Introduction, we know the following facts: (a) $\diamondsuit \Rightarrow ST \neq \emptyset$ (Jensen), (b) $\diamondsuit^* \Rightarrow RE \cap \gamma ST \neq \emptyset$ (Hanazawa [8]) and (c) $\diamondsuit \Rightarrow NCA \cap SAT \neq \emptyset$ (See, Devlin and Shelah [4]). So, we obtain the following:

Proposition 5. (1) $\exists T \in \gamma ST \; \forall \widetilde{T} \lessdot T \, (T \notin ST)$ if $\diamondsuit^*$.

(2) $\exists T \in RE \; \forall \widetilde{T} \lessdot T \, (T \notin SAT)$ if $\diamondsuit^*$.

(3) $\exists T \in NCA \; \forall \widetilde{T} \lessdot T \, (T \notin AST)$ if $\diamondsuit$.

(4) $\exists T \in AT \; \forall \widetilde{T} \lessdot T \, (T \notin NS)$ if $\diamondsuit$.

However, the assumption $\diamondsuit^*$ in (2) can be replaced by a weaker assumption $\diamondsuit$. We show this fact in the next section together with the rest.

§3. Main cases.

In this section, we show the following which answer all of our remaining questions:

Theorem 6. (1) $\exists T \in NS\ \forall \widetilde{T} \lessdot T\ (\widetilde{T} \notin RE)$ if $\diamond$.

(2) $\exists T \in RE\ \forall \widetilde{T} \lessdot T\ (\widetilde{T} \notin SAT)$ if $\diamond$.

(3) $\exists T \in AT\ \forall \widetilde{T} \lessdot T\ (\widetilde{T} \notin NCA)$ if $\diamond$.

(4) $\exists T \in AST\ \forall \widetilde{T} \lessdot T\ (\widetilde{T} \notin \gamma ST)$ if $\diamond^*$.

3.1. Proof of Theorem 6.(1). Assume $\diamond$. Let $\langle Z_\alpha : \alpha < \omega_1 \rangle$ and $\langle e_\alpha : \alpha < \omega_1 \rangle$ be the sequences given in Lemma 3. We give a tree T with the required properties by defining, for each $\alpha < \omega_1$, $T_\alpha \subseteq \mathbb{T}_\alpha$ and also simultaneously a set $A_v(\alpha)$ for every $v \in T\lceil\alpha \cup T_\alpha$ by induction on $\alpha < \omega_1$ so that the following conditions (a)-(i) hold at each α th stage:

(a) $0 \neq |T_\alpha| \leq \aleph_0\ \&\ \forall x \in T_\alpha\ \forall y \in \mathbb{T}\ (y < x \Rightarrow y \in T\lceil\alpha)$;

(b) $A_v(\alpha)$ is an anti-chain contained in $T\lceil\alpha \cup T_\alpha$ for $v \in T\lceil\alpha$;

(c) $x \in A_v(\alpha) \Rightarrow v < x$, for $x, v \in T\lceil\alpha \cup T_\alpha$;

(d) $\beta < \alpha \Rightarrow A_v(\beta) = A_v(\alpha) \cap T\lceil\beta+1$;

(e) $x, v \in T\lceil\alpha\ \&\ x \overset{A}{<} v \Rightarrow \forall y \in A_v(\alpha)\ (x \overset{A}{<} y)$, where $x \overset{A}{<} v$ stands for $(x < v\ \&\ \forall w \leq x\ (A_w(ht(x)) \cap [0, x] = \emptyset \Rightarrow A_w(ht(v)) \cap [0, v] = \emptyset))$;

(f) $\beta < \alpha \Rightarrow \forall x \in T_\beta\ \exists y \in T_\alpha\ (y \overset{A}{>} x)$;

(g) $\alpha \in \Omega\ \&\ Z_\alpha = \{v\} \subset T\lceil\alpha \Rightarrow A_v(\alpha) \cap T_\alpha \neq \emptyset$;

(h) $\alpha \in \Omega\ \&\ \widehat{Z_\alpha} = Z_\alpha\ \&\ \forall x \in Z_\alpha\ \exists y \in T\lceil\alpha\ (y \overset{A}{\geqslant} x\ \&\ y \notin Z_\alpha) \Rightarrow$
$$\Rightarrow \forall u \in T_\alpha\ \exists v < u\ (v \notin Z_\alpha);$$

(i) If $\alpha \in \Omega\ \&\ \widehat{Z_\alpha} = Z_\alpha\ \&\ \exists x \in Z_\alpha\ \forall y \in T\lceil\alpha\ (y \overset{A}{\geqslant} x \Rightarrow y \in Z_\alpha)\ \&$ $e_\alpha : Z_\alpha \to [0, 1)\ \&\ \forall x, y \in Z_\alpha\ (x < y \Rightarrow e_\alpha(x) < e_\alpha(y))$, then there exist $u \in T_\alpha$ and $v \overset{A}{<} u$ such that $\forall y \in T\lceil\alpha\ (y \overset{A}{\geqslant} v \Rightarrow y \in Z_\alpha)\ \&$
$$\forall n \in \mathbb{N}\ \exists z \overset{A}{<} u\ (v \overset{A}{\leqslant} z\ \&\ \forall x \in T\lceil\alpha\ (x \overset{A}{>} z \Rightarrow e_\alpha(x) \leq e_\alpha(z) + 1/n)).$$

Notice that the relation $\overset{A}{<}$ is transitive.

Definition of T_α. I. $T_0 = \{0\}$. $A_0(0) = \emptyset$.

II. $T_{\beta+1} = \{x \cup \{\langle n, \beta \rangle\} : n \in \omega,\ x \in T_\beta\}$.

$$A_x(\beta+1) = \begin{cases} A_x(\beta) & \text{if } x \in T\restriction\beta+1 , \\ \emptyset & \text{if } x \in T_{\beta+1} . \end{cases}$$

We can easily check that Conditions (a) - (f) hold for $\alpha = \beta + 1$.

III. Let $\alpha \in \Omega$. To define T_α, we first define $t_\alpha(x) \in T_\alpha$ for each $x \in T\restriction\alpha$. Let $x \in T\restriction\alpha$ be given. Take a monotone sequence $\langle \alpha_n : n \in \omega \rangle$ such that $\alpha_0 = ht(x)$ and $\lim_{n<\omega} \alpha_n = \alpha$. Put $x_0 = x$ and pick inductively x_n for each $n < \omega$ so that $x_{n-1} \overset{A}{<} x_n$ and $ht(x_n) = \alpha_n$. (Such x_n exists by inductive assumption (f).) We set $t_\alpha(x) = \bigcup_{n<\omega} x_n \in T_\alpha$. For $x \in T\restriction\alpha$, $A^<_x(\alpha)$ will stand for $\bigcup\{A_x(\beta) : ht(x) \leqslant \beta < \alpha\}$ throughout this case III. Observe that the following hold:

(j) $x \in T_\beta$ & $\beta < \alpha \Rightarrow A_x(\beta) = \emptyset$. (By (b) & (c));

(k) $x \overset{A}{\leqslant} y \in T\restriction\alpha \Rightarrow A_x(ht(y)) \cap [0, y] = \emptyset$, since $A_x(ht(x)) \cap [0, x] \subseteq A_x(ht(x)) = \emptyset$ by (j);

(l) $x \in T\restriction\alpha \Rightarrow A^<_x(\alpha) \cap [0, t_\alpha(x)] = \emptyset$. (For, note that $A^<_x(\alpha) = \bigcup_{n<\omega} A_x(\alpha_n)$ by (d). So, by (k), $A_x(\alpha_n) \cap [0, t_\alpha(x)] \subseteq A_x(\alpha_n) \cap [0, x_n] = \emptyset$ since $x \overset{A}{<} x_n$.)

(m) $y \overset{A}{\leqslant} x \in T\restriction\alpha \Rightarrow \forall z \leqslant y$ ($A_z(ht(y)) \cap [0, y] = \emptyset \Rightarrow A^<_z(\alpha) \cap [0, t_\alpha(x)] = \emptyset$). (For, suppose $z \leqslant y$ & $A_z(ht(y)) \cap [0, y] = \emptyset$. As $x \overset{A}{<} x_n$ for $n < \omega$, $y \overset{A}{\leqslant} x_n$, so $A_z(\alpha_n) \cap [0, x_n] = \emptyset$. Hence $A^<_z(\alpha) \cap [0, t_\alpha(x)] = \bigcup_{n<\omega} A_z(\alpha_n) \cap \cap [0, t_\alpha(x)] = \bigcup_{n<\omega} (A_z(\alpha_n) \cap [0, x_n]) = \emptyset$.)

Now we define T_α for $\alpha \in \Omega$. There are four cases to consider.

<u>Case</u> 1. $Z_\alpha = \{v\}$ for some $v \in T\restriction\alpha$. Define:

$$T_\alpha = \{t_\alpha(x) : x \in T\restriction\alpha\} ,$$

$$A_x(\alpha) = \begin{cases} A^<_x(\alpha) \cup \{t_\alpha(v)\} & \text{if } x = v , \\ A^<_x(\alpha) & \text{if } x \neq v \in T\restriction\alpha , \\ \emptyset & \text{if } x \in T_\alpha . \end{cases}$$

Conditions (a) - (i) except (f) follow immediately from the following: (1) $t_\alpha(v)$ is comparable with no element of $A^<_v(\alpha)$ by (l); (2) $v < t_\alpha(v)$; (3) $x \overset{A}{<} v \Rightarrow x \overset{A}{<} t_\alpha(v)$ (This is clear by (m) since $A^<_z(\alpha) = A_z(\alpha)$ for $z \neq v$); (4) $t_\alpha(v) \in A_v(\alpha) \cap T_\alpha$. To check (f), suppose $\beta < \alpha$ and $x \in T_\beta$. Take

$\tilde{x} \in T_{\beta+1}$ such that $x < \tilde{x} \not\lessdot t_\alpha(v)$. Then $A_z(\alpha) \cap [0, t_\alpha(\tilde{x})] = A_z^<(\alpha) \cap [0, t_\alpha(\tilde{x})]$ for any z, since $t_\alpha(\tilde{x}) \neq t_\alpha(v)$. Hence $\tilde{x} \overset{A}{<} t_\alpha(\tilde{x})$ by (m). As $x \overset{A}{<} \tilde{x}$, $x \overset{A}{<} t_\alpha(\tilde{x}) \in T_\alpha$.

Case 2. $\widehat{Z_\alpha} = Z_\alpha \subseteq T\lceil\alpha$ and $\forall x \in Z_\alpha \, \exists y \in T\lceil\alpha \; (y \overset{A}{\geqslant} x \;\&\; y \notin Z_\alpha)$. With each $x \in Z_\alpha$, we associate $\tilde{x} \in T\lceil\alpha$ such that $x \overset{A}{\leqslant} \tilde{x} \notin Z_\alpha$. Define $T_\alpha = \{ t_\alpha(\tilde{x}) : x \in T\lceil\alpha\}$ and $A_x(\alpha) = \begin{cases} A_x^<(\alpha) & \text{if } x \in T\lceil\alpha, \\ \emptyset & \text{if } x \in T_\alpha . \end{cases}$

Conditions (a) - (i) are easily checked. We only show (f). Suppose $x \in T_\beta$ and $\beta < \alpha$. Then $x \overset{A}{<} t_\alpha(\tilde{x})$ by (m) since $x \overset{A}{\leqslant} \tilde{x}$ and $A_z^<(\alpha) = A_z(\alpha)$ for every $z \in T\lceil\alpha$.

Case 3. $\widehat{Z_\alpha} = Z_\alpha \subseteq T\lceil\alpha$, $\exists x \in Z_\alpha \, \forall y \in T\lceil\alpha \,(y \overset{A}{\geqslant} x \Rightarrow y \in Z_\alpha)$ and e_α is a function which embeds Z_α into $[0, 1)$. First take $v \in Z_\alpha$ such that $\forall y \in T\lceil\alpha$ $(y \overset{A}{\geqslant} v \Rightarrow y \in Z_\alpha)$. Let $\langle \alpha_n : n \in \mathbb{N} \rangle$ be a monotone sequence such that $\alpha_1 = ht(v)$ and $\lim_{n\in\mathbb{N}} \alpha_n = \alpha$. Put $v_1 = v$ and take inductively $\tilde{v}_n$ and v_{n+1} for each $n \in \mathbb{N}$ so that (1) $v_n \overset{A}{\leqslant} \tilde{v}_n \in Z_\alpha$ and $e(\tilde{v}_n) > \tilde{e}(v_n) - 1/n$, where $\tilde{e}(v_n) = \sup\{e(x) : v_n \overset{A}{<} x \in Z_\alpha\}$ and (2) $\tilde{v}_n \overset{A}{\leqslant} v_{n+1} \in T\lceil\alpha$ and $\alpha_n \leqslant ht(v_{n+1})$. (As is easily seen, $v_n \overset{A}{\geqslant} v$ for each n and hence $v_n \in Z_\alpha = ran(e_\alpha)$ by the definition of v. So, this inductive procedure does not break down.) Set $u_\alpha = \bigcup_{n\in\mathbb{N}} v_n \in T_\alpha$. Define $T_\alpha = \{u_\alpha\} \cup \{t_\alpha(x) : x \in T\lceil\alpha\}$ and

$$A_x(\alpha) = \begin{cases} A_x^<(\alpha) & \text{if } x \in T\lceil\alpha, \\ \emptyset & \text{if } x \in T_\alpha . \end{cases}$$

Conditions (a) - (h) are easily checked. (For (f), note that $x \overset{A}{<} t_\alpha(x)$ by (m) since $A_z^<(\alpha) = A_z(\alpha)$.) To prove (i), it suffices to show the following: (1) $v \overset{A}{<} u_\alpha$ and (2) $v \overset{A}{\leqslant} \tilde{v}_n \overset{A}{\leqslant} u_\alpha \;\&\; \forall x \in T\lceil\alpha \,(\tilde{v}_n \overset{A}{<} x \Rightarrow e(x) \leqslant e(\tilde{v}_n) + 1/n)$. $v \overset{A}{\leqslant} \tilde{v}_n$ holds obviously by the definition. To prove $\tilde{v}_n \overset{A}{\leqslant} u_\alpha$, suppose $z \leqslant \tilde{v}_n$ and $A_z(ht(\tilde{v}_n)) \cap [0, \tilde{v}_n] = \emptyset$. Then, for all $m \geqslant n$, $A_z(ht(\tilde{v}_m)) \cap [0, \tilde{v}_m] = \emptyset$ as $v_m \overset{A}{\geqslant} v_n$. So, $\bigcup_{m\geqslant n} A_z(ht(\tilde{v}_m)) \cap [0, u_\alpha] = \emptyset$, since $A_z(ht(\tilde{v}_m)) \cap [0, \tilde{v}_m] = A_z(ht(\tilde{v}_m)) \cap [0, u_\alpha]$. Hence $A_z(\alpha) \cap [0, u_\alpha] = A_z^<(\alpha) \cap [0, u_\alpha] = \emptyset$ and thus $v_n \overset{A}{\leqslant} u_\alpha$ holds. $v \overset{A}{<} u_\alpha$ follows from $v \overset{A}{<} v_n \overset{A}{<} u_\alpha$. If $v_n \overset{A}{<} x \in T\lceil\alpha$, then $v \overset{A}{<} x$ and hence $x \in Z_\alpha$ which implies $e(x) \leqslant \tilde{e}(v_n) < e(\tilde{v}_n) + 1/n$. (i) is thus proved.

Case 4. Otherwise. $T_\alpha = \{t_\alpha(x) : x \in T\lceil\alpha\}$.

$$A_x(\alpha) = \begin{cases} A_x^<(\alpha) & \text{if } x \in T\lceil\alpha, \\ \emptyset & \text{if } x \in T_\alpha. \end{cases}$$

Conditions (a) - (i) are easily checked. T_α is thus defined.

Now, we show that the tree $T = \bigcup_{\alpha<\omega_1} T_\alpha$ is as required.

Claim 1. T is an ω_1-tree.

Proof. Obvious by (a), (f) and a property of $\mathbb{T}$.

Claim 2. If $X \subseteq T$ and $\forall v \in \hat{X} \exists x \overset{A}{\geqslant} v\ (x \notin \hat{X})$, then X is bounded in T.

Proof. Suppose $X \subseteq T$ and $\forall v \in \hat{X} \exists x \overset{A}{\geqslant} v\ (x \notin \hat{X})$. Then the set $C = \{\alpha \in \Omega : \forall v \in \hat{X} \cap T\lceil\alpha\ \exists x \in T\lceil\alpha\ (v \overset{A}{\leqslant} x \notin \hat{X})\}$ is club. Put $E = \{\alpha < \omega_1 : Z_\alpha = \hat{X} \cap T\lceil\alpha\} \in \mathcal{T}$. Take $\alpha \in C \cap E$. Then, by (h), $\forall u \in T_\alpha \exists v < u\ (v \notin Z_\alpha)$, which means $T_\alpha \cap \hat{X} = \emptyset$. Hence $X \subseteq \hat{X} \subseteq T\lceil\alpha$.

Claim 3. $\bigcup_{\xi > ht(v)} A_v(\xi)$ is an uncountable anti-chain for each $v \in T$.

Proof. $\alpha \in \Omega \cap \{\alpha : \{v\} = Z_\alpha\} \Rightarrow A_v(\alpha) \cap T_\alpha \neq \emptyset$ by (g). $\Omega \cap \{\alpha : \{v\} = Z_\alpha\} \in \mathcal{T}$. Hence $\bigcup_{\xi < ht(v)} A_v(\xi)$ is uncountable. By (b) and (d), it is an anti-chain.

Claim 4. $T \in NS$.

Proof. Suppose X is an unbounded subset of T. By Claim 2, we obtain $v \in \hat{X}$ such that $\forall x \overset{A}{\geqslant} v\ (x \in \hat{X})$. Pick arbitrarily an element $w \in T$ such that $v \overset{A}{\leqslant} w$. Put $A = \bigcup_{\xi > ht(w)} A_w(\xi)$. if $\xi > ht(w)$ and $x \in A_w(\xi)$, then $v \overset{A}{<} x$ by (e). Hence $A \subseteq \hat{X}$. By Claim 3, A is an uncountable anti-chain. For each $x \in A$, pick an element $\tilde{x} \in X$ such that $x \leqslant \tilde{x} \in X$. Put $\tilde{A} = \{\tilde{x} : x \in A\}$. $\tilde{A}$ is clearly an uncountable anti-chain which is contained in X.

Claim 5. $\forall \tilde{T} \triangleleft T\ (\tilde{T} \notin RE)$.

Proof. If not so, we have an ω_1-subtree $\tilde{T}$ of T with an $\mathbb{R}$-embedding $e: \tilde{T} \to [0, 1)$. Put $C = \{\alpha \in \Omega : \forall x \in \tilde{T}\lceil\alpha\ \forall n \in \mathbb{N}\ (\exists y \in \tilde{T}\ (y \overset{A}{>} x\ \&\ e(y) > e(x) + 1/n) \Rightarrow \exists y \in \tilde{T}\lceil\alpha\ (y \overset{A}{>} x\ \&\ e(y) > e(x) + 1/n))\}$,

$$D = \{\alpha \in \Omega : \forall w \in T\lceil\alpha\ [\exists x \overset{A}{\geqslant} w\ (x \notin \tilde{T}) \Rightarrow \exists x \overset{A}{\geqslant} w\ (x \notin \tilde{T}\lceil\alpha)]\},$$

$$E = \{\alpha < \omega_1 : Z_\alpha = \tilde{T} \cap T\lceil\alpha\ \&\ e_\alpha = e \cap \mathbb{R} \times T\lceil\alpha\}.$$

As $\tilde{T}$ is uncountable and $\forall x \in \tilde{T}\ \forall y < x\ (y \in \tilde{T})$, we obtain $v \in \tilde{T}$ such that $\forall x \overset{A}{\geqslant} v\ (x \in \tilde{T})$ by Claim 2. Since $C, D \in \mathcal{L}$ and $E \in \mathcal{T}$, we can take $\alpha \in \Omega$

such that $ht(v) < \alpha \in C \cap D \cap E$. As $v \in \widetilde{T}\restriction\alpha$ and $Z_\alpha = \widetilde{T} \cap T\restriction\alpha$, the antecedent of (i) is satisfied and hence there are $u \in T_\alpha$ and $w \in T\restriction\alpha$ such that

$$w \overset{A}{<} u \;\&\; \forall y \in T\restriction\alpha\,(y \overset{A}{\geqslant} w \Rightarrow y \in Z_\alpha) \;\&$$

$$\& \;\forall n \in \mathbb{N}\,\exists z \in T\restriction\alpha\,(w \overset{A}{\leqslant} z \overset{A}{<} u \;\&\; \forall x \in T\restriction\alpha\,(x \overset{A}{>} z \Rightarrow e(x) \leqslant e(z) + 1/n)).$$

Take $\tilde{u} \in T_{\alpha+1}$ such that $u \overset{A}{<} \tilde{u}$. As $\alpha \in E$, $\forall y \in T\restriction\alpha\,(y \overset{A}{\geqslant} w \Rightarrow y \in \widetilde{T}\restriction\alpha)$. Hence $u, \tilde{u} \in \widetilde{T} = \mathrm{dom}(e)$, since $\alpha \in D$ and $w \overset{A}{<} u \overset{A}{<} \tilde{u}$. Take $m \in \mathbb{N}$ so that $1/m < e(\tilde{u}) - e(u)$. Take $z \in T\restriction\alpha$ such that $w \overset{A}{\leqslant} z \overset{A}{<} u \;\&\; \forall x \in T\restriction\alpha\,(x \overset{A}{>} z \Rightarrow$ $\Rightarrow e(x) \leqslant e(z) + 1/m)$. $\tilde{u} \overset{A}{>} z$ as $\tilde{u} \overset{A}{>} u \overset{A}{>} z$. $e(\tilde{u}) > e(z) + 1/m$ by the definition of m as $e(\tilde{u}) > e(u) > e(z)$. Hence $\exists y \in \widetilde{T}\restriction\alpha\,(y \overset{A}{>} z \;\&\; e(y) > e(z) + 1/m)$ since $\alpha \in C$ and $\tilde{u} \overset{A}{>} z$. This contradicts to the definition of z.

3.2. <u>Proof</u> of Theorem 6.(2). Let $\langle Z_\alpha : \alpha < \omega_1\rangle$ and $\langle e_\alpha : \alpha < \omega_1\rangle$ be the sequences given in Lemma 3. We define $T_\alpha \subseteq \mathbb{T}_\alpha$ for each $\alpha < \omega_1$ inductively. As define T_α, we also define $r(x) \in \mathbb{R}$ for each $x \in T_\alpha$ and observe inductively that the following conditions (a) - (c) hold:

(a) $\emptyset \neq T_\alpha \;\&\; |T_\alpha| = \aleph_0 \;\&\; \forall x \in T_\alpha \forall y < x\,(y \in T\restriction\alpha)$;

(b) $x \in T\restriction\alpha \;\&\; 0 < q \in \mathbb{Q} \Rightarrow \exists y \in T_\alpha\,(y > x \;\&\; r(y) - r(x) < q)$;

(c) $x < y \in T_\alpha \Rightarrow r(x) < r(y)$.

<u>Definition</u> of T_α. We fix an enumeration $\langle q_n : n < \omega\rangle$ of $\mathbb{Q}^+ = \{q \in \mathbb{Q} : q > 0\}$.

I. $T_0 = \{0\}$, $r(0) = 0$.

II. $T_{\beta+1} = \{x \cup \{\langle n, \beta\rangle\} : x \in T_\beta,\ n \in \omega\}$,

$r(x \cup \{\langle n, \beta\rangle\}) = r(x) + q_n$.

III. Let $\alpha \in \Omega$. To define T_α, we preliminarily define $t_\alpha(x, \varepsilon) \in \mathbb{T}_\alpha$ and $\tilde{r}_\alpha(x, \varepsilon) \in \mathbb{R}$ for each $x \in T\restriction\alpha$ and each $\varepsilon \in \mathbb{Q}$ with $r(x) < \varepsilon$. Let $x \in T\restriction\alpha$ and $r(x) < \varepsilon \in \mathbb{Q}$. Take monotone sequence $\langle \alpha_n : n < \omega\rangle$ such that $\alpha_0 = ht(x)$ and $\lim_{n<\omega}\alpha_n = \alpha$. Put $x_0 = x$ and $\varepsilon_0 = \varepsilon$ and by inductive assumption (b) we can pick x_{n+1} and ε_{n+1} so that $x_n < x_{n+1}$, $ht(x_{n+1}) = \alpha_{n+1}$, $r(x_{n+1}) < \varepsilon_n$ and $\varepsilon_{n+1} = (\varepsilon_n + r(x_{n+1}))/2$. We set $t_\alpha(x, \varepsilon) = \bigcup_{n<\omega} x_n \in \mathbb{T}_\alpha$ and $\tilde{r}_\alpha(x, \varepsilon) = \lim_{n<\omega} r(x_n) < \varepsilon$. Now, we define T_α for $\alpha \in \Omega$. There are three cases to consider.

Case 1. $\forall x \in T\lceil\alpha\ \forall \varepsilon \in \mathbb{Q}^+\ \exists y \in T\lceil\alpha\ (x \leqslant y \notin Z_\alpha\ \&\ r(y) - r(x) < \varepsilon)$. By this assumption, for each $x \in T\lceil\alpha$ and each $\varepsilon \in \mathbb{Q}$ with $r(x) < \varepsilon$, we can take $x_\varepsilon \in T\lceil\alpha$ such that $x \leqslant x_\varepsilon \notin Z_\alpha\ \&\ r(x_\varepsilon) < \varepsilon$. We set:

$$T_\alpha = \{t_\alpha(x_\varepsilon, \varepsilon) : x \in T\lceil\alpha\ \&\ \varepsilon \in \mathbb{Q}\ \&\ \varepsilon > r(x)\},$$

$$r(t_\alpha(x_\varepsilon, \varepsilon)) = \tilde{r}_\alpha(x_\varepsilon, \varepsilon).$$

Case 2. $\exists u \in T\lceil\alpha\ \exists \varepsilon \in \mathbb{Q}^+\ \forall x \in T\lceil\alpha\ (u \leqslant x\ \&\ r(x) - r(u) < \varepsilon \Rightarrow x \in Z_\alpha)$ & $e_\alpha : Z_\alpha \to \mathbb{Q}^+\ \&\ \forall x, y \in Z_\alpha\ (x < y \Rightarrow e_\alpha(x) < e_\alpha(y))$. Pick such $u \in T\lceil\alpha$ and $\varepsilon \in \mathbb{Q}^+$. Clearly $u \in Z_\alpha$. Let $\langle \alpha_n : n < \omega \rangle$ be a monotone sequence such that $\alpha_0 = ht(u)\ \&\ \lim_{n<\omega} \alpha_n = \alpha$. Put $u_0 = u$ and $\varepsilon_0 = \varepsilon$. By inductive assumptions (b) and (c), we can pick $u_{n+1} \in Z_\alpha$ and $\varepsilon_{n+1} \in \mathbb{Q}^+$ so that $u_n < u_{n+1}$ & $ht(u_{n+1}) > \alpha_n\ \&\ r(u_{n+1}) - r(u_n) < \varepsilon_n\ \&\ [\exists y \in Z_\alpha\ (u_n < y\ \&\ r(y) - r(u_n) < \varepsilon_n\ \&\ q_n \leqslant e(y)) \Rightarrow q_n \leqslant e(u_{n+1})]\ \&\ r(u_{n+1}) + \varepsilon_{n+1} < r(u_n) + \varepsilon_n$. We set:

$$u_\alpha = \bigcup_{n<\omega} u_n \in \mathbb{T}_\alpha,$$

$$T_\alpha = \{u_\alpha\} \cup \{t_\alpha(x, \varepsilon) : x \in T\lceil\alpha,\ r(x) < \varepsilon \in \mathbb{Q}\},$$

$$r(u_\alpha) = \lim_{n<\omega} u_n \in \mathbb{R},$$

$$r(t_\alpha(x, \varepsilon)) = \tilde{r}_\alpha(x, \varepsilon).$$

Case 3. Otherwise. We set:

$$T_\alpha = \{t_\alpha(x, \varepsilon) : x \in T\lceil\alpha,\ r(x) < \varepsilon \in \mathbb{Q}\},$$

$$r(t_\alpha(x, \varepsilon)) = \tilde{r}_\alpha(x, \varepsilon).$$

T_α is thus defined.

Put $T = \bigcup_{\alpha < \omega_1} T_\alpha$. Clearly $r : T \to \mathbb{R}$ is an order-embedding. It remains to show that $\tilde{T} \notin SAT$ for any $\tilde{T} \lhd T$. To the contrary, assume $\tilde{T} \in SAT$. We have then a $\mathbb{Q}$-embedding $e : \tilde{T} \to \mathbb{Q}^+$.

Claim. $\exists u \in T\ \exists \varepsilon \in \mathbb{Q}^+\ \forall x \in T\ (x \geqslant u\ \&\ r(x) - r(u) < \varepsilon \Rightarrow x \in \tilde{T})$.

Suppose not. Then the following set is club:

$$C = \{\alpha \in \Omega : \forall x \in T\lceil\alpha\ \forall \varepsilon \in \mathbb{Q}^+\ \exists x \in T\lceil\alpha\ (u \leqslant x \in \tilde{T}\lceil\alpha\ \&\ r(x) - r(u) < \varepsilon).$$

Since $\{\alpha : \tilde{T} \cap T\lceil\alpha = Z_\alpha\} \in \mathcal{T}$, we can take $\alpha \in C$ such that $\tilde{T} \cap T\lceil\alpha = Z_\alpha$. For such α, T_α must be defined in Case 1. However, then, every element $t_\alpha(x_\varepsilon, \varepsilon)$ of T_α does not belong to $\tilde{T}$ since $x_\varepsilon < t_\alpha(x_\varepsilon, \varepsilon)$ and $x_\varepsilon \notin Z_\alpha =$

$\widetilde{T}\lceil\alpha$. This means $T_\alpha \cap \widetilde{T} = \emptyset$ which contradicts to $\widetilde{T} \lessdot T$. Claim is thus proved.

By the claim, we take $\widetilde{u} \in T$ and $\widetilde{\varepsilon} \in \mathbb{Q}^+$ such that $\forall x \in T\ (\widetilde{u} \leqslant x\ \&\ r(x) - r(\widetilde{u}) < \widetilde{\varepsilon} \Rightarrow x \in \widetilde{T})$. Put:

$$C = \{\alpha \in \Omega :\ \forall x \in T\lceil\alpha\ \forall q \in \mathbb{Q}\ \forall n \in \omega\ [\exists y \in \widetilde{T}\ (y > x\ \&\ r(y) < q\ \&\ e(y) \geqslant q_n) \Rightarrow$$
$$\Rightarrow \exists y \in \widetilde{T}\lceil\alpha\ (y > x\ \&\ r(y) < q\ \&\ e(y) \geqslant q_n)]\},$$
$$D = \{\alpha \in \Omega :\ \forall u \in \widetilde{T}\lceil\alpha\ \forall \varepsilon \in \mathbb{Q}^+\ [\forall x \in T\lceil\alpha\ (x > u\ \&\ r(x) < r(u) + \varepsilon \Rightarrow x \in \widetilde{T}) \Rightarrow$$
$$\Rightarrow \forall x \in T\ (x > u\ \&\ r(x) < r(u) + \varepsilon \Rightarrow x \in \widetilde{T})]\}.$$

As $C, D \in \mathcal{L}$ and $\{\alpha : \widetilde{T}\lceil\alpha = Z_\alpha\ \&\ e_\alpha = e\lceil(T\lceil\alpha)\} \in \mathcal{T}$, we can take $\alpha \in C \cap D$ such that $\alpha > \mathrm{ht}(\widetilde{u})$, $\widetilde{T}\lceil\alpha = Z_\alpha$ and $e_\alpha = e\lceil(T\lceil\alpha)$. For such α, T_α must have been defined in Case 2. Recall the definition of T_α in Case 2. Let u, ε, $\{u_n : n < \omega\}$ and $\{\varepsilon_n : n < \omega\}$ be the ones given there. We observe that $u_\alpha \in T_\alpha$ and $r(u_\alpha) < r(u) + \varepsilon$ as $r(u_{n+1}) + \varepsilon_{n+1} < r(u_n) + \varepsilon_n$ for every $n < \omega$. So, as $\alpha \in D$, by choice of u and ε, $u_\alpha \in \widetilde{T} = \mathrm{dom}(e)$ and hence there must be $m \in \omega$ such that $q_m = e(u_\alpha)$. Since $u_\alpha > u_m\ \&\ r(u_\alpha) < r(u_m) + \varepsilon_m\ \&\ q_m \leqslant e(u_\alpha)$, $\exists y \in T\lceil\alpha\ (y > u_m\ \&\ r(y) < r(u_m) + \varepsilon_m\ \&\ q_m \leqslant e(y))$ as $\alpha \in C$. So, by choice of u_{m+1}, $q_m \leqslant e(u_{m+1})$ which is absurd since $u_{m+1} < u_\alpha$.

3.3. <u>Proof</u> of Theorem 6.(3). Assume $\Diamond$. Take $\langle Z_\alpha : \alpha < \omega_1\rangle$ that was given in Lemma 3. We define $T_\alpha \subseteq \mathbb{T}_\alpha$ by induction. As we define T_α, we also define simultaneously $e(x) \in \mathbb{Q}$ for each $x \in T_\alpha$. We will easily see that the following hold at each α-th stage:

(a) $T_\alpha \neq \emptyset\ \&\ |T_\alpha| = \aleph_0\ \&\ \forall x \in T_\alpha\ \forall y \in \mathbb{T}\ (y < x \Rightarrow y \in T\lceil\alpha)$,

(b) $x < y \in T_\alpha \Rightarrow e(x) < e(y)$,

(c) $x \in T_\beta\ \&\ \beta < \alpha\ \&\ e(x) < q \in \mathbb{Q} \Rightarrow \exists y \in T_\alpha\ (y > x\ \&\ e(y) = q)$.

<u>Definition</u> of T_α. Fix $\{q_n : n < \omega\}$, an enumeration of $\{q \in \mathbb{Q} : q > 0\}$.

I. $T_0 = \{0\}$, $e(0) = 0$.

II. $T_{\beta+1} = \{x \cup \{\langle n, \beta\rangle\} : x \in T_\beta,\ n \in \omega\}$,
$e(x \cup \{\langle n, \beta\rangle\}) = e(x) + q_n$.

III. Let $\alpha \in \Omega$. We preliminarily define $t_\alpha(x, q) \in \mathbb{T}_\alpha$ for each $x \in T\lceil\alpha$

and each $q \in \mathbb{Q}$ with $q > e(x)$. Let $x \in T\lceil\alpha$, $q \in \mathbb{Q}$ and $q > e(x)$. Take a monotone sequence $\langle\alpha_n : n > \omega\rangle$ such that $\alpha_0 = ht(x)$ and $\lim_{n<\omega} \alpha_n = \alpha$. Take a monotone sequence $\{p_n : n<\omega\} \subset \mathbb{Q}$ such that $p_0 = e(x)$ and $\lim_{n<\omega} p_n = q$. We inductively take x_n for each $n < \omega$ as follows: We put $x_0 = x$. $x_n \in T\lceil\alpha$ is taken so that $ht(x_n) = \alpha_n$, $x_n > x_{n-1}$ and $e(x_n) = p_n$. (This is possible by (c).) Set $t_\alpha(x, q) = \bigcup_{n<\omega} x_n \in \mathbb{T}_\alpha$. Now, we define T_α, $\alpha \in \Omega$. The following two cases are distinguished:

Case 1. $\widetilde{Z}_\alpha = Z_\alpha$ and $\forall x \in T\lceil\alpha \, \forall q \in \mathbb{Q} \, (q > e(x) \Rightarrow \exists y \in T\lceil\alpha \, (y>x \,\&\, e(y)<q \,\&\, y \notin Z_\alpha)$. With each $x \in T\lceil\alpha$ and each $q \in \mathbb{Q}$ with $q > e(x)$, we associate $x_q \in T\lceil\alpha$ such that $x_q > x \,\&\, e(x_q) < q \,\&\, x_q \notin Z_\alpha$. We set:

$$T_\alpha = \{t_\alpha(x_q, q) : x \in T\lceil\alpha,\ e(x) < q \in \mathbb{Q}\},$$
$$e(t_\alpha(x_q, q)) = q.$$

Case 2. Otherwise. Set:

$$T_\alpha = \{t_\alpha(x, q) : x \in T\lceil\alpha,\ e(x) < q \in \mathbb{Q}\},$$
$$e(t_\alpha(x, q)) = q.$$

T_α is thus defined.

We show that the tree $T = \bigcup_{\alpha<\omega_1} T_\alpha$ is as required. The function $e: T \to \mathbb{Q}$ is clearly a $\mathbb{Q}$-embedding and so $T \in AT$.

Claim. If a subset X of T satisfies that $\widehat{X} = X$ and $\forall x \in T \, \forall q \in \mathbb{Q} \, (q > e(x) \Rightarrow \exists y \in T \, (x < y \notin X \,\&\, e(y) < q))$, then X is bounded in T.

For, by the routine argument, we have $\alpha \in \Omega$ such that $X \cap T\lceil\alpha = Z_\alpha$ and $\forall x \in T\lceil\alpha \, \forall q \in \mathbb{Q} \, (q > e(x) \Rightarrow \exists y \in T\lceil\alpha \, (x < y \notin X \,\&\, e(y) < q))$. Then by the definition in Case 1, every element of T_α has the form $t_\alpha(x_q, q)$, an extension of x_q which is not in $Z_\alpha = X \cap T\lceil\alpha \subseteq \widehat{X}$, and hence does not belong to $\widehat{X}$. So $\widehat{X} \cap T_\alpha = \emptyset$, which implies $X \subset T\lceil\alpha$. Claim is thus proved.

Suppose $\widetilde{T} \lhd T$. Then, by the claim, we obtain $x \in T$ and $q \in \mathbb{Q}$ such that $q > e(x)$ and $\forall y \in T \, (y > x \,\&\, e(y) < q \Rightarrow y \in \widetilde{T})$. Clearly $x \in \widetilde{T}$. Take $\tilde{q} \in \mathbb{Q}$ so that $e(x) < \tilde{q} < q$. Put $A = \{y \in T : y > x \text{ and } e(y) = \tilde{q}\}$. Then A is clearly an anti-chain by (b). By the definition of x, q and $\tilde{q}$, $A \subseteq \widetilde{T}$. By (c), $\{ht(z) : z \in A\} = \{\alpha : \alpha > ht(x)\} \in \mathcal{L}$.

3.4. Proof of Theorem 6.(4). Assume $\diamondsuit^*$. Take $\langle Z_\alpha : \alpha < \omega_1\rangle$, $\langle e_\alpha : \alpha < \omega_1\rangle$ and $\langle W_\alpha : \alpha < \omega_1\rangle$ that were given in Lemmas 3 and 4. By Ω_0 and Ω_1, we denote the sets $\{\omega\cdot(\beta + 1) : \beta < \omega_1\}$ and $\{\omega^2\cdot\beta : \beta < \omega_1\}$ respectively. We give a required tree T by defining $T_\alpha \subseteq \mathbb{T}_\alpha$ inductively for each $\alpha < \omega_1$. We give also closed subsets A_n of T for all $n < \omega$ such that $A_n \subset T\lceil\Omega_0$ and $A_n \cap A_m = \emptyset$ $(n \neq m)$, by defining the sets $A_n \cap T_\alpha$ simultaneously when T_α is defined for $\alpha \in \Omega_0$. We will observe inductively that the following conditions (a) - (f) hold at each α-th stage:

(a) $x \in T\lceil\alpha \Rightarrow \forall n < \omega\, \exists y \in T_\alpha\, (x \overset{n}{<} y)$, where $x \overset{n}{<} y$ stands for $x < y$ & & $(x, y] \cap \bigcup_{i<n} A_i = \emptyset$.

(b) $x \in T\lceil\alpha$ & $\alpha \in \Omega_0 \Rightarrow \forall n\, \exists y\, (x \overset{n}{<} y \in A_n)$.

(c) $x \in T_\alpha$ & $\alpha \in \Omega_1 \Rightarrow \forall n\, \exists y \in T\lceil\alpha\, (y \overset{n}{<} x)$.

(d) $\alpha \in \Omega_1$ & $\widehat{Z}_\alpha = Z_\alpha$ & $\forall x \in Z_\alpha\, \forall n < \omega\, \exists y \in T\lceil\alpha\, (x \overset{n}{<} y \notin Z_\alpha) \Rightarrow \forall u \in T_\alpha$ $\exists y < u\, (y \notin Z_\alpha)$.

(e) $\alpha \in \Omega_1$ & $e_\alpha = \{\langle n, x\rangle\}$ & $n < \omega \subseteq \mathbb{R}$ & $x \in T\lceil\alpha$ & (Z_α is a nbd of $\{y \in T\lceil\alpha : y \in A_n$ & $y \overset{n}{>} x\}$) & $0 \notin Z_\alpha \Rightarrow \exists t \in T_\alpha\, (x \overset{n}{<} t$ & $\forall y < t\, \exists z < t\, (y < z \in Z_\alpha$)).

(f) $\alpha \in \Omega_1 \Rightarrow \forall t \in T_\alpha\, \forall X \in W_\alpha\, \exists n < \omega\, \exists x \overset{n}{<} t[\exists y \in T\lceil\alpha\, (x \overset{n}{<} y \in X) \Rightarrow$ $\Rightarrow \exists y < t\, (x \overset{n}{<} y \in X)]$.

Definition of T_α. I. $T_0 = \{0\}$.

II. $T_{\beta+1} = \{x \cup \{\langle n, \beta\rangle\} : x \in T_\beta,\ n \in \omega\}$.

III. Let $\alpha = \omega\beta + \omega \in \Omega_0$. Take a bijective function $f\colon T\lceil(\alpha \setminus \omega\beta) \times \omega \to \omega$. We preliminarily define $t_\alpha(x, n) \in \mathbb{T}_\alpha$ for each $x \in T\lceil(\alpha \setminus \omega\beta)$ and each $n \in \omega$ by induction on $f(x, n)$. Let $x \in T\lceil(\alpha \setminus \omega\beta)$ and $n \in \omega$. Take $\widetilde{x} \in T_{ht(x)+1}$ so that $\widetilde{x} > x$ and $\widetilde{x} \not< t_\alpha(x', n')$ for any x', n' with $f(x', n') < f(x, n)$. Put $t_0 = \widetilde{x}$. Take $t_{k+1} \in T\lceil\alpha$ such that $t_{k+1} > t_k$. Put $t_\alpha(x, n) = \bigcup_{k<\omega} t_k$. We set:

$$T_\alpha = \{t_\alpha(x, n) : x \in T\lceil(\alpha \setminus \omega\beta),\ n < \omega\},$$

$$A_n \cap T_\alpha = \{t_\alpha(x, n) : x \in T\lceil(\alpha \setminus \omega\beta)\} \quad \text{for } n < \omega.$$

Clearly $x \overset{n}{<} t_\alpha(x, n) \in A_n$.

IV. Let $\alpha \in \Omega_1$. Let $\langle X_k : k < \omega \rangle$ enumerate W_α. We first define $u_\alpha(x, n) \in T_\alpha$ for $x \in T\restriction\alpha$ and $n < \omega$ preliminarily. Let $x \in T\restriction\alpha$ and $n \in \omega$. Take a monotone sequence $\langle \alpha_n : n < \omega \rangle$ such that $\alpha_0 = ht(x)$ and $\lim_{n<\omega} \alpha_n = \alpha$. Inductively take u_k and $\tilde{u}_k$, for $k < \omega$, as follows: Put $u_0 = x$. Take $\tilde{u}_k \in T\restriction\alpha$ so that $u_k \overset{k+n}{<} \tilde{u}_k$ and $[\exists y \in T\restriction\alpha\ (u_k \overset{k+n}{<} y \in X_k) \Rightarrow \tilde{u}_k \in X_k]$. Take $u_{k+1} \in T\restriction\alpha$ so that $\tilde{u}_k \overset{k+n}{<} u_{k+1}$ & $ht(u_{k+1}) > \alpha_k$. Now we set $u_\alpha(x, n) = \bigcup_{k<\omega} u_k$. By $\tilde{u}_k(x, n)$ and $u_k(x, n)$ we will refer respectively to $\tilde{u}_k$ and u_k in the above. Now we define T_α for $\alpha \in \Omega_1$. There are three cases to consider.

Case 1. $\hat{Z}_\alpha = Z_\alpha$ & $\forall x \in Z_\alpha \forall n < \omega \exists y \in T\restriction\alpha\ (x \overset{n}{<} y \notin Z_\alpha)$. With each $x \in T\restriction\alpha$ and each $n < \omega$, associate $\tilde{x}$ such that $x \overset{n}{\leqslant} \tilde{x} \notin Z_\alpha$. We set:

$$T_\alpha = \{u_\alpha(\tilde{x}, n) : \tilde{x} \in T\restriction\alpha,\ n < \omega\}.$$

Conditions (a) - (f) are easily checked by observing the following: (1) $\tilde{x} < u_\alpha(\tilde{x}, n)$ & $\tilde{x} \notin Z_\alpha$; (2) $x \overset{n}{<} u_\alpha(\tilde{x}, n)$; (3) $k + n > i \Rightarrow \tilde{u}_k(\tilde{x}, n) \overset{i}{<} u_\alpha(\tilde{x}, n)$; (4) $\exists y \in T\restriction\alpha\ (u_k(\tilde{x}, n) \overset{k+n}{<} y \in X_k) \Rightarrow \tilde{u}_k(\tilde{x}, n) \in X_k$; (5) $u_k(\tilde{x}, n) \overset{k+n}{<} \tilde{u}_k(\tilde{x}, n) \overset{k+n}{<} u_\alpha(\tilde{x}, n)$.

Case 2. $0 \notin Z_\alpha$ and there are $n \in \omega$ and $x \in T\restriction\alpha$ such that $e_\alpha = \{\langle n, x \rangle\}$ and Z_α is a nbd of $\{y \in T\restriction\alpha : x \overset{n}{<} y \in A_n\}$. Take such $n \in \omega$ and $x \in T\restriction\alpha$. Let $\{\alpha_k : k < \omega\}$ be a monotone sequence such that $\alpha_0 = ht(x)$ & $\lim_{k<\omega} \alpha_k = \alpha$. Inductively pick s_k, v_k, a_k and z_k for each $k < \omega$ as follows: Put $s_0 = x$. Pick v_k so that $\alpha_k \leqslant ht(v_k)$ and $s_k \overset{n+k}{\leqslant} v_k$. (It is possible by (a).) Pick a_k so that $v_k < a_k \in A_n$ & $ht(a_k) = ht(v_k) + \omega$. (It is possible by (b).) Notice here that $x \overset{n}{<} v_k$ (which is shown by induction) and $v_k \overset{n}{<} a_k$ (trivially) and hence $x \overset{n}{<} a_k$ which implies $a_k \in Z_\alpha$. Pick z_k so that $\emptyset \neq (z_k, a_k] \subseteq (v_k, a_k] \cap Z_\alpha$. $v_k \overset{n+k}{\leqslant} z_k \in Z_\alpha$ holds trivially. Take s_{k+1} such that $z_k \overset{n+k}{<} s_{k+1}$ & $[\exists y \in T\restriction\alpha\ (z_k \overset{n+k}{<} y \in X_k) \Rightarrow s_{k+1} \overset{n+k}{<} X_k]$. Now put $s_\alpha = \bigcup_{k<\omega} s_k$. We set: $T_\alpha = \{s_\alpha\} \cup \{u_\alpha(x, n) : x \in T\restriction\alpha,\ n < \omega\}$.

Conditions (a) - (f) are easily checked by the following: (1) $k + n > i \Rightarrow$ $\Rightarrow s_k \overset{i}{<} s_\alpha$ since $s_k \overset{k+n}{<} v_k \overset{k+n}{\leqslant} z_k \overset{k+n}{<} s_{k+1}$; (2) $\hat{Z}_\alpha \neq Z_\alpha$ as $0 \notin Z_\alpha$; (3) $x \overset{n}{<} s_\alpha$ & $\forall y < s_\alpha \exists k\ (y < s_k < z_k \in Z_\alpha)$; (4) $z_k \overset{k+n}{<} s_\alpha$ & $[\exists y \in T\restriction\alpha\ (z_k \overset{k+n}{<} y \in$

$\in X_k) \Rightarrow s_{k+1} < s_\alpha$ & $z_k \overset{n+k}{<} s_{k+1} \in X_k]$.

Case 3. Otherwise. We set:

$$T_\alpha = \{u_\alpha(x, n) : x \in T\lceil\alpha,\ n < \omega\}.$$

T_α is thus defined for each $\alpha < \omega_1$.

We set $T = \bigcup_{\alpha<\omega_1} T_\alpha$, which is an ω_1-subtree of $\mathbb{T}$ obviously.

Claim 1. If X is an uncountable subset of T, then

$$\exists x \in \hat{X}\, \exists n < \omega\ \{y \in T : x \overset{n}{<} y\} \subset \hat{X}.$$

Proof. To the contrary, suppose $\forall x \in \hat{X}\, \forall n < \omega\ \{y \in T : x \overset{n}{<} y\} \not\subset \hat{X}$. Put $C = \{\alpha \in \Omega_1 : \forall x \in \hat{X}\lceil\alpha\ \forall n < \omega\ \{y \in T\lceil\alpha : x \overset{n}{<} y\} \not\subset \hat{X}\} \in \mathcal{L}$ and $E = \{\alpha \in \omega_1 : Z_\alpha = \hat{X} \cap T\lceil\alpha\} \in \mathcal{T}$. Pick $\alpha \in C \cap E$. Then, by (d), $\forall u \in T_\alpha\, \exists y < u\ (y \notin Z_\alpha = \hat{X} \cap T\lceil\alpha)$. Hence $T_\alpha \cap \hat{X} = \emptyset$ which implies $\hat{X} \subset T\lceil\alpha$, contradicting to $|X| > \aleph_0$. The claim is thus proved.

$T \in AT$ follows immediately from this claim.

Claim 2. $T \in AST$.

Proof. Suppose A is an antichain of T. Put:

$C = \{\alpha \in \Omega_1 : \forall x \in T\lceil\alpha\ \forall n < \omega[\exists y \in A\ (x \overset{n}{<} y) \Rightarrow \exists y \in A \cap T\lceil\alpha\ (x \overset{n}{<} y)]\}$.

Let D be a club subset of $\{\alpha : A \cap T\lceil\alpha \in W_\alpha\}$. We show $A \cap T\lceil(C \cap D) = \emptyset$. To the contrary, suppose that there are α and t such that $\alpha \in C \cap D$ and $t \in A \cap T_\alpha$. By (f), we can take $n \in \omega$ and $x \overset{n}{<} t$ such that $\exists y \in T\lceil\alpha\ (x \overset{n}{<} y \in A \cap T\lceil\alpha) \Rightarrow \exists y < t\ (x \overset{n}{<} y \in A \cap T\lceil\alpha)$. As $x \overset{n}{<} t \in A$, by $\alpha \in C$, we obtain $y \in A \cap T\lceil\alpha$ such that $x \overset{n}{<} y$. Hence, by choice of n and x, $\exists y < t\ (y \in A)$, which is absurd since A is an anti-chain. Claim 2 is thus proved.

Claim 3. $\forall \tilde{T} \lhd T\ (\tilde{T} \notin \gamma ST)$.

Proof. Suppose $\tilde{T} \lhd T$. It suffices to prove that $\tilde{T}$ is not normal w.r.t. tree topology. By Claim 1, we have $n < \omega$ and $x \in \tilde{T}$ such that $\{y \in T : x \overset{n}{<} y\} \subset \tilde{T}$. Put $X = \{y \in T : x \overset{n}{<} y \in A_n\}$. X is closed in $\tilde{T}$. (For, suppose $y \in \tilde{T}$ and $y \notin X$. If $ht(y) \in \Omega \setminus \Omega_1$, then clearly $\exists z < y\ ((z, y] \cap X = \emptyset)$. Let $y \in T_\alpha$ with $\alpha \in \Omega_1$. By (c), for some $z < y$, $z \overset{n+1}{<} y$, which implies $(z,y] \cap X = \emptyset$.) $\tilde{T}\lceil\Omega_1$ is closed in $\tilde{T}$ clearly. Also clear is $X \cap \tilde{T}\lceil\Omega_1 = \emptyset$. Suppose that U and V are open subsets of $\tilde{T}$ such that $X \subseteq U$ and $\tilde{T}\lceil\Omega_1 \subseteq V$. We show $U \cap V \neq \emptyset$. We may assume $0 \notin U$. Put $E = \{\alpha \in \omega_1 : \{\langle n, x\rangle\} = e_\alpha$ & $U \cap T\restriction\alpha = Z_\alpha\} \in \mathcal{V}$. Pick $\alpha \in E$ such that $\alpha > ht(x)$ & $\alpha \in \Omega_1$. Then, by (e), there is $t \in T_\alpha$ such that $x \overset{n}{<} t$ and $\forall y < t \, \exists z < t \, (y < z \in Z_\alpha)$. By choice of n and x, $t \in \widetilde{T}$. Hence $t \in \widetilde{T}\restriction\Omega_1$ $(\subset V)$. Since V is open, there is $y < t$ such that $(y, t] \subset V$. By the original property of t, we obtain $z < t$ such that $y < z \in Z_\alpha$, which implies $z \in (y, t] \cap U$ as $\alpha \in E$. Hence $z \in U \cap V$. Claim 3 is thus proved.

This completes the proof of Theorem 6.(4).

§4. Appendix.

This section is divided into three subsections. In 4.1, we remark on non-Souslin trees. In 4.2, we prove $RE \cap \gamma ST \neq \emptyset$ under $\diamondsuit^*$. In 4.3, we would like to write shortly a recent topics concerning our interest.

4.1. Obviously a non-Souslin tree does not contain a Souslin subtree. But conversely, any Aronszajn tree that is not non-Souslin does contain a Souslin subtree in the sense which may be recognized in the proof of the following:

Proposition 7. $AT \setminus NS \neq \emptyset \implies ST \neq \emptyset$.

Proof. Suppose $T \in AT \setminus NS$. Then there is an uncountable subset X of T such that X contains no uncountable antichain. We may assume $\widehat{X} = X$. (For, if A is an uncountable antichain contained in $\widehat{X}$, then for any choice function $f : \mathbb{P}(\widehat{X}) \setminus \{0\} \to \widehat{X}$, the set $\{f(\{x \in X : a \leqslant x\}) : a \in A\}$ is an uncountable antichain contained in X.) We can take $x \in X$ such that $\forall z \in X \, (x \leqslant z \implies \forall \alpha < \omega_1 \, \exists y \in X \, (z < y \,\&\, ht(y) > \alpha))$. (For, suppose not. Then we can inductively take an element $x_\alpha \in X$ for $\alpha < \omega_1$ so that $\forall y \in X \, (x_\beta < y \implies ht(y) < ht(x_\alpha))$ if $\beta < \alpha < \omega_1$. Clearly $\{x_\alpha : \alpha < \omega_1\}$ is an uncountable antichain contained in X. A contradiction.) For that x, $\forall y \in X \, (x < y \implies \exists \alpha > ht(y) \, (|\{z \in X : y < z\} \cap T_\alpha| \geqslant 2))$, since $\{z \in X : y < z\}$ is unbounded in T and is not a branch as $T \in AT$. Put $\widetilde{T} = \{y \in X : x \leqslant y\}$ and $C = \{\alpha \in \Omega : \forall y \in T\restriction\alpha \cap \widetilde{T} \; \exists w, z \in T\restriction\alpha \cap \widetilde{T} \, (w, z > y \,\&\, w \nleqslant z \,\&\, z \nleqslant w)\} \in \mathcal{L}$. Then clearly $\widetilde{T}\restriction C$ is an ω_1-tree w.r.t. the ordering of T and is Souslin.

4.2. We show in sketched form that $\diamondsuit^*$ implies $RE \cap \gamma ST \neq \emptyset$. Assuming $\diamondsuit^*$, take $\langle W_\alpha : \alpha < \omega_1 \rangle$ that is given in Lemma 4.

Lemma 8. Let $T \leqslant \mathbb{T}$. Suppose that a function f embeds T into $\mathbb{R}$. Then $T \in \gamma ST$, if the following holds for all $\alpha \in \Omega$:

(1) If X is an antichain of $T\lceil\alpha$ such that $X \in W_\alpha$, then for all $x \in T_\alpha$, there exist a $y < x$ and a $q \in \mathbb{Q}$ with $q > 0$ such that

$$f(x) < f(y) + q \quad \& \quad \forall z \in X \; (y < z \Rightarrow f(y) + 2q \leqslant f(z)).$$

We refer the reader to [9] for the proof of this lemma.

Construction of $T \in RE \cap \gamma ST$ and $f: T \to \mathbb{R}$. We define $T_\alpha \subseteq \mathbb{T}_\alpha$ and $f\lceil T_\alpha$ by induction on α. As we define them, we ensure the following conditions:

(a) $\emptyset \neq T_\alpha$ & $|T_\alpha| \leqslant \aleph_0$ & $\forall x \in T_\alpha \forall y < x \; (y \in T\lceil\alpha)$,

(b) $x < y \in T_\alpha \Rightarrow f(x) < f(y)$,

(c) $x \in T\lceil\alpha$ & $q > f(x) \Rightarrow \exists y \in T_\alpha \; (y > x$ & $f(y) < q)$.

I. $T_0 = \{\emptyset\}$, $f(\emptyset) = 0$.

II. $T_{\beta+1} = \{x^\frown\langle n\rangle : x \in T_\beta, \; n \in \omega\}$, $f(x^\frown\langle n\rangle) = f(x) + q_n$, where $\langle q_n : n \in \omega\rangle$ enumerates $\mathbb{Q}^+$.

III. $\alpha \in \Omega$. Let $\langle X_k : k \in \omega\rangle$ enumerate W_α. Let $x \in T\lceil\alpha$ and $q \in \mathbb{Q}$ with $q > f(x)$. Let $\alpha_0 < \alpha_1 < \dots < \alpha_n < \dots$ converge to α. Define $\{x_n : n \in \omega\} \subseteq T\lceil\alpha$ and $\{q_n : n \in \omega\}$ as follows: (1) $x_0 = x$ and $q_0 = q$;

(2) If there is $y \in X_k$ such that $x_k < y$ and $f(y) < q_k$, then take $x_{k+1} \in T\lceil\alpha$ so that $y < x_{k+1}$ & $f(x_{k+1}) < q_k$ & $\alpha_k < ht(x_{k+1})$, and put $q_{k+1} = q_k$; Otherwise, put $q_{k+1} = (f(x_k) + q_k)/2$ and take $x_{k+1} \in T\lceil\alpha$ so that $x_k < x_{k+1}$ & $f(x_{k+1}) < q_{k+1}$ & $\alpha_k < ht(x_{k+1})$.

Put $x^* = \bigcup\{x_n : n \in \omega\}$, $T_\alpha = \{x^* : x \in T\lceil\alpha\}$ and $f(x^*) = \sup\{f(x_n) : n \in \omega\}$.

The tree thus defined satisfies the condition in Lemma 8 and so is as required.

4.3. One of the main interests to Aronszajn trees is a topological character of them. For topological terms and knowledge, we refer the reader to Rudin [11]. Devlin and Shelah investigate the tree topology in [4] and characterize some topological properties as we described in Introduction. Their work is closely connected with the normal Moore space conjecture (NMC) which asserts that a normal Moore space is metrizable. Put NT = the family of all normal trees. The following are the best so far concerning NMC as far as I know:

(a) MA + $\neg$CH $\Rightarrow$ SAT $\subseteq$ NT (Fleissner [6]); this means that MA + $\neg$CH refutes NMC.

(b) CH (or even $2^{\aleph_0} < 2^{\aleph_1}$) $\Rightarrow$ SAT $\cap$ NT = $\emptyset$ (Devlin and Shelah[5]).

(c) However, NMC can not be proved in ZFC + GCH (Shelah, see [5]).

On the direction of the consistency of NMC, P. Nyikos has proved that:

(d) Cons(ZFC + $\exists\kappa$: a strong compact cardinal) $\Rightarrow$ Cons(ZFC + NMC).

It is of course an interesting open question, whether the assumption of this can be weakened. Meanwhile, since MA + $\neg$CH implies $2^{\aleph_0} = 2^{\aleph_1}$, it may be reasonable to ask the following:

(e) Does $2^{\aleph_0} = 2^{\aleph_1}$ imply SAT $\subseteq$ NT ?

Also interesting is:

(f) Is it possible, that there is a pair of two special Aronszajn trees such that one is normal and the other is not normal ?

(e) and (f) are not both true of course. Recently Fleissner [7] refined a result of [4] and besides proved the following:

(g) ($\diamondsuit^+$) There is an Aronszajn tree which is not countably metacompact,

(h) No tree can be a Dowker space.

He raised also a question there, that is the following:

(h) Suppose the tree topology on T is such that every closed subset of T is G_δ. Must T be a special Aronszajn tree ?

(We would like to remark here that $\diamondsuit$ suffices for the assertion (g).) On Souslin trees, many problems has already been solved. As far as I know, there are no new results on Souslin trees after Devlin and Johnsbråten [3]. However, it seems to remain still not a few interests to them; e.g. the position of the hypothesis of the existence of Souslin trees seems still unclear among other hypotheses independent of ZFC.

REFERENCES

[1] J. Baumgartner, Decompositions and embeddings of trees, Notices A.M.S. 17 (1970), 967

[2] K. J. Devlin, Note on a theorem of J. Baumgartner, Fund. Math. 76 (1972) 255-260.

[3] K. J. Devlin and H. Johnsbråten, The Souslin problem, Lecture Notes in Mathematics 405 (Springer, Berlin, 1974).

[4] K. J. Devlin and S. Shelah, Souslin properties and tree topologies, Proc. London Math. Soc. 39 (1979), 237-252.

[5] K. J. Devlin and S. Shelah, A note on the normal Moore space conjecture, Canad. J. Math. 31 (1979) 241-251.

[6] W. G. Fleissner, When is Jones' space normal?, Proc. A.M.S. 50 (1975) 375-378.

[7] W. G. Fleissner, Remarks on Souslin properties and tree topologies, Proc. A.M.S. 80 (1980) 320-326.

[8] M. Hanazawa, On a classification of Aronszajn trees, Tsukuba J. Math. (to appear).

[9] M. Hanazawa, On a classification of Aronszajn trees II, (to appear).

[10] E. W. Miller, A note on Souslin's problem, Amer. J. Math. 65 (1943) 673-678.

[11] M. E. Rudin, Lectures on set-theoretic topology, American Mathematical Society Regional Conference Series in Mathematics 23 (Providence, R.I., 1975).

Department of mathematics,
Faculty of science, Saitama university,
Urawa, 338 Japan

ON SET THEORIES IN TOPOSES

Susumu HAYASHI
Department of mathematics, University of Tsukuba
Sakura-mura, Ibaraki, Japan

In this note, we introduce Kripke-Joyal semantics for first order set theory in any elementary topos. It is a generalization of the models of set-objects in well-pointed toposes in Osius [7]. Osius's model is constructed with all set-objects, however, our semantics is defined for each preuniverse which is an appropriate subclass of transitive set objects. For any Grothendieck topos E, we will construct a complete preuniverse, which is called the von Neumann universe in E and the intuitionistic ZF-set theory is valid in it in the sense of our semantics. We will show that the von Neumann universe in the Grothendieck topos over a small complete Heyting algebra Ω is the Heyting-valued model of Scott-Solovay for Ω. (Cf.[3,4].)

After finishing this work, the auther learned from Dr. R. J. Grayson that the cumulative hierarchy, i.e. von Neumann universe, in sheaves are known already. Fourman [1] defined an interpretation of the intuitionistic ZF-set theory allowing urelements for any locally small cocomplete elementary topos. (He assumed also the existence of small limits, however, it will be superfluous.) He showed that Scott-Solovay's Boolean valued models, Mostowski-Fraenkel's permutation models and Cohen-Scott's symmetric extensions are particular cases of his interpretation. We explain the relation between our semantics and his interpretation. Firstly, it is easy to generalize our semantics to the set theory allowing urelements. Secondly, it is easy to check that all of our results for Grothendieck toposes are valid even for locally small cocomplete toposes without any change. After these preparations, it is easy to see that Fourman's interpretation may be identified with our semantics for the generalized von Neumann universe allowing urelements in locally small cocomplete toposes.

Let E be an elementary topos. We use the terminologies and notations of Johnstone [5].

Def. 1. Let E_{tr} be the preordered class of transitive objects of E (cf. [5]). A nonempty subclass U of E_{tr} is a preuniverse of E_{tr} iff

(i) $\forall A,B\in U\ \exists C\in U(A\cup B\subset C)$,

(ii) $\forall A\in U\ \exists B\in U(PA\subset B)$.

A preuniverse is a universe iff

(iii) $\forall S \subset U(S \text{ is small} \longrightarrow \exists A \in U(\forall B \in S(B \subset A)))$.

We now introduce Kripke-Joyal semantics for each preuniverse.

Def. 2. Let E be an elementary topos. Let $X \in Ob(E)$, $(A,r) \in E_{tr}$. We write $X \xrightarrow{a} (A,r)$ for the pair $(X \xrightarrow{a} A, r)$. Let U be a preuniverse of E_{tr}. We set

$$\Gamma(X,U) = \{X \xrightarrow{a} (A,r) : (A,r) \in U\},$$

and the elements of $\Gamma(X,U)$ are called sections of U over X or (generalized) elements of U at stage X. Let $\phi(x_1,\ldots,x_n)$ be a formula of the first order ZF-set theory with its free variables among $x_1,\ldots,x_n$. Let $a_1,\ldots,a_n$ be a sequence of sections of U over X. Then we define $U \models_X \phi(a_1,\ldots,a_n)$ by induction on the complexity of ϕ, and read it '$\phi(a_1,\ldots,a_n)$ is valid over X (or at stage X)'.

(i) $U \models_X a_1 = a_2$ iff $i_1 a_1 = i_2 a_2$, where a_1, a_2 are $X \xrightarrow{a_1} (A_1,r_1)$, $X \xrightarrow{a_2} (A_2,r_2)$ and i_1, i_2 are inclusions from A_1, A_2 to $A_1 \cup A_2$, respectively,

(ii) $U \models_X a_1 \in a_2$ iff

$$X \xrightarrow{\langle i_1 a_1, i_2 a_2 \rangle} A_1 \cup A_2 \times A_1 \cup A_2 \xrightarrow{\langle id \times r_1 \cup r_2 \rangle} A_1 \cup A_2 \times P(A_1 \cup A_2) \xrightarrow{ev} \Omega = true_X,$$

(iii) the definitions for $\vee$, $\wedge$, $\Rightarrow$ are the same as the usual Kripke-Joyal semantics (see [9]),

(iv) $U \models_X \forall y \phi(a_1,\ldots,a_n)$ iff for any $t: Y \to X$, $b \in \Gamma(Y,U)$,
$U \models_X \phi(a_1 t,\ldots,a_n t, b)$,

(v) $U \models_X \exists y \phi(a_1,\ldots,a_n)$ iff there exists $t: Y \to X$, $b \in \Gamma(Y,U)$ such that $U \models_X \phi(a_1 t,\ldots,a_n t, b)$.

Lemma 1. For any $t: Y \to X$, if $U \models_X \phi(a_1,\ldots,a_n)$ then $U \models_X \phi(a_1 t,\ldots,a_n t)$.

Lemma 2. For any epimorphic family $\{X_i \xrightarrow{t_i} X\}_{i \in I}$, if $U \models_X \phi(a_1 t_i,\ldots,a_n t_i)$ for any i, then $U \models_X \phi(a_1,\ldots,a_n)$.

Lemma 3. Let $X \xrightarrow{a} (A,r) = X \xrightarrow{b} (B,s) \xrightarrow{i} (A,r)$ and let i be an inclusion. Then $U \models_X \phi(a_1,\ldots,a_n,a)$ iff $U \models_X \phi(a_1,\ldots,a_n,b)$.

Lemma 4. (a) Let a_i be $X \to (A_i,r_i)$. Then, $U \models_X \forall x(x \in a_1 \Rightarrow \phi(x,a_1,\ldots,a_n))$ iff for any $Y \xrightarrow{t} X$, $Y \xrightarrow{b} (A_1,r_1)$, if $U \models_X b \in a_1 t$ then $U \models_X \phi(b,a_1 t,\ldots,a_n t)$.
(b) $U \models_X \exists x(x \in a_1 \wedge \phi(x,a_1,\ldots,a_n)$ iff there exists $Y \xrightarrow{t} X$, $Y \xrightarrow{b} (A_1,r_1)$ such that

$U \models_X b\epsilon a_1 t$ and $U \models_X \phi(b,a_1t,\ldots,a_nt)$.

Lemma 5 (Corollary of Lemma 4). Let a_i be $X \to (A_i,r_i)$ and let $\phi(x_1,\ldots,x_n)$ be a bounded formula, i.e. formula has only bounded quantifiers. We define a formula $\tilde{\phi}(x_1,\ldots,x_n)$ of Mitchell-Bénabou language (see [5]) as follows: the type of any free or bound variables of $\tilde{\phi}$ is $A=A_1\cup\ldots\cup A_n$ and ; if ϕ is $x_i \epsilon x_j$, $x_i = x_j$ then $\tilde{\phi}$ is $x_i \epsilon_A r(x_j)$, $x_i =_A x_j$, where $r=r_1\cup\ldots\cup r_n$, and if ϕ is $\phi_1 \circ \phi_2$ then $\tilde{\phi}$ is $\tilde{\phi}_1 \circ \tilde{\phi}_2$, where $\circ$ is $\wedge$, $\vee$ or $\Rightarrow$, if ϕ is $\forall y \epsilon x_j \phi_1(y)$ then $\tilde{\phi}$ is $\forall y (y \epsilon_A r(x_j) \Rightarrow \tilde{\phi}_1(y))$, and similarly for existential quantifier. Let $\tilde{a}_i$ be $X \xrightarrow{a_i} A_i \xrightarrow{I_i} A$, where I_i is the inclusion. Then

$$U \models_X \phi(a_1,\ldots,a_n) \quad \text{iff} \quad \models_X \tilde{\phi}(\tilde{a}_1,\ldots,\tilde{a}_n),$$

where the right hand side is the usual Kripke-Joyal semantics for Mitchell-Bénabou language.

Lemma 6. Let E be an elementary topos whose subobject lattices are small complete Heyting algebras. Let $a_i \epsilon \Gamma(X,U)$. Set

$$S = \sup\{Y \overset{s}{\rightarrowtail} X;\ U \models_Y \phi(a_1 s,\ldots,a_n s)\}.$$

Then for any $Z \xrightarrow{t} X$, $U \models_Z \phi(a_1t,\ldots,a_nt)$ iff $\mathrm{Im}(t) \le S$ in $\mathrm{Sub}(X)$.

Lemma 7. Let E be a Grothendieck topos, and let $C=\{(A_i,r_i)\}_{i\epsilon I}$ be a small full and filtered subcategory of E_{tr}. Set $A_\infty = \mathrm{Colim}(C)$. Let $I_i : A_i \to A_\infty$ be the cannonical injections. Let

$$B_i \overset{e_i}{\rightarrowtail} A_i \times A_i \underset{\mathrm{true}_{A_i\times A_i}}{\overset{ev(id\times r_i)}{\rightrightarrows}} \Omega$$

be an equalizer diagram. Let $E \overset{e_\infty}{\rightarrowtail} A_\infty \times A_\infty$ be the supremum of the subobjects $\{I_i\times I_i e_i\}_{i\epsilon I}$ of $A_\infty \times A_\infty$. Then there exists $A_\infty \xrightarrow{r_\infty} P(A_\infty)$ such that

$$\begin{array}{ccc} E & \overset{e_\infty}{\rightarrowtail} & A_\infty \times A_\infty \\ \downarrow & & \downarrow{\scriptstyle id\times r_\infty} \\ S & \overset{f}{\rightarrowtail} & A_\infty \times P(A_\infty) \end{array}$$

is a pullback diagram, where f is the monomorphism classified by the evaluation $A_\infty \times P(A_\infty) \to \Omega$. Then (A_∞, r_∞) is a transitive object of E, and (A_∞, r_∞) is the supremum of $\{(A_i,r_i)\}_{i\epsilon I}$ in E_{tr}.

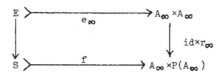

Def. 3. We define the intuitionistic set theory IZ to be the first order

intuitionistic theory in the language of ZF-set theory with the following axioms:

(i) (extensionality) $\forall x(x\in u \Leftrightarrow x\in v) \Rightarrow u=v$,

(ii) (empty) $\exists x\forall y(\neg y\in x)$,

(iii) (unordered pair) $\exists x(u\in x \wedge v\in x)$,

(iv) (power set) $\exists x\forall y(\forall z(z\in y \Rightarrow z\in u) \Rightarrow y\in x)$,

(v) (bounded separation) for any bounded formula ϕ,

$$\exists x\forall y(y\in x \Leftrightarrow (y\in u \wedge \phi(y))),$$

(vi) (transitive closure) $\exists x(u\in x \wedge \forall yz(y\in z\in x \Rightarrow y\in x))$,

(vii) (regularity) $\forall x\in u(x\cap u\subset v \Rightarrow x\in v) \Rightarrow u\subset v$.

Note that the axiom of union is a consequence of (vi). The intuitionistic ZF-set theory IZF is obtained by adding the following three axioms to IZ:

(viii) (infinity) $\exists x(\emptyset\in x \wedge \forall y(y\in x \Rightarrow y\cup\{y\}\in x))$,

(ix) (separation) $\exists x\forall y(y\in x \Leftrightarrow (y\in u \wedge \phi(y)))$,

(x) (collection) $\forall x\in u\exists y\ \phi(x,y) \Rightarrow \exists z\forall x\in u\exists y\in z\ \phi(x,y)$.

Prop. 1. Let E be an elementary topos, and let U be a preuniverse. Then IZ is valid under $U \models_X$, i.e. if ϕ is a theorem in IZ and $a_1,\ldots,a_n\in\Gamma(X,U)$, then $U \models_X \phi(a_1,\ldots,a_n)$. If E has a natural number object, then the axiom of infinity is valid, too. If E is Boolean, the axiom of excluded middle is valid for any <u>bounded</u> formula.

Prop. 2. If E is a well-powered elementary topos and U is a universe, then the axiom of collection is valid. If E is an elementary topos whose subobject lattices are complete Heyting algebras and U is a preuniverse, then the axiom of separation is valid, furthermore, if E is Boolean then the axiom of excluded middle is valid. Thence, if E is a Grothendieck topos and U is a universe, then the intuitionistic ZF-set theory IZF is valid, furthermore, if E is Boolean then ZF- set theory is valid.

Proofs of Propostions 1, 2.

(A) The validity of axioms and inference rules of the Heyting calculas is a trivial consequence of Lemmas 1, 2. The additional validity excluded middle in Boolean cases is a consequence of Lemma 5, 6.

(B) The validity of set theoretic axioms.

(i) (extensionality) This axiom is an immediate consequence of the extensionality principle for the usual Kripke-Joyal semantics (4.15 in [8]) by Lemma 5.

(ii) (power set, etc.) Let $a:X \to (A,r)$ be a section of U. Let $f:X \to PPA$ be the exponential adjoint of $true_{X\times PA}$. Take an inclusion $g:PP(A,r) \rightarrowtail (B,s)$ such that $(B,s)\in U$. Then $U \models_X \forall x(x\subset a \Rightarrow x\in gf)$. Similarly for the axioms (ii), (iii), (vi) in Def. 3.

(iii) (bounded separation) Let $\phi(x_1,\ldots,x_n,y)$ be a bounded formula, and let ψ be the formula $\phi \wedge y\in z$. Define a formula $\tilde{\psi}$ of Mitchell-Bénabou language as in Lemma 5 such that the only type of the variables of $\tilde{\psi}$ is $A_1\cup\ldots\cup A_n\cup B$. Set

$$f_i = B\times X \xrightarrow{pr_2} X \xrightarrow{a_i} A_i \xrightarrow{I_i} A_1\cup\ldots A_n\cup B,$$
$$g = B\times X \xrightarrow{pr_1} B \xrightarrow{J} A_1\cup\ldots A_n\cup B,$$
$$h = B\times X \xrightarrow{pr_2} X \xrightarrow{b} B \xrightarrow{K} A_1\cup\ldots\cup A_n\cup B,$$

where I_i, J and K are inclusions. We recognize f_i, g and h as function symbols of Mitchell-Bénabou language. Let $k:B\times X \to \Omega$ be the internal interpretation of $\tilde{\psi}(f_1(x),\ldots,f_n(x),g(x),h(x))$. Let $m:X \to PB$ be the exponential adjoint of k. Take an inclusion $n:(PB,\exists s) \to (C,t)$ such that $(C,t)\in U$. Then, by Lemmas 3, 4, 5,

$$U \models_X \forall y(y\in nm \Leftrightarrow (\ \phi(a_1,\ldots,a_n,y) \wedge y\in b\ )).$$

(iv) (regularity) By Lemma 4 and the definition of transitive objects.

(ix) (separation) Let E be an elementary topos whose subobject lattices are small complete Heyting algebras, and let U be a preuniverse of it. Let $a_i:X \to (A_i,r_i)$ $(i=1,\ldots,n)$, $b:X \to (B,s)$ be sections of U. Set

$$S = \{Y \xrightarrow{<x,t>} B\times X\ :\ U \models_X \phi(a_1t,\ldots,a_nt,x) \wedge x\in bt\}.$$

Set $S^* = \{Im(f) : f\in S\}$. Let $m:Sup(S^*) \rightarrowtail B\times X$ be the supremum of S^*. Let $f:X \to PB$ be the exponential adjoint of the classifing map of the monomorphism m. Take an inclusion $g:(PB,\exists s) \to (C,u)$ such that $(C,u)\in U$. Then

$$U \models_X \forall x(x\in gf \Leftrightarrow \phi(a_1,\ldots,a_n,x) \wedge x\in b).$$

(vi) (collection) Let E be a well-powered elementary topos, and let U be a universe. Let $a_i:X \to (A_i,r_i)$ $(i=1,\ldots,n)$ be sections of U. Assume that $U \models_X \forall x\in a\ \exists y\ \phi(a_1,\ldots,a_n,x,y)$. Set

$$S = \{Y \xrightarrow{<x,t>} A\times X\ :\ U \models_X x\in at\}\ .$$

Set $S^* = \{Im(f) : f\in S\}$. Since E is well-powered, S^* is small. Set $S^* = \{<x_i,t_i>: Z_i \rightarrowtail A\times X\}_{i\in I}$ such that I is small. By Lemma 1, 2, $U \models_{Z_i} x_i\in at_i$. Hence

$$U \models_{Z_i} \exists y\phi(a\ t_i,\ldots,a_nt_i,x_i,y).$$

Thence, for each i there exists $w_i:W_i \to Z_i$, $b_i:W_i \to (B_i,s_i)$ such that

$$U \models_X \phi(a_1t_iw_i,\ldots,a_nt_iw_i,x_iw_i,b_i).$$

Since U is a universe, there exists an upper bound $(C,u)\in U$ of $\{(B_i,r_i)\}_{i\in I}$. Let $f:X\to PC$ be the exponential adjoint of $\mathrm{true}_{X\times C}$. Take an inclusion $g:(PC,\exists u)\to(D,v)$ such that $(D,v)\in U$. Then

$$U\models_X \forall x\in a\,\exists y\in gf\ \phi(a_1,\ldots,a_n,x,y).$$

Example. (i) Let E be an elementary topos. Then E_{tr} is a preuniverse. Let U_f be $\{(0,0),\ P(0,0),\ PP(0,0),\ldots\}$. Then U_f is the preuniverse of finite sets. Assume that E has a natural number object N, and let (N,r) be the corresponding transitive object. Then we define the Zermelo preuniverse $U_z=\{(N,r),\ P(N,r),\ PP(N,r),\ldots\}$. By Prop. 2, $U_z\models_X$ IZ+(infinity). (ii) Let E be a Grothendieck topos. By Lemma 7, we can construct the von Neumann universe U_n in E:

$$U_n=\{V_0,V_1,\ldots,V_\alpha,\ldots\}\qquad(\alpha\in On),$$

where

$$V_0=(0,0),\qquad V_{\alpha+1}=PV_\alpha,$$

$$V_\alpha=\sup_{\beta<\alpha}V_\beta\quad(\ \alpha\text{ is a limit number }).$$

Let Ω be a small complete Heyting algebra, and let $(V_\alpha^{(\Omega)},[\![\ -\ ]\!])$ be the Heyting valued model of Scott-Solovay for Ω. We will show that it may be identified with our semantics for the von Neumann universe in the topos of Ω-sets (see [2, 6] for Ω-sets).

Prop. 3. Set $V_\alpha=(A_\alpha,r_\alpha)$. Let S be the subobject of $A_\alpha\times PA_\alpha$ classified by the evaluation. Let R be the following binary relation on A_α :

$$(id\times r_\alpha)^{-1}(S)\rightarrowtail A_\alpha\times A_\alpha\ .$$

Then A_α is isomorphic to the Ω-set $(V_\alpha^{(\Omega)},[\![\ =\ ]\!])$ and the relation R on it is given by $[\![\ \in\ ]\!]$.

Def. 4. (i) In the topos of Ω-sets, a terminal object is given by $(\{*\},\delta(*,*)=1)$. Let $a:1\to V_\alpha$. We define $j(a)\in V_\alpha^{(\Omega)}$ by

$$j(a)(x)=\bigvee_{y\in V_\alpha^{(\Omega)}}[x=y]\wedge a(*,y),$$

$$\mathrm{dom}(j(a))=V_\alpha^{(\Omega)}.$$

(ii) Let $a_1,\ldots,a_n\in\Gamma(1,U_n)$. Let $\|\phi(a_1,\ldots,a_n)\|$ be the following open object:

$$\bigvee\{Y\overset{t}{\rightarrowtail}1:\ U_n\models_Y\phi(a_1t,\ldots,a_nt)\}.$$

Prop. 4. (i) Let $a_1,\ldots,a_n\in\Gamma(1,U_n)$. Then

$$\|\phi(a_1,\ldots,a_n)\| = [\![\phi(j(a_1),\ldots,j(a_n))]\!].$$

(ii) For any $u\in V^{(\Omega)}$, there exists $a\in\Gamma(1,U_n)$ such that $[\![j(a)=u]\!] = 1$. Hence the Ω-classes $(\Gamma(1,U_n),\| = \|)$ and $(V^{(\Omega)},[\![=]\!])$ is isomorphic by the Ω-class isomorphism given by $[\![j(\ )=]\!]$.

REFERENCES

1. M.P. Fourman, Sheaf models for set theory, preprint.

2. M.P. Fourman and D.S. Scott, Sheaves and logic. "Applications of Sheaves", Springer Lecture Notes in Math. 753 (1979) ,302-401.

3. R.J. Grayson, A sheaf approach to models of set theory (M. Sc. thesis, Oxford, 1975).

4. R.J. Grayson, Heyting-valued models for intuitionistic set theory. "Applications of Sheaves", Springer Lecture Notes in Math. 753 (1979) , 402-414.

5. P.T. Johnstone, Topos theory (Academic Press, London, 1977).

6. M. Makkai and G.E. Reyes, First Order Categorical Logic, Springer Lecture Notes in Math. 611 (1977).

7. G. Osius, Categorical Set Theory: a characterization of the category of the sets. J. Pure and Applied Algebra 4 (1974), 79-119.

8. G. Osius, Logical and set-theoretical tools in elementary topoi. "Model Theory and Topoi", Springer Lecture Notes in Math. 445 (1975), 297-346.

9. G. Osius, A note on Kripke-Joyal semantics for the internal language of topoi, Ibid. , 349-354.

A REPRESENTATION FOR SPECTOR SECOND ORDER CLASSES IN COMPUTATION THEORIES ON TWO TYPES

by

Ken Hirose and Fujio Nakayasu

Waseda University, Tokyo, Japan

§0. INTRODUCTION

In the generalized recursion theories, the notion of a Spector second order class is introduced in Moschovakis [1], and a general development of computation theories on the domain of two types was first presented by Moldestad [2].

We shall be interested in the relationship between Spector second order classes and computation theories on two types. The purpose of this lecture is to give a computation theoretic representation for a Spector second order class.

In [5], Moschovakis showed:

A first order class Γ on S is a Spector first order class if and only if there is a finite computation theory on S such that

$$en(\Theta) = \Gamma.$$

And Fenstad [4] constructed such a Θ inductively.

We shall try to lift up the above result from a Spector first order class to second order class, following Fenstad [4, Remark 4.4.2].

In §2, following Kechris [3], we shall explain the notion of a Spector second order class. In §3, we shall develop the computation theory on two types which is associated with signatures. This is obtained from the theory of Moldestad, by adding the notion of signatures. And we shall define the interpretation * between second order relations and relations on two types.

And then, in §7, we shall show

THEOREM

Let Γ be a Spector second order class which is closed under $\forall^{P(S)}$ and a coding of finite sequences from $S \cup (\bigcup_{n\in\omega} P(S^n))$. Then, for each signature ν, there is a computation Θ_ν^Γ such that

$$en^*(\Theta_\nu^\Gamma)_\nu = \Gamma_\nu,$$

where

$$\Gamma_\nu = \{Q \in \Gamma \mid Q \text{ has signature } \nu\}$$

$$en^*(\Theta) = \{R^* \mid R \in en(\Theta)\}.$$

(The meaning of a coding is given in §4.)

§1. PRELIMINARIES

Let $m = \langle S, R_1, R_2, \ldots, R_\ell \rangle$ be a structure. In the following lines, we shall assume that m contains a copy $\langle N, \leq^N \rangle$ of $\langle \omega, \leq \rangle$ and that functions M, K, L and $+1$ are elementary on m, where M is an injection $S \times S \longrightarrow S$, K and L are functions $S \longrightarrow S$ such that $K(M(r,s)) = r$ and $L(M(r,s)) = s$. In addition, N is closed under K, L and M. From these functions, we can define the following functions and relations :

$$\langle\ \rangle_0 = 0,$$
$$\langle r_1, \ldots, r_{n+1} \rangle_{n+1} = M(\langle r_1, \ldots, r_n \rangle_n, r_{n+1}),$$
$$\langle r_1, \ldots, r_n \rangle = M(n, \langle r_1, \ldots, r_n \rangle_n),$$
$$Seq : \text{the image of } \langle\ \rangle, \text{ i.e.}$$
$$Seq(r) \Leftrightarrow \exists n \, \exists r_1, \ldots, \exists r_n \quad r = \langle r_1, \ldots, r_n \rangle,$$
$$lh(\langle r_1, \ldots, r_n \rangle) = n$$
$$(\langle r_1, \ldots, r_n \rangle)_i = r_i.$$

Let $I = S \cup {}^S\omega$, where ${}^S\omega$ is the set of functions from S to ω. In general we shall use lower letters x, y, z as variables over S, capital letters P, X, Y, Z, W as variables over relations on S and a, b, c for members of I. We shall reserve e, i, j, k, l, m, n for members of ω. Finally, barred letters will denote finite sequences, as for example

$$\bar{x} = (x_1, \ldots, x_n) \text{ or } \bar{P} = (P_1, \ldots, P_m).$$

By a *signature*, we mean a finite sequence from ω, and also say a signature ν has length n, if ν is in the form of $(k_1, \ldots, k_n)$. We shall denote $\langle \nu \rangle$ the coding of signature ν, and $\langle \nu \rangle$ is often identified with ν.

A second order relation on S is a relation of the form

$$Q(x_1, \ldots, x_n, P_1, \ldots, P_m)$$

where P_i varies over k_i-ary relations on S. And we call $(\underbrace{0, \ldots, 0}_{n}, k_1, \ldots, k_m)$ the signature of Q. An n-ary relation R on S is also a second order relation which has signature $(0, \ldots, 0)$. We also say that the sequence $(x_1, \ldots, x_n, P_1, \ldots, P_m)$ has signature $(0, \ldots, 0, k_1, \ldots k_m)$.

A 2-class or second order class is a collection of second order relations on S.

§2. SPECTOR SECOND ORDER CLASS

The concept of Spector 2-class was introduced by Moschovakis [1].

Following Kechris [3], for a structure m, we say a 2-class Γ is a Spector 2-class on m, if Γ satisfies properties (i)-(v) :

(i) Γ contains $=$, $R_1, \dots, R_l$ and their negations.

(ii) Γ is closed under $\wedge$, $\vee$, $\exists^S$, $\forall^S$.

(iii) Γ is normed, i.e., if $Q(\bar{x},\bar{P})$ is in Γ, there is a function σ: $Q(\{(\bar{x},\bar{P})|Q(\bar{x},\bar{P})\}) \longrightarrow O_n$ such that corresponding relations $(\bar{x},\bar{P}) <_\sigma (\bar{x}',\bar{P}')$ and $(\bar{x},\bar{P}) \leq_\sigma (\bar{x}',\bar{P}')$ are in Γ, where $<_\sigma$ and $\leq_\sigma$ defined by

$$(\bar{x},\bar{P}) \in Q \wedge [(\bar{x}',\bar{P}') \in Q \longrightarrow \sigma(\bar{x},\bar{P}) < \sigma(\bar{x}',\bar{P}')]$$

and $(\bar{x},\bar{P}) \in Q \wedge [(\bar{x}',\bar{P}') \in Q \longrightarrow \sigma(\bar{x},\bar{P}) \leq \sigma(\bar{x}',\bar{P}')]$

respectively. Such a σ is called Γ-norm on Q.

(iv) Γ is ω-parametrized, i.e., if ν is a signature, there is a relation $U(\in\Gamma)$ of the signature $(0,\nu)$ which is universal for second order relations of signature ν: in other words, if $U_e = \{(\bar{x},\bar{P})|U(e,\bar{x},\bar{P})\}$, then

$$\{U_e | e \in \omega\} = \{Q \in \Gamma | Q \text{ has signature } \nu\}.$$

(v) If $Q \in \Gamma$, then Q *is* Γ *on* Δ. That is, for any R_i, P_i, $1 \leq i \leq k$, in Γ there is $S \in \Gamma$ such that :
for every $\bar{x}_1, \bar{Y}_1, \dots, \bar{x}_k, \bar{Y}_k$ such that

$$\forall i \leq k\ \forall \bar{z}_i [R_i(\bar{x}_i,\bar{z}_i,\bar{Y}_i) \Leftrightarrow \neg P_i(\bar{x}_i,\bar{z}_i,\bar{Y}_i)]$$

we have

$$S(\bar{x},\bar{x}_1,\bar{Y}_1,\dots,\bar{x}_k,\bar{Y}_k) \Leftrightarrow Q(\bar{x}, \{\bar{z}_1|R_1(\bar{x}_1,\bar{z}_1,\bar{Y}_1)\},\dots,\{\bar{z}_k|R_k(\bar{x}_k,\bar{z}_k,\bar{Y}_k)\}).$$

This last property (v) is equivalent in the presence of (i) - (iv) to the following one:

(v*) Γ is closed under Δ-bounded existential quantification.

In detail this means the following:

Let $\check{\Gamma} = \{\neg Q | Q \in \Gamma\}$, $\Delta = \Gamma \cap \check{\Gamma}$. For each 2-class Γ and each $\bar{x},\bar{Y}$ let $\Gamma(\bar{x},\bar{Y}) = \{Q(\bar{u},\bar{V}) | \textit{There is } R \in \Gamma \textit{ such that } Q(\bar{u},\bar{V}) \Leftrightarrow R(\bar{x},\bar{Y},\bar{u},\bar{V})\}$ and $\Delta(\bar{x},\bar{Y}) = \Gamma(\bar{x},\bar{Y}) \cap \check{\Gamma}(\bar{x},\bar{Y})$. In this notation (v*) asserts that if Q is in Γ so is

$$S(\bar{x},\bar{Y}) \Leftrightarrow \exists z \in \Delta(\bar{x},\bar{Y}) Q(\bar{x},\bar{Y},z).$$

§3. COMPUTATIONS ON TWO TYPES ASSOCIATED WITH SIGNATURES

In [2] J. Moldestad defined a set of computations on two types inductively. In this section, we shall define analogously a set of computations Θ which is associated with the notion of *signatures*.

A set Θ is called a *computation set* (or a set of computations) on two types if it satisfies

(i) $\Theta \subset \bigcup_{n \in \omega} I^n \quad (I = S \cup S_\omega)$.

(ii) if $(e,\bar{a},x) \in \Theta$, then $e \in N$ and $x \in S$.

(iii) if $(e,\bar{a},x)$, $(e,\bar{a},x') \in \Theta$, then $x = x'$.

Let Θ be a computation set on two types. For $e \in N$, let $\{e\}_\Theta$ be the partial function defined by

$$\{e\}_\Theta(\bar{a}) \simeq x \quad \textit{iff}\ (e,\bar{a},x) \in \Theta.$$

A relation R on I is Θ-semicomputable, if there is $e \in N$ such that

$$R(\bar{a}) \Leftrightarrow \exists x \{e\}_\Theta(\bar{a}) \simeq x.$$

Such an e is called the index of R.

We set

$$en(\Theta) = \{R \mid R \text{ is } \Theta\text{-semicomputable}\},$$

$en(\Theta)$ is called the envelope of Θ.

We shall construct a computation set Θ for a Spector 2-class Γ and represent Γ by $en(\Theta)$. For this purpose, we shall define interpretations between $en(\Theta)$ and second order relations on S.

DEFINITION 3.1

The signature of $R \in en(\Theta)$ is defined by $(e)_2$, where e is the index of R.

DEFINITION 3.2

Let * be the interpretation mapping

$$\begin{array}{ccc} en(\Theta) & \longrightarrow & \text{second order relations on } S \\ \Psi & & \Psi \\ R & \longmapsto & R^* \end{array}$$

defined as follows:

If the signature of R is $(0,\ldots 0, k_1,\ldots, k_m)$, then R^* is the second order relations on S of signature $(0,\ldots 0, k_1,\ldots, k_m)$ such that, for some $\bar{\alpha}$,

$$R^*(\bar{x},\bar{Y}) \text{ if and only if}$$
$$R(\bar{x},\bar{\alpha}) \wedge [(y_1,\ldots, y_{ki}) \in Y_i \Leftrightarrow \alpha_i(\langle y_1,\ldots, y_{ki}\rangle)=0 \quad \textit{for } 1 \le i \le m]$$

Conversely, ^ is the mapping

$$\begin{array}{ccc} \text{second order relations on } S & \longrightarrow & \text{relations on } I \\ \Psi & & \Psi \\ Q & \longmapsto & Q^{\wedge} \end{array}$$

If the signature of Q is $(0,\ldots, 0, k_1,\ldots, k_m)$, then $Q^\wedge$ is the relation on I such that, for some $\overline{Y}$,

$$Q^\wedge(\overline{x},\overline{\alpha}) \text{ if and only if } Q(\overline{x},\overline{Y}) \wedge [(y_1,\ldots, y_{ki}) \in Y_i \Leftrightarrow \alpha_i(\langle y_1,\ldots, y_{ki}\rangle)=0]$$

DEFINITION 3.3

Now we define a computation set Θ on two types associated with signatures. The computation Θ is defined inductively by the monotone operator Φ which is defined by the following set of clauses, where the clauses [I] - [IX] introduce the basic initial functions and clauses [X] - [XIV] are the closure properties for the operations.

Let $X \subset \bigcup_{n\in\omega} I^n$. $\Phi(X)$ is the subset of $\bigcup_{n\in\omega} I^n$ defined by:

For all $n \in N$, $\lambda \in \{\lambda | \lambda$ is a signature of length $n\}$ and $\overline{a} \in I^n$,

[I] $(\langle 1, \langle 0,\lambda\rangle\rangle,c,\overline{a},0) \in \Phi(X)$ if $c \in N$
$(\langle 1, \langle 0,\lambda\rangle\rangle,c,\overline{a},1) \in \Phi(X)$ otherwise

[II] $(\langle 2, \langle 0,\lambda\rangle\rangle,c,\overline{a},0) \in \Phi(X)$ if $c \in S$
$(\langle 2, \langle 0,\lambda\rangle\rangle,c,\overline{a},1) \in \Phi(X)$ otherwise

[III] $(\langle 3, \langle 0,\lambda\rangle\rangle,c,\overline{a},c) \in \Phi(X)$ if $c \in S$
$(\langle 3, \langle 0,\lambda\rangle\rangle,c,\overline{a},0) \in \Phi(X)$ otherwise

[IV] $(\langle 4, \langle 0,\lambda\rangle\rangle,c,\overline{a},c+1) \in \Phi(X)$ if $c \in N$
$(\langle 4, \langle 0,\lambda\rangle\rangle,c,\overline{a},0) \in \Phi(X)$ otherwise

[V] $(\langle 5, \langle\lambda\rangle,m\rangle,\overline{a},m) \in \Phi(X)$ $m \in N$

[VI] $(\langle 6, \langle 0,0,\lambda\rangle\rangle ,c,b,\overline{a}, M(c,b)) \in \Phi(X)$ if $c,b \in S$
$(\langle 6, \langle 0,0,\lambda\rangle\rangle ,c,b,\overline{a},0) \in \Phi(X)$ otherwise

[VII] $(\langle 7, \langle 0,\lambda\rangle\rangle,c,\overline{a}, K(c)) \in \Phi(X)$ if $c \in S$
$(\langle 7, \langle 0,\lambda\rangle\rangle,c,\overline{a},0) \in \Phi(X)$ otherwise

[VIII] $(\langle 8, \langle 0,\lambda\rangle\rangle,c,\overline{a}, L(c)) \in \Phi(X)$ if $c \in S$
$(\langle 8, \langle 0,\lambda\rangle\rangle,c,\overline{a},0) \in \Phi(X)$ otherwise

[IX] $(\langle 9, \langle i,0,\lambda\rangle\rangle,c,b,\overline{a}, c(b)) \in \Phi(X)$ if $c \in {}^S\omega$, $b \in S$ and $b=\langle (b)_1,\ldots,(b)_i\rangle$
$(\langle 9, \langle i,0,\lambda\rangle\rangle,c,b,\overline{a},1) \in \Phi(X)$ otherwise

[X] $\exists y \in S\ [(e,\overline{a},y) \in X \wedge (e',y,\overline{a},x) \in X]$
$\Rightarrow (\langle 10, \langle\lambda\rangle,e,e'\rangle,\overline{a},x) \in \Phi(X)$, where $(e)_2=\langle\lambda\rangle$ and $(e')_2= \langle 0,\lambda\rangle$

[XI] $(e,\overline{a},x) \in X \Rightarrow (\langle 11, \langle 0,\lambda\rangle,e,e'\rangle,0,\overline{a},x) \in \Phi(X)$
$\exists y \in S\ [(\langle 11, \langle 0,\lambda\rangle,e,e'\rangle, m,\overline{a},y) \in X \wedge (e',y,m,\overline{a},x) \in X]$
$\Rightarrow (\langle 11, \langle 0,\lambda\rangle, e,e'\rangle, m+1,\overline{a},x) \in \Phi(X)$,
where $(e)_2= \langle\lambda\rangle$ and $(e')_2= \langle 0,0,\lambda\rangle$

[XII] $(e,\overline{a}',x) \in X \Rightarrow (\langle 12, \langle\lambda\rangle, e,i\rangle, \overline{a},x) \in \Phi(X)$, where $\overline{a}'$ and $(e)_2$ are obtained from $\overline{a}$ and λ respectively by moving $(i+1)$-st object in a and in λ to the front of the list

[XIII] $(e,\bar{a},x) \in X \Rightarrow (<13, <0,\lambda>>, e,\bar{a},x) \in \Phi(X)$, where $(e)_2 = <\lambda>$

[XIV] $\forall x \in S\ \exists y \in N(e,x,\bar{a},y) \in X \wedge (e',\alpha,\bar{a},u) \in X$

$\Rightarrow (<14, <\lambda>,e,e'>,\bar{a},u) \in \Phi(X)$,

where $\alpha(<y_1,\ldots, y_i>) = y \Leftrightarrow (e, <y_1,\ldots, y_i>, \bar{a},y) \in X$,

$(e)_2 = <0,\lambda>$ and $(e')_2 = <i,\lambda>$ for $i \neq 0$

Furthermore, for each Spector 2-class Γ, we define Φ^Γ by the following scheme [XVν]. That is, Φ^Γ is defined by [I] - [XIV] for all signatures ν. Φ^Γ is defined by [I] - [XIV] and [XVν] for a fixed signature ν:

[XV$_\nu$] Let U_ν be a universal relation for relations in Γ of signature ν, and let σ_ν be a Γ-norm on U_ν.

If $(e,\bar{x},\bar{Y}) \in U_\nu \wedge$

$\forall e' \in N\ \forall \bar{x}'\ \forall \bar{Y}'\ [(e',\bar{x}',\bar{Y}') <_{\sigma\nu} (e,\bar{x},\bar{Y})$

$\longrightarrow (<15, <\nu,\lambda>,e'>, \bar{x}',\bar{\alpha}',\bar{a},0) \in X]$,

then $(<15,<\nu,\lambda>,e>, \bar{x},\bar{\alpha},\bar{a},0) \in \Phi(X)$, where $\bar{\alpha}$ and $\bar{\alpha}'$ are the representing functions of $\bar{Y}$ and $\bar{Y}'$, i.e.

$(y_1,\ldots, y_{ki}) \in Y_i \Leftrightarrow \alpha_i(<y_1,\ldots, y_{ki}>) = 0.$

We set

Θ = the least fixed point of Φ (Φ^Γ or Φ^Γ_ν).

That is,

$$\Phi^\xi = \Theta^\xi = \Phi(\bigcup_{\eta<\xi}\Theta^\eta) \quad \text{and}$$

$$\Phi^\infty = \Theta = \bigcup_{\xi\in On}\Theta^\xi .$$

Let $|\,|_\Theta$ be the function

$|\,|_\Theta : \Theta \longrightarrow On$, where

$|e,\bar{a},x|_\Theta$ = the least ξ such that $(e,\bar{a},x) \in \Theta^\xi$.

DEFINITION 3.4

A partial function φ on I is called Θ-computable with the signature λ, if $\varphi = \{e\}_\Theta$ and $(e)_2 = <\lambda>$.

Now we collect the basic properties of the pair $(\Theta,|\,|_\Theta)$ as follows:

(i) Θ is a computation set on two types.

(ii) For all $n \in N, \lambda \in \{\lambda | \lambda$ is a signature of length $n\}$, and $\bar{a} \in I^n$, the following functions are;

Θ-computable with signature $(0,\lambda)$

$$f(c,\bar{a}) = \begin{cases} 0 & \text{if } c \in N \\ 1 & \text{otherwise} \end{cases}$$

$$f(c,\bar{a}) = \begin{cases} 0 & \text{if } c \in S \\ 1 & \text{otherwise} \end{cases}$$

$$f(c,\bar{a}) = \begin{cases} c & \text{if } c \in S \\ 0 & \text{otherwise} \end{cases}$$

$$f(c,\bar{a}) = \begin{cases} c+1 & \text{if } c \in N \\ 0 & \text{otherwise} \end{cases}$$

$$f(c,\bar{a}) = \begin{cases} K(c) & \text{if } c \in S \\ 0 & \text{otherwise} \end{cases}$$

$$f(c,\bar{a}) = \begin{cases} L(c) & \text{if } c \in S \\ 0 & \text{otherwise} \end{cases}$$

Θ-computable with signature $(0,0,\lambda)$

$$f(c,b,\bar{a}) = \begin{cases} M(c,b) & \text{if } c,b \in S \\ 0 & \text{otherwise} \end{cases}$$

Θ-computable with signature λ

$$f(\bar{a}) = m \qquad m \in N$$

and Θ-computable with signature $(i,0,\lambda)$

$$f(c,b,\bar{a}) = \begin{cases} c(b) & \text{if } c \in {}^{S}\omega,\ b \in S \text{ and } b = \langle (b)_1,..(b)_i\rangle \\ 0 & \text{otherwise} \end{cases}$$

for $i \neq 0$.

(iii) The following operations are allowed: substitution, primitive recursion, permutations of the list of arguments of a function, adding dummy arguments, substitution of a function for an element in ${}^{S}\omega$, diagonalization and the S^n_λ-property (λ is a signature). To make precise what is meant by "an operation is allowed", let us regard substitution, diagonalization and S^n_λ-property.

Substitution : There is a Θ-computable mapping g_1 with signature $(0,0,0)$ such that for all $e,f,\bar{a},x$:

$$\{g_1(e,f,\langle\lambda\rangle)\}_\Theta(\bar{a}) \simeq x \Leftrightarrow \exists u[\{e\}_\Theta\ (\bar{a}) \simeq u \wedge \{f\}_\Theta\ (u,\bar{a}) \simeq x]$$

and

$$|g_1(e,f,\langle\lambda\rangle),\bar{a},x|_\Theta \geq sup\{|e,\bar{a},u|_\Theta+1,\ |f,u,\bar{a},x|_\Theta+1\}$$

when

$$\{e\}_\Theta(\bar{a}) \simeq u \text{ and } \{f\}_\Theta(u,\bar{a}) \simeq x,$$

where $(g_1(e,f,\langle\lambda\rangle))_2 = \langle\lambda\rangle$ and $(f)_2 = \langle 0,\lambda\rangle$.

Diagonalization : There is a Θ-computable mapping g_2 with signature $(0,0)$ such that for all $e,\bar{a},\bar{b},x$:

$$\{g_2(\langle\lambda\rangle,\ \langle\nu\rangle)\}_\Theta\ (e,\bar{a},\bar{b}) \simeq x \Leftrightarrow \{e\}_\Theta(\bar{a}) \simeq x$$

and

$$|g_2(\langle\lambda\rangle,\langle\nu\rangle),\ e,\bar{a},\bar{b},x|_\Theta\ \geq |e,\bar{a},x|_\Theta+1 \quad \text{when} \quad \{e\}_\Theta(\bar{a}) \simeq x,$$

where $(g_2(\langle\lambda\rangle,\langle\nu\rangle))_2 = \langle 0,\lambda,\nu\rangle$ and $(e)_2 = \langle\lambda\rangle$.

S^n_λ-property : There is a Θ-computable mapping g_3 with signature $(0,0)$ such that $g_3(n,\langle\lambda\rangle)$ is the index for the mapping S^n_λ with the following property :

For all $e \in N$, $x_1,\ldots, x_n \in N$, $\bar{a}, z$:

$$\{S^n_\lambda(e,x_1,\ldots, x_n)\}_\Theta(\bar{a}) \simeq z \iff \{e\}_\Theta(x_1,\ldots, x_n,\bar{a}) \simeq z$$

and

$$|S^n_\lambda(e,x_1,\ldots, x_n),\bar{a},z|_\Theta \geq |e,x_1,\ldots, x_n,\bar{a},z|_\Theta +1$$

when $\{e\}_\Theta(x_1,\ldots, x_n,\bar{a}) \simeq z$, where $(S^n_\lambda(e,x_1,\ldots, x_n))_2 = \langle\lambda\rangle$ and $(e)_2 = \langle\underbrace{0,\ldots, 0}_{n},\lambda\rangle$

(The mappings above are totally defined)

DEFINITION 3.5

We call a pair $(T, ||_T)$ a computation theory on two types associated with signatures, if it satisfies the properties (i) - (iii) above.

DEFINITION 3.6

Let $\mathfrak{F}$ be a partial monotone functional. $\mathfrak{F}$ is called weakly T-computable with signature (λ,ν), if there is a T-computable mapping g with signature $(0,0)$ such that for all $e,\bar{a},\bar{b}$:

(i) $\mathfrak{F}(\varphi,\bar{a}) \simeq \{g(e,\langle\pi\rangle)\}_T(\bar{b},\bar{a})$,
where $\varphi = \lambda\bar{c}\{e\}_T(\bar{c},\bar{b})$, $(e)_2 = \langle\lambda,\pi\rangle$ and $(g(e,\langle\pi\rangle))_2 = \langle\pi,\nu\rangle$.

(ii) If $\{g(e,\langle\pi\rangle)\}_T(\bar{b},\bar{a}) \simeq r$, then there is a subfunction ψ of $\lambda\bar{c}\{e\}_T(\bar{c},\bar{b})$ such that $\mathfrak{F}(\psi,\bar{a}) \simeq r$
and

$$|\{g(e,\langle\pi\rangle)\}_T(\bar{b},\bar{a})|_T > |\{e\}_T(\bar{c},\bar{b})|_T \text{ for all } \bar{c} \in dom\ \psi .$$

THEOREM 1 (THE FIRST RECURSION THEOREM)

Suppose $\mathfrak{F}$ is a partial monotone functional which is weakly T-*computable with signature (λ,λ). Then there is the least solution φ to the equality*

$$\forall\bar{a}\ [\mathfrak{F}(\varphi,\bar{a}) \simeq \varphi(\bar{a})]$$

and this solution is T-*computable with signature λ.*

THEOREM 2 (THE SECOND RECURSION THEOREM)

$$\forall e\, \exists x\, \forall\bar{a}\ [\{e\}_T(x,\bar{a}) \simeq \{x\}_T(\bar{a})],$$

where $(e)_2 = \langle 0,\lambda\rangle$ *and* $(x)_2 = \langle\lambda\rangle$.

The proof of theorem 1 and 2 are quite similar to those of the first recursion theorem and the second recursion theorem in J. E. Fenstad [4].

§4. RELATIONSHIPS BETWEEN SPECTOR 2-CLASSES AND COMPUTATION THEORIES

In this section, by using the interpretation mappings * and ^, we shall see relationships between Spector 2-class and computation theories on two types associated with signatures.

In the following lines, we restrict $S\mathrm{U}^S\omega$ to $S\mathrm{U}^S 2$. That is, we put

$$I = S\mathrm{U}^S 2$$

Of course, the least fixed point of Φ in this case also satisfies the properties (i) - (iii) in definition 3.5. And we also say a computation set T on $S\mathrm{U}^S 2$ is a computation theory on two types associated with signatures, if it satisfies the properties in definition 3.5. In the following sections, for each Spector 2-class Γ, we set

$$\Theta^\Gamma_\nu = \text{the least fixed point of } \Phi^\Gamma_\nu,$$
$$\Theta^\Gamma = \text{the least fixed point of } \Phi^\Gamma.$$

THEOREM 3

Let Γ be a Spector 2-class. Then there exists a computation theory T *on two types associated with signatures such that*

$$Q \in \Gamma \Rightarrow Q^\wedge \in en(\mathrm{T}).$$

That is, if we identify $\{Q^\wedge | Q \in \Gamma\}$ with Γ, then we have

$$\Gamma \subset en(\mathrm{T}).$$

Proof.
Let $\Theta^\Gamma = \mathrm{T}$, then the scheme [XV$\nu$] allows us to prove that for each universal relation U_ν,

$$(e,\bar{x},\bar{Y}) \in U_\nu \Leftrightarrow (<15,<\lambda>,e>,\bar{x},\bar{\alpha},0) \in \Theta^\Gamma,$$

where $\bar{\alpha}$ are the representing functions of $\bar{Y}$ respectively. That is

$$Q \in \Gamma \Rightarrow Q^\wedge \in en(\Theta^\Gamma).$$

COROLLARY 3.1

Let Γ be a Spector 2-class. Then

$$\Gamma \subset \{R^* | R \in en(\Theta^\Gamma)\}.$$

COROLLARY 3.2

Let Γ be a Spector 2-class. Then for each signature ν

$$\{Q \in \Gamma | Q \text{ has signature } \nu\} \subset \{R^* | R \in en(\Theta^\Gamma_\nu)\}.$$

DEFINITION 4.1

A coding $<< \quad >>$ of finite sequences from $S \cup (\bigcup_{n\in\omega} P(S^n))$ is an injection

$$\begin{array}{cccc} << \quad >> : & \bigcup_{n\in\omega} J^n & \longrightarrow & P(S) \\ & \Psi & & \Psi \\ & (\bar{x},\bar{Y}) & \longmapsto & <<\bar{x},\bar{Y}>> \end{array}$$

where $$J = S \cup (\bigcup_{n\in\omega} P(S^n)).$$

We say 2-class Γ is closed under $<< \quad >>$ and $\forall^{P(S)}$ if Γ satisfies (a), (b) and (c).

(a) Γ is closed under $\forall^{P(S)}$

(b) For each signature λ and for each $(\bar{x},\bar{Y})$ which has signature λ, if $A \in \Gamma$, then $A' \in \Gamma$, where

$$A'(X,\bar{w},\bar{W}) \Leftrightarrow \exists \bar{x},\bar{Y}\ [A(\bar{x},\bar{Y},\bar{w},\bar{W}) \wedge X = <<\bar{x},\bar{Y}>>]$$

and if $B \in \Gamma$, then $B' \in \Gamma$, where

$$B'(\bar{x},\bar{Y},\bar{w},\bar{W}) \Leftrightarrow B(<<\bar{x},\bar{Y}>>,\bar{w},\bar{W}).$$

To express the property (c) we need the following notions:

Let L be a 2-class. We say a second order relation Q is *second order definable from L*, if Q is defined by a second order formula φ, where atomic formulas are enlarged by the elements in L and quantification of the relation variables are allowed.

For each $A(x,\bar{w},\bar{W})$, $B_i(\bar{x}_i,\bar{Y}_i) \in \Gamma$ we define

$$\Gamma^{A(B_1,\ldots,B_l)}(\bar{w},\bar{W})$$

as follows:

$$\Gamma^{A(B_1,\ldots,B_l)}(\bar{w},\bar{W})$$
$$= \{Q \in \Gamma(\bar{w},\bar{W}) \mid Q \textit{ is second order definable from } L\}.$$

Where L is a 2-class such that

$Q \in L$ if and only if
Q is B_i for some $1 \le i \le l$
or for some signature λ,
Q is $\lambda\bar{x}, \bar{Y}A^{\lambda}(\bar{x},\bar{Y},\bar{w},\bar{W})$ such that
$\lambda\bar{x}, \bar{Y}A^{\lambda}(\bar{x},\bar{Y},\bar{w},\bar{W}) \Leftrightarrow A(<<\bar{x},\bar{Y}>>,\bar{w},\bar{W})$

(c) C is in Γ, if for each signature λ,

$$C^{\lambda}(\bar{x},\bar{Y},\bar{w},\bar{W}) \Leftrightarrow C(<<\bar{x},\bar{Y}>>, \bar{w},\bar{W})$$

is in $\Gamma^{A(B_1,\ldots,B_l)}(\bar{w},\bar{W})$, where $(\bar{x},\bar{Y})$ has signature λ, for some fixed A,

$B_1, \ldots, B_l \in \Gamma$. And if $C(X,\overline{w},\overline{W})$ then

$$X = \langle\langle \overline{x},\overline{Y}\rangle\rangle \quad \text{for some } (\overline{x},\overline{Y}).$$

THEOREM 4

Let Γ be a Spector 2-class which is closed under $\forall^{P(S)}$ and a coding of finite sequences from $S \cup (\bigcup_{n\in\omega} P(S^n))$. Then

$$R \in en(\Theta_\nu^\Gamma) \Rightarrow R^* \in \Gamma.$$

That is, if we identify $\{R^* | R \in en(\Theta_\nu^\Gamma)\}$ with $en(\Theta_\nu^\Gamma)$, then we have

$$en(\Theta_\nu^\Gamma) \subset \Gamma.$$

§5. LEMMATA

In this section, we prepare the proof of Theorem 4 in a series of lemmata.

DEFINITION 5.1

** is the mapping of the form

$$\begin{array}{ccc} P(\bigcup_{n\in\omega} I^n) & \longrightarrow & P(\bigcup_{n\in\omega} J^n) \\ \Psi & & \Psi \\ T & \longmapsto & T^{**} \end{array}$$

and T** is defined as follows:
for all $n, m \in N$

$$(e,x_1,\ldots, x_n, Y_1,\ldots, Y_m,y) \in T^{**}$$

if and only if

$$(e,x_1,\ldots, x_n, \alpha_1,\ldots, \alpha_m,y) \in T$$

and $\quad (e)_2 = (0,\ldots, 0, k_1,\ldots, k_m)$, where

$$(y_1,\ldots, y_{ki}) \in Y_i \Leftrightarrow \alpha_i(\langle y_1,\ldots,y_{ki}\rangle) = 0.$$

LEMMA 1

Let Γ be a 2-class and T be a computation set on two types. Then

$$\{R^* | R \in en(T)\} = en(T^{**}),$$

*where en(T**) is obtained from the same definition for en(T).*

Proof.

$R \in en(T^{**})$ if and only if there is $e \in N$ such that

$$R(\overline{x},\overline{Y}) \Leftrightarrow \exists y(e,\overline{x},\overline{Y},y) \in T^{**}$$

Therefore the proof of this lemma is immediate from the definition

3.1 and 5.1.

We define monotone operator Φ^{**} on $\bigcup_{n\in\omega} J^n$ $(J = S \cup (\bigcup_{n\in\omega} P(S^n)))$ such that

$$\Theta^{**} = \text{the least fixed point of } \Phi^{**},$$

where Θ is the least fixed point of Φ.

DEFINITION 5.2

The operator Φ^{**} is defined by the following set of clauses which are obtained from the clauses in the definition of Φ.

Let $X \subset \bigcup_{n\in\omega} J^n$. $\Phi^{**}(X)$ is the subset of $\bigcup_{n\in\omega} J^n$ defined by:

for all $\lambda = (0,\dots,0, k_1,\dots, k_m)$ and $(x_1,\dots, x_n, Y_1,\dots, Y_m)$ where Y_i is k_i-ary relation

(I**) $(\langle 1, \langle 0,\lambda\rangle\rangle, x, \bar{x}, \bar{Y}, 0) \in \Phi^{**}(X)$ if $x \in N$

$(\langle 1, \langle 0,\lambda\rangle\rangle, x, \bar{x}, \bar{Y}, 1) \in \Phi^{**}(X)$ otherwise

(II**) $(\langle 2, \langle 0,\lambda\rangle\rangle, x, \bar{x}, \bar{Y}, 0) \in \Phi^{**}(X)$

(III**) $(\langle 3, \langle 0,\lambda\rangle\rangle, x, \bar{x}, \bar{Y}, x) \in \Phi^{**}(X)$

(IV**) $(\langle 4, \langle 0,\lambda\rangle\rangle, x, \bar{x}, \bar{Y}, x+1) \in \Phi^{**}(X)$ if $x \in N$

$(\langle 4, \langle 0,\lambda\rangle\rangle, x, \bar{x}, \bar{Y}, 0) \in \Phi^{**}(X)$ otherwise

(V**) $(\langle 5, \langle\lambda\rangle, m\rangle, \bar{x}, \bar{Y}, m) \in \Phi^{**}(X)$ $m \in N$

(VI**) $(\langle 6, \langle 0,0,\lambda\rangle\rangle, x, y, \bar{x}, \bar{Y}, M(x,y)) \in \Phi^{**}(X)$

(VII**) $(\langle 7, \langle 0,\lambda\rangle\rangle, x, \bar{x}, \bar{Y}, K(x)) \in \Phi^{**}(X)$

(VIII**) $(\langle 8, \langle 0,\lambda\rangle\rangle, x, \bar{x}, \bar{Y}, L(x)) \in \Phi^{**}(X)$

(IX**) $(\langle 9, \langle i,0,\lambda\rangle\rangle, Y, x, \bar{x}, \bar{Y}, 0) \in \Phi^{**}(X)$ if $((x)_1,\dots,(x)_i) \in Y \wedge x = \langle (x)_1,\dots,(x)_i\rangle$

$(\langle 9, \langle i,0,\lambda\rangle\rangle, Y, x, \bar{x}, \bar{Y}, 1) \in \Phi^{**}(X)$ otherwise

($i \neq 0$ and Y is a n-ary relation)

(X**) $\exists y[(e,\bar{x},\bar{Y},y) \in X \wedge (e',y,\bar{x},\bar{Y},x) \in X]$

$\Rightarrow (\langle 10, \langle\lambda\rangle, e, e'\rangle, \bar{x}, \bar{Y}, x) \in \Phi^{**}(X)$, where $(e)_2 = \langle\lambda\rangle$ and $(e')_2 = \langle 0,\lambda\rangle$

(XI**) $(e,\bar{x},\bar{Y},x) \in X \Rightarrow (\langle 11, \langle 0,\lambda\rangle, e, e'\rangle, 0, \bar{x}, \bar{Y}, x) \in \Phi^{**}(X)$

$\exists y[(\langle 11, \langle 0,\lambda\rangle, e, e'\rangle, m, \bar{x}, \bar{Y}, y) \in X \wedge (e', y, m, \bar{x}, \bar{Y}, x) \in X]$

$\Rightarrow (\langle 11, \langle 0,\lambda\rangle, e, e'\rangle, m+1, \bar{x}, \bar{Y}, x) \in \Phi^{**}(X)$, where $(e)_2 = \langle\lambda\rangle$ and $(e')_2 = \langle 0,0,\lambda\rangle$

(XII**) $(e, (\bar{x},\bar{Y})', x) \in X \Rightarrow (\langle 12, \langle\lambda\rangle, e, i\rangle, \bar{x}, \bar{Y}, x) \in \Phi^{**}(X)$, where $(\bar{x},\bar{Y})'$ and $(e)_2$ are obtained from $(\bar{x},\bar{Y})$ and λ respectively by moving the $(i+1)$-st object in $(\bar{x},\bar{Y})$ and in λ to the front of the list

(XIII**) $(e,\bar{x},\bar{Y},x) \in X \Rightarrow (\langle 13, \langle 0,\lambda\rangle\rangle, e, \bar{x}, \bar{Y}, x) \in \Phi^{**}(X)$, where $(e)_2 = \langle\lambda\rangle$

(XIV**) $\forall x\, \exists y \in N\ (e,x,\bar{x},\bar{Y},y) \in X \wedge (e',Z,\bar{x},\bar{Y},u) \in X$
$\Rightarrow (<14,\ <\lambda>,e,e'>,\bar{x},\bar{Y},u) \in \Phi^{**}(X),$
where $(y_1,\ldots,\ y_i) \in Z \Leftrightarrow (e,<y_1,\ldots,\ y_i>,\bar{x},\bar{Y},0) \in X,$
$(e)_2 = <0,\lambda>$ and $(e')_2 = <i,\lambda>$ for $i \neq 0$

We defined Φ^{Γ}_{ν} and Φ^{Γ} in § 3. We define $\Phi^{\Gamma}_{\nu}{}^{**}$ and $\Phi^{\Gamma **}$ in an analogous way by the following scheme (XVν**).

(XVν**) $(<15,\ <\nu,\lambda>,e\ >,\bar{z},\bar{Z},\bar{x},\bar{Y},0) \in \Phi^{**}(X),$
if $\forall e' \in N\, \forall \bar{z}'\, \forall \bar{Z}'\, [(e',\bar{z}',\bar{Z}') <_{\sigma\nu} (e,\bar{z},\bar{Z})$
$\Rightarrow (<15,\ <\nu,\lambda>,e'>,\bar{z}',\bar{Z}',\bar{x},\bar{Y},0) \in X]$ and $(e,\bar{z},\bar{Z}) \in U_{\nu},$
where U_{ν} is a universal relation and σ_{ν} is a Γ-norm on U_{ν}.

LEMMA 2

$(\Theta^{\Gamma})^{**}$ = the least fixed point of $\Phi^{\Gamma **}$ and
$(\Theta^{\Gamma}_{\nu})^{**}$ = the least fixed point of $\Phi^{\Gamma}_{\nu}{}^{**}$.

Proof.

We can easily see from the definition of Φ^{**} and ** that

$$(\Phi(X))^{**} = \Phi^{**}(X^{**}).$$

From the above, we show by induction ξ that

$$(\Theta^{\xi})^{**} = (\Phi^{**})^{\xi},$$

where $(\Phi^{**})^{\xi} = \Phi^{**}(\bigcup_{\eta<\xi} (\Phi^{**})^{\eta}).$

$$\begin{aligned}(\Theta^{\xi})^{**} &= (\Phi(\bigcup_{\eta<\xi} \Theta^{\eta}))^{**} \\ &= \Phi^{**}((\bigcup_{\eta<\xi} \Theta^{\eta})^{**}) \\ &= \Phi^{**}(\bigcup_{\eta<\xi} (\Theta^{\eta})^{**}) \\ &= \Phi^{**}(\bigcup_{\eta<\xi} (\Phi^{**})^{\eta}) \\ &= (\Phi^{**})^{\xi}.\end{aligned}$$

Thus we have $\Theta^{**} = (\Phi^{**})^{\infty}$.

DEFINITION 5.3

Let << >> be a coding of finite sequences from $S \cup (\bigcup_{n\in\omega} P(S^n))$. The monotone operator $<<\Phi^{**}>>$ is defined by the set of clauses <<I**>> - <<XV**>> which is obtained from (I**) - (XV**) by using << >>.

$$\begin{array}{rccc} <<\Phi^{**}>> : & P(P(S)) & \longrightarrow & P(P(S)) \\ & \Psi & & \Psi \\ & X & \longmapsto & <<\Phi^{**}>>(X) \end{array}$$

For example:

$<<II^{**}>>$ $\quad << <2,\ <0,\lambda>>,\ x,\bar{x},\bar{Y},0>> \in <<\Phi^{**}>>$

$<< X^{**}>>$ $\quad \exists y[<< e,\bar{x},\bar{Y},y>> \in X \wedge << e',y,\bar{x},\bar{Y},x>> \in X]$

$\Rightarrow << <10,\ <\lambda>,e,e'>,\bar{x},\bar{Y},x >> \in <<\Phi^{**}>>(X),$

where $(e)_2 = <\lambda>$ and $(e')_2 = <0,\lambda>$.

LEMMA 3

$<<(\Theta^{\Gamma})^{**}>>$ = the least fixed point of $<<\Phi^{\Gamma**}>>$, and

$<<(\Theta^{\Gamma}_{\nu})^{**}>>$ = the least fixed point of $<<\Phi^{\Gamma}_{\nu}{}^{**}>>$,

where $\quad <<\Theta^{**}>> = \{<<\bar{x},\bar{Y}>> \mid (\bar{x},\bar{Y}) \in \Theta^{**}\}$

Proof.

From the difinition, we can easily see that

$$<<\Phi^{**}>>(<<X>>) = <<\Phi^{**}(X)>>.$$

From the above,

$$<<\Phi^{**}>>(<<\Theta^{**}>>) = <<\Phi^{**}(\Theta^{**})>> = <<\Theta^{**}>>,$$

therefore $<<\Theta^{**}>>$ is a fixed point of $<<\Phi^{**}>>$.

And if $<<\Phi^{**}>>(<<T>>) = <<T>>$, then

$$\Phi^{**}(T) = T.$$

Hence $\quad <<\Theta^{**}>> \subset <<T>>$.

LEMMA 4

Let Γ be a Spector 2-class which is closed under $\forall^{P(S)}$ and a coding $<<\ >>$ of finite sequences from $S \cup (\bigcup_{n \in \omega} P(S^n))$. Then $<<\Phi^{\Gamma}_{\nu}{}^{**}>>$ is Γ on Γ, that is, if for each $A \in \Gamma$, the relation B defined by

$$B(X,\bar{w},\bar{W}) \Leftrightarrow X \in <<\Phi^{\Gamma}_{\nu}{}^{**}>>(\{X' \mid A(X',\bar{w},\bar{W})\})$$

is also in Γ.

Proof.

B is defined by the set of clauses,

(I**) $\quad (<< <1,\ <0,\lambda>>,x,\bar{x},\bar{Y},0>>,\bar{w},\bar{W}) \in B \quad$ if $x \in N$

$(<< <1,\ <0,\lambda>>,x,\bar{x},\bar{Y},1>>,\bar{w},\bar{W}) \in B \quad$ otherwise

(II**) $\quad (<< <2,\ <0,\lambda>>,x,\bar{x},\bar{Y},0>>,\bar{w},\bar{W}) \in B$

(III**), (IV**), (V**), (VI**), (VII**), and (VIII**) are similar to the above.

(IX**) $\quad (<< <9,\ <i,0,\lambda>>,Y,x,\bar{x},\bar{Y},0>>,\bar{w},\bar{W}) \in B$

if $((x)_1,\ldots,(x)_i) \in Y \wedge x = <(x)_1,\ldots,(x)_i>$

$(\ll <9,\ <i,0,\lambda>>, Y, x, \bar{x}, \bar{Y}, 1>>, \bar{w}, \bar{W}) \in B$ otherwise. $(i \neq 0)$

(X**) $\exists y[(\ll e, \bar{x}, \bar{Y}, y>>, \bar{w}, \bar{W}) \in A \wedge (\ll e', y, \bar{x}, \bar{Y}, x>>, \bar{w}, \bar{W}) \in A]$
$\Leftrightarrow (\ll <10,\ <\lambda>, e, e'>, \bar{x}, \bar{Y}, x>>, \bar{w}, \bar{W}) \in B$

(XI**), (XII**) and (XIII**) are similar to the above.

(XIV**)

From the uniqueness of Θ**, we can put clause (XIV**) as follows:

$\forall x \exists ! y \in N\ (\ll e, x, \bar{x}, \bar{Y}, y>>, \bar{w}, \bar{W}) \in A \wedge (\ll e', Z, \bar{x}, \bar{Y}, u>>, \bar{w}, \bar{W}) \in A$
$\Leftrightarrow (\ll <14,\ <\lambda>, e, e'>, \bar{x}, \bar{Y}, u>>, \bar{w}, \bar{W}) \in B,$
where $(e')_2 = <i, \lambda>$ and
$(y_1, \ldots, y_i) \in Z \Leftrightarrow (\ll e,\ <y_1, \ldots, y_i>,\ \bar{x}, \bar{Y}, 0>>, \bar{w}, \bar{W}) \in A.$

From the Δ-selection theorem (Moshovakis [5]), we can see that

Z is in $\Delta(\bar{x}, \bar{Y}, \bar{w}, \bar{W})$

Therefore the property (V) of Spector classes holds.

(XVν**) $\forall e' \in N \forall \bar{z}' \forall \bar{Z}'\ [(e, \bar{z}, \bar{Z}) \leq_{\sigma\nu} (e', \bar{z}', \bar{Z}') \vee$
$(\ll <15,\ <\nu, \lambda>, e'>, \bar{z}', \bar{Z}', \bar{x}, \bar{Y}, 0>>, \bar{w}, \bar{W}) \in A]$
$\Leftrightarrow (\ll <15,\ <\nu, \lambda>, e>, \bar{z}, \bar{Z}, \bar{x}, \bar{Y}, 0>>, \bar{w}, \bar{W}) \in B$

Thus, we can check $B(X, \bar{w}, \bar{W})$ satisfies the property (c) in definition 4.1. That is, $B \in \Gamma$.

§6. THE PROOF OF THEOREM 4

In this section, we give the proof of Theorem 4.

From lemma 1, it is enough to say that

$$\ll \Theta_\nu^\Gamma ** >> \in \Gamma.$$

We shall show $\ll \Theta_\nu^\Gamma ** >> \in \Gamma$ by the First Recursion Theorem for Spector 2-classes (Moschovakis [6]) and lemmata.

Choose U which is universal for relations in Γ of signature $(0,1)$. Let

$$\sigma : \longrightarrow On$$

be a Γ-norm on U and we put

$$Q(m, X) \Leftrightarrow X \in \ll \Phi_\nu^\Gamma ** >>\ (\{x' \mid (m, m, X') <_\sigma (m, m, X)\})$$

From the lemma 4, the relation Q is in Γ. It follows that for a fixed $e\# \in \omega$,

$$Q(m, X) \Leftrightarrow U(e\#, m, X)$$

We put

$$R(X) \Leftrightarrow Q(e\#, X)\ (\Leftrightarrow U(e\#, e\#, X)),$$

then, from the definition,

$$R(X) \Leftrightarrow X \in \langle\langle \Phi_\nu^\Gamma ** \rangle\rangle \; (\{X' | (e\#,e\#,X') <_\sigma (e\#,e\#,X)\}).$$

If we can show

$$R(X) \Leftrightarrow X \in \langle\langle \Theta_\nu^\Gamma ** \rangle\rangle,$$

the proof will be completed.

First we show

$$R(X) \Rightarrow X \in \langle\langle \Theta_\nu^\Gamma ** \rangle\rangle$$

by induction on $\sigma(e\#,e\#,X)$.

We suppose

$$R(X') \Rightarrow X' \in \langle\langle \Theta_\nu^\Gamma ** \rangle\rangle$$

for an X' such that

$$\sigma(e\#,e\#,X') < \sigma(e\#,e\#,X).$$

Then, if $R(X)$,

$$\{X' | (e\#,e\#,X') <_\sigma (e\#,e\#,X)\} \subset \langle\langle \Theta_\nu^\Gamma ** \rangle\rangle.$$

So by the monotonicity of $\langle\langle \Phi_\nu^\Gamma ** \rangle\rangle$,

$$X \in \langle\langle \Phi_\nu^\Gamma ** \rangle\rangle \; (\langle\langle \Theta_\nu^\Gamma ** \rangle\rangle).$$

And from lemma 3

$$\langle\langle \Phi_\nu^\Gamma ** \rangle\rangle \; (\langle\langle \Theta_\nu^\Gamma ** \rangle\rangle) = \langle\langle \Theta_\nu^\Gamma ** \rangle\rangle$$

That is,

$$X \in \langle\langle \Theta_\nu^\Gamma ** \rangle\rangle.$$

Next we show

$$X \in \langle\langle \Theta_\nu^\Gamma ** \rangle\rangle \Rightarrow R(X)$$

by induction on ξ, that is

$$X \in (\langle\langle \Phi_\nu^\Gamma ** \rangle\rangle)^\xi \Rightarrow R(X)$$

It will be enough to obtain a contradiction from the hypothesis,

$$X \in (\langle\langle \Phi_\nu^\Gamma ** \rangle\rangle)^\xi, \quad \bigcup_{\eta<\xi} (\langle\langle \Phi_\nu^\Gamma ** \rangle\rangle)^\eta \subset R \text{ and } \neg R(X).$$

From the difinition of $<_\sigma$, if $\neg R(X)$, then

$$\{X' | (e\#,e\#,X') <_\sigma (e\#,e\#,X)\} = R$$

Thus, from $\neg R(X)$ and the above we obtain

$$X \notin \langle\langle \Phi_\nu^\Gamma ** \rangle\rangle (R)$$

And from the definition of $\langle\langle \Phi_\nu^\Gamma ** \rangle\rangle^\xi$, we have

$$X \in \langle\langle \Phi_\nu^\Gamma ** \rangle\rangle \; (\bigcup_{\eta<\xi} (\langle\langle \Phi_\nu^\Gamma ** \rangle\rangle)^\eta).$$

But those contradict the monotonicity of $<<\Phi_\nu^\Gamma**>>$.

Hence $<<\Phi_\nu^\Gamma**>>^\infty \subset R$, that is, $<<\Theta_\nu^\Gamma**>> \subset R$.

§7. A LOCAL REPRESENTATION OF A SPECTOR 2-CLASS

In this section, we give a computation theoretic local representation of a Spector 2-class.

For a 2-class Γ and a signature ν, we put

$$\Gamma_\nu = \{Q \in \Gamma \mid Q \textit{ has signature } \nu\}$$

and

$$en^*(\Theta_\nu^\Gamma) = \{R^* \mid R \in en(\Theta_\nu^\Gamma)\}.$$

From corollary 3.2 and Theorem 4, we have

THEOREM 5

Let Γ be a Spector 2-class which is closed under $\forall^{P(S)}$ and a coding of finite sequences from $S \cup (\bigcup_{n \in \omega} P(S^n))$. Then for each signature ν

$$en^*(\Theta_\nu^\Gamma)_\nu = \Gamma_\nu.$$

REFERENCES

[1] Y. N. Moschovakis, *Descriptive Set Theory* (North-Holland, Amsterdam, 1980).

[2] J. Moldestad, Computations in Higher Types, *Lecture Notes in Mathematics*, No. 574.

[3] A. S. Kechris, Spector Second Order Classes and Reflection, in: J. E. Fenstad, R. O. Gandy and G. E. Sacks (eds.), *Generalized Recursion Theory II* (North-Holland, Amsterdam, 1978), pp.147-183.

[4] J. E. Fenstad, General Recursion Theory, *Perspectives in Mathematical Logic* (Springer-Verlag, 1980).

[5] Y. N. Moschovakis, *Elementary Induction on Abstract Structures* (North-Holland, Amsterdam, 1974).

[6] Y. N. Moschovakis, Structural Characterizations of Classes of Relations, in: J. E. Fenstad and P. G. Hinman (eds.), *Generalized Recursion Theory* (North-Holland, Amsterdam, 1974), pp.53-79.

Precipitousness of the ideal of thin sets on a measurable cardinal

Yuzuru Kakuda

Kobe University

1. **Introduction.** In [7], which seems to be the first publication for approach toward the problem of saturation of ideals by modern set-theoretical methods, Solovay proved that if a regular uncountable cardinal κ has a κ^+-saturated κ-complete non-trivial ideal then we can form the well-founded ultrapower of the universe V inside the Boolean-valued model. Expressed in the terminology introduced by Jech-Prikry [2], a κ^+-saturated κ-complete non-trivial ideal on a regular uncountable cardinal κ is precipitous. On the other hand, there is a well- known problem among set-theoreticians whether the ideal of thin sets on a regular uncountable κ can be κ^+-saturated. In [6], Namba showed that the ideal of thin sets on κ is not κ^+-saturated if κ has a κ-saturated κ-complete non-trivial ideal, and Baumgartner-Taylor-Wagon [1] (also, independently, Kakuda [3]) improved Namba's result so that the ideal of thin sets on κ is not κ^+-saturated for large cardinals κ in a broad class.

On the analogy of the ablove results, it is natural to ask whether the ideal of thin sets on large cardinals can be precipitous. In this paper, we get the following result concerning this problem.

Theorem. *Let* κ *be a measurable cardinal. If the ideal of thin sets on* κ *is precipitous, then, for any ordinal* θ*, there is a transitive model of* ZFC *with* θ *measurable cardinals.*

2. Preliminaries. *M-ultrafilters:* Let M be a transitive model of ZFC, and ρ be an ordinal in M. A subset U of $P(\rho)\cap M$ is *an M-ultrafilter on* ρ if;

i) $\{\xi\}\notin U$ for $\xi<\rho$,

ii) $\forall x\in U\ \forall y\in P(\rho)\cap M\ (x\subset y\rightarrow y\in U)$,

iii) $\forall x\in P(\rho)\cap M\ (x\in U\vee \rho-x\in U)$,

iv) if $\xi<\rho$, $\langle x_\zeta:\zeta<\xi\rangle\in M$ and each $x_\zeta\in U$, then $\bigcap\{x_\zeta:\zeta<\xi\}\in U$.

Let U be an M-ultrafilter on ρ. As usual, we can form the ultrapower $\mathrm{Ult}(M,U)$ of M ($\mathrm{Ult}(M,U)$ is the collection of equivalence classes of functions *in* M from ρ to M), and the elementary embedding $j:M\rightarrow \mathrm{Ult}(M,U)$.

Precipitous ideals: Let κ be a regular uncountable cardinal, and I be a κ-complete no-trivial ideal on κ. Let $I^+=\{x\subset\kappa : x\notin I\}$. We consider I^+ as the set of forcing conditions such that x has more information than y if $x\subset y$. Let G be an I^+-generic filter over the universe V. Then, we can see that G is a V-ultrafilter on κ. We say that I is precipitous if $\Vdash$ "$\mathrm{Ult}(V,G)$ is well-founded".

Iterable M-ultrafilters: Let U be an M-ultrafilter on ρ. We say that U is *iterable* if it satisfies the following condition:

v) if the sequence $\langle x_\xi:\xi<\rho\rangle\in M$, then $\{\xi : x_\xi\in U\}\in M$.

Due to Kunen [5], if U is an iterable M-ultrafilter, we can form the iterated ultrapower $\mathrm{Ult}_\alpha(M,U)$ and the elementary embedding $i_{\alpha\beta}:\mathrm{Ult}_\alpha(M,U)$

$\to \mathrm{Ult}_\beta(M, U)$ for $\alpha<\beta$. Much of the proof of our theorem is due to Kunen's proof of the same conclusion from the existence of a strongly compact cardinal and also to his proof that any normal κ^+-saturated filter on a regular uncountable cardinal κ is strong. The reader should refer to Kunen [5] for detailed exposition of iterated ultrapowers. Here we quote only the following results from [5] which will be essentially used in our proof.

Fact 1. *If arbitrary countable intersetion of elements of U are non--empty, then* $\mathrm{Ult}(M, U)$ *is well-founded for all* α.

Fact 2. i) *If β is any cardinal greater than $2^{\rho(M)}$ and $\mathrm{Ult}_\beta(M, U)$ is well--founded, then $i_{0\beta}(\rho) = \beta$.*

ii) *Suppose $\gamma \geq 1$ and $\mathrm{Ult}_\gamma(M, U)$ is well-founded. If $\mathrm{cf}^{(M)}(\delta) \neq \rho$, δ is a cardinal $> \gamma$, and for all $\xi < \delta$, $(\xi^{\rho(M)})^{=} < \delta$, then $i_{0\gamma}(\delta) = \delta$.*

Normal ideal of thin sets: Let κ be a regular uncountable cardinal. A subset C of κ is closed unbounded in κ if ;

i) $\forall \xi < \kappa \exists \zeta \in C\ (\xi < \zeta)$,

ii) for all limit ordinal $\alpha < \kappa$, $\alpha = \sup(\alpha \cap C)$ impleis $\alpha \in C$.

A subset A of κ is *thin* (or *non-stationary*) if $A \cap C = \emptyset$ for some closed unbounded subset C of κ. It is a well-known fact that the set of all thin subsets of κ is the least normal ideal on κ.

A lemma for the theorem: To prove the theorem, we need the following lemma which was shown, in Kakuda [4], under the weaker condition that κ has a κ-saturated κ-complete non-trivial ideal.

Lemma. *Let κ be a measurable cardinal and U a normal ultrafilter on κ.*

If the ideal of thin sets on κ *is precipitous,* $\{\rho < \kappa : \rho$ *is a regular uncountable cardinal and the ideal of thin sets on* ρ *is precipitous*$\}$ *is in* U.

Proof. For each regular cadinal ρ, I_ρ denotes the ideal of thin sets on ρ. Suppose that I_κ is precipitous. Let M be the transitive collapse of $\mathrm{Ult}(V, U)$. Since M is closed under κ-sequences, we have,

$$M \models \text{"}I_\kappa \textit{ is the ideal of thin sets on } \kappa\text{"}.$$

Hence it suffices to show that,

$$M \models \text{"}I_\kappa \textit{ is precipitous }\text{"}.$$

Let S be an arbitrary set of positive I_κ-measure. Let G be a V-generic ultrafilter with respect to I_κ such that $S \in G$. Clearly, G is an M-generic ultrafilter with respect to I_κ. Since I_κ is precipitous, $\mathrm{Ult}(V, G)$ is well-founded. Let N be its transitive collapse and $j_G ; V \to N$ be the canonical elementary embedding. Since $j_G(U)$ is a normal ultrafilter on $j_G(\kappa)$ in N, $\mathrm{Ult}(N, j_G(U))$. is well--founded. Let M' be the transitve collapse of $\mathrm{Ult}(N, j_G(U))$. Then, the restriction of j_G to M is an elementary embedding of M into M'. It is easily seen that $G = \{x \in P(\kappa) \cap M : \kappa \in j_G(x)\}$. Define $k : \mathrm{Ult}(M, G) \to M'$ as follows: Let $f \in {}^\kappa M \cap M$. Set $k([f]_G) = j_G(f)(\kappa)$. Then, k is anelementary embedding. Therefore, $\mathrm{Ult}(M, G)$ is well-founded. We have shown that for any set S of positive I_κ-measure there exists an M-generic ultrafilter G such that $S \in G$ and $\mathrm{Ult}(M, G)$ is well-founded. Therfore, we have,

$$M \models \text{"}I_\kappa \textit{ is precipitous}\text{"}.$$

3. **Proof of the theorem.** Suppose that the ideal I_κ of thin sets on a measurable cardinal κ is precipitous. Then, by our lemma, we have that

$$P = \{\rho < \kappa : \rho \text{ is a regular uncountable cardinal and } I_\rho \text{ is precipitous}\}$$

is unbounded in κ.

Let θ be any orcinal less than κ. We can define cardinals ρ_μ, $\gamma_{\mu,n}$, λ_μ for $\mu < \theta$, $n < \omega$ as followos: Let Ω be the set of inaccessible cardinals less than κ.

(i) $\langle \gamma_{\mu,n} : n < \omega \rangle$ is an increasing sequence of elements in Ω such that $(\gamma_{\mu,n} \cap \Omega)^= = \gamma_{\mu,n}$ and $\gamma_{\mu,n} > \rho$,

(ii) $\lambda_\mu = \sup\{\gamma_{\mu,n} : n < \omega\}$,

(iii) $\rho_\mu \in P$, $\rho_\mu > \sup\{\lambda_\nu : \nu < \mu\}$ and $\rho_0 > \theta$.

Set $a = \{\gamma_{\mu,n} : \mu \leq \theta,\ n < \omega\}$ and $W_\mu = \{x \in P(\rho_\mu) \cap L[a] : \rho_\mu - x \in I_{\rho_\mu}\}$ for $\mu < \theta$.

Claim I. W_μ is a normal and iterable $L[a]$-ultrafilter on ρ_μ. Moreover, the intersection of any countable subsets of W_μ is non-empty.

Proof. Set $K_\mu = \rho_\mu \cup \{\gamma\ \ \Omega : \gamma > \rho_\mu\} \cup \{\eta : \eta$ is a strongly limit cardinals such that $cf(\eta) > \kappa\}$. We first show that, for each $x \in P(\rho_\mu) \cap L[a]$, there exists a formula ϕ and a finite subset F of K_μ such that

$$x = \{\zeta < \rho : L[a] \models \phi(\zeta, F, a)\}. \qquad (1)$$

Let A be the class of all sets definable in $L[a]$ from elements of $K_\mu \cup \{a\}$. Then, $A \prec L[a]$. Let π be the transitive collapsing of A. Since $(\gamma_{\nu,n} \cap K_\mu)^= = \gamma_{\nu,n}$, $\pi(\gamma_{\nu,n}) = \gamma_{\nu,n}$ for all $\nu < \theta$, $n < \omega$. Hence, $\pi(a) = a$, and so A is isomorphic to $L[a]$. Since π is identity on ρ_μ, $\pi^{-1}(x) = x$, and so $x \in A$ for each $x \in P(\rho_\mu) \cap L[a]$. Therefore, for each $x \in P(\rho_\mu) \cap L[a]$, there exists a formula ϕ and a finite subset F of K_μ satisfying (1).

Let G_μ be a V-generic ultrafilter with respect to I_{ρ_μ}. Since $\rho_\mu \in P$, $\mathrm{Ult}(V, G_\mu)$ is well-founded. We identify $\mathrm{Ult}(V, G_\mu)$ with its transitive collapse. Let $j_\mu : V \to \mathrm{Ult}(V, G_\mu)$ be the canonical embedding. It is clear that $j_\mu(\eta) = \eta$ for all

$\eta \in K_\mu$, in particular, $j_\mu(\gamma_{\nu,n}) = \gamma_{\nu,n}$ for all $\nu < \theta$, $n < \omega$. It follows that $j_\mu(a) = a$ since a has the length of $\theta \cdot \omega < \rho \leq \rho_\mu$.

To show that W_μ is an $L[a]$-ultrafilter, let $x \in P(\rho_\mu) \cap L[a]$ be such that $x \in I_{\rho_\mu}$. Let ϕ and F be a formula and a finite subset of K_μ satisfying (1). Since $x \Vdash x \in G_\mu$, $x \Vdash \kappa \in j_\mu(x)$. Since $j_\mu(a) = a$ and $j_\mu(F) = F$, $x \Vdash$"$L[a] \models \phi(\kappa, F, a)$". Since $\| L[a] \models \phi(\kappa, F, a) \| = 0$ or 1, $\Vdash$ "$L[a] \models \phi(\kappa, F, a)$". Therefore, $\Vdash \kappa \in j_\mu(x)$, and so $\Vdash x \in G_\mu$. This means that $\rho_\mu - x \in I_{\rho_\mu}$. Thus, W_μ is an $L[a]$-ultrafilter.

Since I_{ρ_μ} is non-trivial and normal, it is clear that W_μ is non-principal and normal. For the iterability of W_μ, it is from the fact that $W_\mu = G_\mu \cap L[a]$ and $j_\mu(a) = a$. Since I_{ρ_μ} is ρ_μ-complete, the intersection of any countable subset of W_μ is non-empty.

By Claim I and Fact 1, $\mathrm{Ult}_\alpha(L[a], W_\mu)$ is well-founded for all α. Throughout the proof, $i^\mu_{0\alpha}$ is $i^{W_\mu}_{0\alpha} : L[a] \to \mathrm{Ult}_\alpha(L[a], W_\mu)$.

Claim II. Let ν, $\mu < \theta$.

(i) $i^\mu_{0\alpha}(\gamma_{\nu,n}) = \gamma_{\nu,n}$ if $\alpha < \gamma_{\mu,n}$.

(ii) $i_{0\alpha}(\lambda_\nu) = \lambda_\nu$ if $\alpha < \lambda_\mu$.

Proof. For (i), if $\nu < \mu$, it is clear since $\gamma_{\nu,n} < \lambda_\nu < \rho_\mu$. If $\nu \leq \mu$, it is from the fact that $\alpha < \gamma_{\mu,n} \leq \gamma_{\nu,n}$ and $\gamma_{\nu,n}$ is stronglyinaccessible. For (ii), $i_{0\alpha}(\lambda_\nu) = \sup\{ i_{0\alpha}(\gamma_{\nu,n}) : n < \omega\} = \sup\{ \gamma_{\nu,m} : m \geq n\}$ if $\alpha < \gamma_{\mu,n}$ since $cf(\lambda_\nu) = \omega < \rho_\mu$. Therefore, $i^\mu_{0\alpha}(\lambda_\nu) = \lambda_\nu$.

We define a filter F_μ on λ_μ defined by

$$x \in F_\mu \quad \text{iff} \quad m \quad n \geq m(\gamma_{\mu,n} \in x).$$

Let $\vec{F} = \langle F_\mu : \mu < \theta \rangle$. Since $\vec{F}$ is definable from a, $L[\vec{F}] \subset L[a]$.

Claim III. Let $\mu<\theta$, $\alpha<\lambda_\mu$ and $\nu<\theta$.

(i) Let $x \in P(\lambda_\nu) \cap L[a] \cap \mathrm{Ult}_\alpha([L[a], \mathcal{W}_\mu)$. Then,

$x \in F_\nu$ iff $x \in i^\mu_{0\alpha}(F_\nu \cap L[a])$.

(ii) $i^\mu_{0\alpha}(L[\vec{F}]) = L[\vec{F}]$.

(iii) $i^\mu_{0\alpha}(\vec{F} \cap L[\vec{F}]) = \vec{F} \cap L[\vec{F}]$.

Proof. (i): Since a has the length of $\omega\cdot\theta$, $i^\mu_{0\alpha}(a) = \{i^\mu_{0\alpha}(\gamma_{\nu,n}) ; \nu<\theta, n<\omega\}$. If $x \in F_\nu$, then $n \geq m(\gamma_{\nu,n} \in x)$ for some $m<\omega$. We may suppose that $\alpha<\gamma_{\mu,m}$. Then, $i_{0\alpha}(\gamma_{\nu,n}) \in x)$ for all $n \geq m$ since $i_{0\alpha}(\gamma_{\nu,n}) = \gamma_{\nu,n}$ for all $n \geq m$ by (i) of Claim II. Thus, $x \in i^\mu_{0\alpha}(F_\nu \cap L[a])$. The converse can be proved by the same argument.

(ii): We can show that $L_\eta[\vec{F}] = L_\eta[i_{0\alpha}(\vec{F} \cap L[a])]$ by induction on η by using (i). Since $i_{0\alpha}$ $(L[\vec{F}]) = L[i_{0\alpha}(\vec{F} \cap L[a])]$, we can have the required equality.

(iii): It is clear from (ii).

The following claim can be proved exactly by the same argument of the proof of Lemma 10.10 in Kunen [2].

Claim IV. $F_\mu \cap L[\vec{F}]$ is a normal ultrafilter on λ_μ in $L[\vec{F}]$ for each $\mu<\theta$.

Thus, we have seen that for any ordinal θ less than κ there exists a transitive model of ZFC with θ measurable cardinals. For $\theta \geq \kappa$, we take an $\mathrm{Ult}_\alpha(V, \mathcal{U})$ such that $i_{0\alpha}(\kappa) > \theta$, and do the same discussion inside $\mathrm{Ult}_\alpha(V, \mathcal{U})$.

Note. T. Jech pointed out that our lemma can be also seen by the Π^2_1 indescribability of measurable cardinals and the game theoretic characterization of precipitous ideals.

References

[1] J. Baumgartner, A. Taylor, and S. Wagon, On splitting stationary subsets of large cardinals, Jour. Symbolic Logic, Vol 42 (1977), 203-214.

[2] T. Jech and K. Prikry, Ideal of sets and power set operation, Bull. Amer. Math. Soc. 82 (1976), 593-595.

[3] Y. Kakuda, Saturation of ideals and pseudo-Boolean algebras of ideals of sets, Math. Sem. Notes, Kobe Univ., Vol. 6 (1978), 269-321.

[4] Y. Kakuda, On a condition for Cohen extensions which preserve precipitous ideals, Jour. Symbolic Logic, to appear.

[5] K. Kunen, Some application of iterated ultrapowers in set theory, Ann. Math. Logic 1 (1970), 179-227.

[6] K. Namba, On the closed unbounded ideals of ordinal numbers, Comment. Math. Univ. St. Pauli, XXII-2 (1973), 33-56.

[7] R, M. Solovay, Real-Valued measurable cardinals, Axiomatic set theory (D. Scott Editor), Proceedings of Symposia in Pure Mathematics, Vol. 13, Part I, i971, 397-428.

AXIOM SYSTEMS OF NONSTANDARD SET THEORY

Toru Kawai
Kagoshima University

§1. Introduction.

We propose two similar axiom systems of nonstandard set theory, which can be used as frameworks for nonstandard mathematics.

A theory with an axiom system, which we write NST, is an extension of *internal set theory* IST which Nelson [1] has given. The theory NST deals with external sets directly while IST does not. The axiom system of the theory NST is similar to that of a theory $\mathfrak{R}\mathfrak{S}_2$ which Hrbacek [2] has given. We give a more elementary proof of the conservation theorem for NST. All classes are regarded as external sets in NST, but not in $\mathfrak{R}\mathfrak{S}_2$.

We also define a theory WNST which is weaker than NST. The proof of the conservation theorem for WNST is more elementary, because the ultralimit construction is not used in it. The theory WNST suffices for many applications of nonstandard set theory. For example, countable saturation principle, which is useful in nonstandard probability theory, holds within WNST, as seen in §4.

In this note we replace predicates S (standard) and I (internal) in [3] by two corresponding constants. This enables us to give the axiom schema of saturation in a more convenient form and to deal with any class as an external set.

In §4 we give an outline of the way to approach nonstandard mathematics on the basis of NST or WNST.

§2. Axioms.

Let ZFC denote Zermelo-Fraenkel set theory with the axiom of choice. Variables of ZFC are denoted by letters A,B,$\cdots$,a,b,$\cdots$.

We define a nonstandard set theory NST. The nonlogical symbols of NST are binary predicate $\in$, constant S and constant I. Capital letters A,B,$\cdots$ denote variables of NST. We consider that they range over the "universe of discourse" of NST consisting of external sets. $A \in S$ means that A is a standard set. Intuitively, S can be identified with the "universe of discourse" of ZFC. Lower-case letters a,b,$\cdots$ denote variables ranging over standard sets. $A \in I$ means that A is an internal set. Variables ranging over internal sets are denoted by Greek letters $\alpha,\beta,\gamma,\delta,\zeta,\eta$. A formula ϕ is called an S-formula if ϕ is obtained from a formula of ZFC by replacing all variables by variables ranging over standard sets. A formula obtained by replacing all variables of an S-formula ϕ by variables ranging over internal sets (external sets, respectively) is called an I-formula (E-formula, respectively) and is denoted by ${}^{I}\phi$ (${}^{E}\phi$, respectively). In other words, E-formulas are formulas of NST which do not contain constants S and I. For an S or I-formula ψ, we use $\bar{\psi}$ to denote the formula of ZFC from which ψ is obtained. An S-formula which has no free variables is called an S-sentence.

The nonlogical axioms of NST are the following [A.1]-[A.12].

[A.1] An S-sentence ϕ is an axiom of NST whenever $\bar{\phi}$ is an axiom of ZFC.

[A.2] (Axiom of Extensionality)

$$(\forall A,B)[A = B \equiv (\forall X)[X\in A \equiv X\in B]].$$

[A.3] (Axiom of Pairing)

$$(\forall A,B)(\exists C)(\forall X)[X\in C \equiv X=A \vee X=B].$$

[A.4] (Axiom of Union)

$$(\forall A)(\exists B)(\forall X)[X\in B \equiv (\exists Y)[X\in Y \wedge Y\in A]].$$

[A.5] (Axiom of Power Set)

$$(\forall A)(\exists B)(\forall X)[X\in B \equiv X\subset A].$$

[A.6] (Axiom Schema of Comprehension)

Let $\Omega(X)$ be a formula of NST with free variable X and possibly other free variables. Then

$$(\forall A)(\exists B)(\forall X)[X\in B \equiv X\in A \wedge \Omega(X)].$$

We write $B = \{X\in A : \Omega(X)\}$.

[A.7] (Well-ordering Principle)

$$(\forall A)(\exists B)[B \text{ well-orders } A].$$

The axiom of choice for external sets is deduced from this.

[A.8] $(\forall a)\ [a\in I]$.

All standard sets are internal.

[A.9] $(\forall \alpha)(\forall A)\ [A\in \alpha \rightarrow A\in I]$.

I is transitive.

[A.10] (Transfer Principle)

Let $\phi(a_1,\cdots,a_n)$ be an S-formula with free variables $a_1,\cdots,a_n$ and no other free variables. Then

$$(\forall a_1,\cdots,a_n)\ [\phi(a_1,\cdots,a_n) \equiv {}^{I}\phi(a_1,\cdots,a_n)].$$

[A.11] (Axiom of standardization)

$$(\forall A)[(\exists s)\ A\subset s \rightarrow (\exists a)(\forall x)[x\in A \equiv x\in a]].$$

The standard set a having the same standard elements as A is denoted by ${}^{*}A$.

Definition. D is A-finite $\equiv$ $(\exists\, n;$ natural number$)(\exists F)[F : n \to D$ (1:1, onto)$]$ (where $n = \{0,1,2,\cdots,n-1\}$).

Definition. D has S-size $\equiv$ $(\exists F)[F : S \to D$ (onto)$]$.

For example, S has S-size.

[A.12] (Axiom Schema of Saturation)

Let $\Phi(\alpha,\beta,\gamma_1,\cdots,\gamma_n)$ be an I-formula with free variables $\alpha,\beta,\gamma_1,\cdots,\gamma_n$ and no other free variables. Then

$$(\forall D;\ D \text{ has S-size})\ (\forall \gamma_1,\cdots,\gamma_n)$$
$$\left[\begin{array}{l} (\forall \delta)[\delta \text{ is A-finite} \wedge \delta \subset D \to (\exists \beta)(\forall \alpha \in \delta)\ \Phi(\alpha,\beta,\gamma_1,\cdots,\gamma_n)] \\ \to (\exists \beta)(\forall \alpha \in D)\ \Phi(\alpha,\beta,\gamma_1,\cdots,\gamma_n) \end{array} \right].$$

Remark. Our axiom [A.11] is weaker than an axiom (CO) in [2]. In NST we can not standardize every external set. For example, suppose that $(\forall x)[x \in S \equiv x \in {}^*S]$. Taking $x = {}^*S$, we have ${}^*S \in {}^*S$. This contradicts the axiom of regularity in [A.1]. On the other hand, our axiom schema [A.12] is stronger than ($B4^+$) in [2]. The definition of S-size in NST differs from that of *standard size* in $\mathfrak{NS}_2$.

Under the above axioms of NST, we have the following
Strong Extension Principle:

$$(\forall A)(\forall \alpha)(\forall \beta)(\forall F) \left[\begin{array}{l} A \subset \alpha \wedge A \text{ has S-size} \wedge [F : A \to \beta \text{ (map)}] \\ \to (\exists \gamma)[[\gamma : \alpha \to \beta \text{ (map)}] \wedge (\forall X \in A)[\gamma(X) = F(X)]] \end{array} \right].$$

A weak nonstandard set theory WNST is defined from NST by replacing [A.12] by the following [A.12E] and [A.12W]; that is to say, the nonlogical axioms of WNST are [A.1]-[A.11], [A.12E], and [A.12W].

[A.12E] (Axiom Schema of Enlarging)

Let $\phi(a,b,x_1,\cdots,x_n)$ be an S-formula with free variables $a,b,x_1,\cdots,x_n$ and no other free variables. Then

$$(\forall x_1,\cdots,x_n) \left[\begin{array}{l} (\forall d)[d \text{ is finite} \to (\exists b)(\forall a \in d)\ \phi(a,b,x_1,\cdots,x_n)] \\ \to (\exists \beta)(\forall a)\ {}^I\phi(a,\beta,x_1,\cdots,x_n) \end{array} \right].$$

Remark. A standard set d is finite iff d is A-finite (see §4).

For every external set A, we write ${}^\circ A = \{x \in S : x \in A\}$.

[A.12W] (Weak Extension Principle)

$$(\forall a)(\forall \beta)(\forall F)$$
$$[[F : {}^\circ a \to \beta \text{ (map)}] \to (\exists \gamma)[[\gamma : a \to \beta \text{ (map)}] \wedge (\forall x \in {}^\circ a)[\gamma(x) = F(x)]]].$$

It is easily seen that [A.12E] and [A.12W] are deduced from axioms of NST.

§3. The conservation theorem.

The following conservation theorem asserts that an S-sentence proved in NST or WNST holds true as a theorem of standard mathematics. This shows that NST and WNST can be used for study of standard mathematics. It suffices to prove it for NST only; however, we supply an elementary proof for WNST.

Conservation Theorem for NST (WNST, respectively). *If an S-sentence ψ is a theorem of NST (WNST, respectively), then $\overline{\psi}$ is a theorem of ZFC.*

Proof. Only finitely many of axioms from [A.1], say $\psi_1,\cdots,\psi_m$, and axioms from [A.2]-[A.12] are used in the proof of ψ within NST. Let $R(0) = 0$ and $R(\kappa) = \bigcup_{\sigma<\kappa} P(R(\sigma))$ for every ordinal κ. By reflection principle, there exists a limit ordinal κ such that $(\overline{\psi} \equiv \overline{\psi}^R) \wedge \overline{\psi}_1{}^R \wedge \cdots \wedge \overline{\psi}_m{}^R$, where $\overline{\psi}^R$ and others are the relativizations of $\overline{\psi}$ and others to $R = R(\kappa)$, respectively. Let J be the set of all finite subsets of R, and let $t(j) = \{i \in J : j \subset i\}$ $(j \in J)$. Since $\{t(j) : j \in J\}$ has the finite intersection property, there exists an ultrafilter $\mathfrak{F}$ which includes it. Let $\mathcal{W}_0 = [W_0, \varepsilon_0]$ be the structure defined by $W_0 = R$ and $\varepsilon_0 = \{(a,b) : a,b \in R,\ a \in b\}$.

(*)

Let λ be a limit ordinal such that $cf(\lambda) > |R|$, where $|R|$ is the cardinal of R. For example, it suffices to take $\lambda = 2^{|R|}$. We define structures $\mathcal{W}_\mu = [W_\mu, \varepsilon_\mu]$ $(\mu \leqq \lambda)$ and injections $k_\nu^\mu : W_\nu \to W_\mu$ $(\nu \leqq \mu \leqq \lambda)$ by transfinite induction. Suppose that $\mathcal{W}_\nu$ and $\{k_\rho^\nu\}_{\rho \leqq \nu}$ have been defined for all ν such that $\nu < \mu$. For a successor ordinal $\mu = \nu+1$, let $\mathcal{W}_\mu$ be the ultrapower of $\mathcal{W}_\nu$ over $\mathfrak{F}$, and let $k_\nu^\mu : W_\nu \to W_\mu$ be the canonical embedding. Let $k_\mu^\mu : W_\mu \to W_\mu$ be the identity mapping, and for $\rho < \nu$ let $k_\rho^\mu = k_\nu^\mu \circ k_\rho^\nu$. For a limit ordinal μ, let $(W_\mu, \{k_\nu^\mu\}_{\nu<\mu})$ be the direct limit of $(\{W_\nu\}_{\nu<\mu}, \{k_\rho^\nu\}_{\rho \leqq \nu < \mu})$, and let $k_\mu^\mu : W_\mu \to W_\mu$ be the identity mapping.

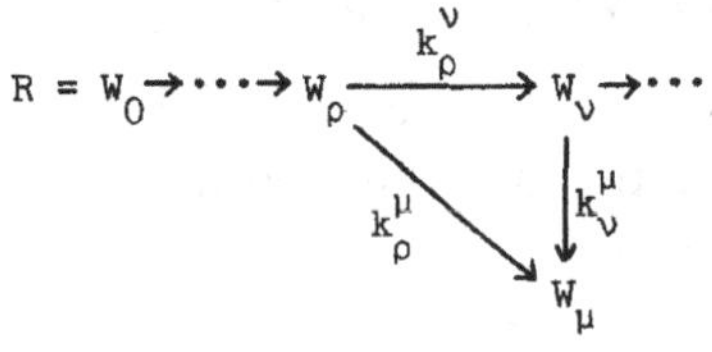

We define the binary relation ε_μ on W_μ as follows: $(x,y) \in \varepsilon_\mu$ iff there are $\nu < \mu$ and $a,b \in W_\nu$ such that $x = k_\nu^\mu(a)$, $y = k_\nu^\mu(b)$, and $(a,b) \in \varepsilon_\nu$. Thus $\{\mathcal{W}_\mu\}_{\mu \leqq \lambda}$ and $\{k_\nu^\mu\}_{\nu \leqq \mu \leqq \lambda}$ have been defined by transfinite induction.

(**)

Let $\nu \leq \mu \leq \lambda$. Then for any formula ϕ of ZFC and any $x_1, \cdots, x_n$ in W_ν,

$$\mathcal{W}_\nu \models \phi[x_1, \cdots, x_n] \quad \text{iff} \quad \mathcal{W}_\mu \models \phi[k_\nu^\mu(x_1), \cdots, k_\nu^\mu(x_n)].$$

That is, k_ν^μ is an elementary embedding of $\mathcal{W}_\nu$ into $\mathcal{W}_\mu$. Furthermore, we proceed by induction. Let $\mathcal{V}_0 = [V_0, e_0]$ be the structure defined by $V_0 = W_\lambda$ and $e_0 = \varepsilon_\lambda$. Since R is transitive and k_0^λ is an elementary embedding of $\mathcal{W}_0$ into $\mathcal{V}_0$, it follows that

$$\mathcal{V}_0 \models (\forall\, a,b)[a = b \equiv (\forall\, c)[c \in a \equiv c \in b]].$$

We define structures $\mathcal{V}_n = [V_n, e_n]$ and injections $p_{n+1} : V_n \longrightarrow V_{n+1}$ for all natural numbers n as follows:

$$V_{n+1} = P(V_n) \text{ (power set);}$$

$$p_{n+1}(a) = \{b \in V_n : (b,a) \in e_n\} \quad \text{for } a \in V_n;$$

$$e_{n+1} = \{(a,b) \in V_{n+1} \times V_{n+1} : (\exists\, c \in V_n)[a = p_{n+1}(c) \wedge c \in b]\};$$

$$\mathcal{V}_{n+1} \models (\forall\, a,b)[a = b \equiv (\forall\, c)[c \in a \equiv c \in b]].$$

Let $(U, \{q_n\}_0^\infty)$ be the direct limit of $(\{V_n\}_0^\infty, \{p_n\}_1^\infty)$. Let $\mathcal{U}_n = [U_n, E_n]$ be structures which correspond to $\mathcal{V}_n = [V_n, e_n]$ by injections $q_n : V_n \longrightarrow U$, and let $E = \bigcup_{n=0}^{\infty} E_n$. We write $h_\mu = q_0 \circ k_\mu^\lambda$ for all $\mu \leq \lambda$. Let $\mathcal{U}^\mu = [U^\mu, E^\mu]$ be structures defined by $U^\mu = h_\mu[W_\mu]$ and $E^\mu = E \cap (U^\mu \times U^\mu)$. Then $\mathcal{W}_\mu$ is isomorphic to $\mathcal{U}^\mu$ for any $\mu \leq \lambda$.

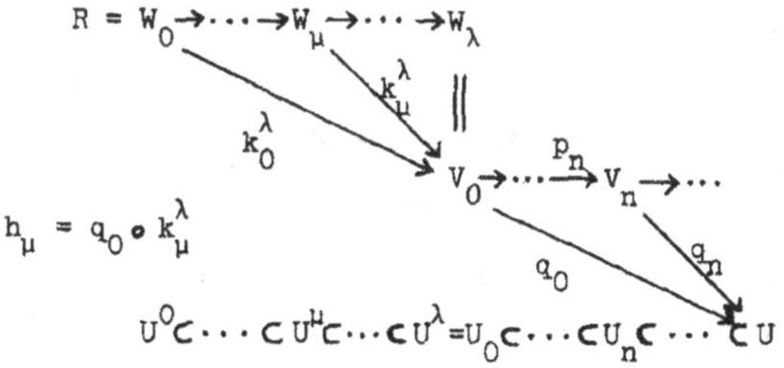

Let $\mathcal{U} = [U, E, S, I]$ be the structure with two constants $S = q_1(k_0^\lambda[R])$ and $I = q_1(V_0)$. We interpret formulas of NST by $\mathcal{U}$ and claim that $\mathcal{U}$ satisfies axioms $\psi_1, \cdots, \psi_m$ and [A.2]-[A.12]. It follows from $\bar{\psi}_1{}^R, \cdots, \bar{\psi}_m{}^R$ that $\mathcal{U}$ satisfies $\psi_1, \cdots, \psi_m$. Let $A \in U$, $n \geq 0$, and $B \in U_{n+1}$. Then $(A,B) \in E$ iff $A \in U_n$ and $q_n^{-1}(A) \in q_{n+1}^{-1}(B)$. This implies that $\mathcal{U}$ satisfies [A.2]-[A.9]. Since $\mathcal{U}_0$ is an elementary extension of $\mathcal{U}^0$, $\mathcal{U}$ satisfies [A.10]. The axiom [A.11] follows from the fact that any subset of an element of R is an element of R. To prove that $\mathcal{U}$ satisfies [A.12], let D be a subset of U_0, let F be a mapping of U^0 onto D, and let $x_1, \cdots, x_n \in U_0$. Since $|D| \leq$

$|R| < cf(\lambda)$, there exists $\rho < \lambda$ such that $D \subset U^{\rho}$. Let τ be an ordinal such that $\rho < \tau < \lambda$ and $x_1, \cdots, x_n \in U^{\tau}$. If $j = \{r_1, \cdots, r_k\} \in J$; that is, j is a finite subset of R, then

$$\mathcal{U}^{\lambda} \models (\exists b)[\bar{\Phi}[F(h_0(r_1)), b, x_1, \cdots, x_n] \wedge \cdots \wedge \bar{\Phi}[F(h_0(r_k)), b, x_1, \cdots, x_n]].$$

Since $\mathcal{U}^{\lambda}$ is an elementary extension of $\mathcal{U}^{\tau}$, the structure $\mathcal{U}^{\tau}$ satisfies the same formula. Hence there exists $b \in W_{\tau}{}^{J}$ such that for all $j \in J$ and $r \in j$,

$$\mathcal{U}^{\tau} \models \bar{\Phi}[F(h_0(r)), h_{\tau}(b(j)), x_1, \cdots, x_n].$$

Put $B = h_{\tau+1}(b/\mathfrak{F})$, where $b/\mathfrak{F}$ is the equivalence class to which b belongs. Let $A = F(h_0(r)) \in D$. Since $t(\{r\})$ is in the filter $\mathfrak{F}$, the larger set

$$\{j \in J : \mathcal{U}^{\tau} \models \bar{\Phi}[A, h_{\tau}(b(j)), x_1, \cdots x_n]\}$$

is in $\mathfrak{F}$. By Łoś's theorem, we have $\mathcal{U}^{\tau+1} \models \bar{\Phi}[A, B, x_1, \cdots, x_n]$. Therefore the same is true in the structure $\mathcal{U}_0$. This shows that $\mathcal{U}$ satisfies [A.12]. Thus the claim has been established. Now the proof of ψ from $\psi_1, \cdots, \psi_m$, [A.2]-[A.12] gives a proof of $\mathcal{U} \models \psi$ from the fact that $\mathcal{U}$ satisfies $\psi_1, \cdots, \psi_m$, [A.2]-[A.12]. The sentence $\bar{\psi}^R$ follows, and so we have $\bar{\psi}$. This gives a proof of $\bar{\psi}$ within ZFC, and the proof of the conservation theorem is complete.

Remark. We have constructed the ultralimit in the proof. Using a good ultrafilter, we can replace it by the only ultrapower construction, but the proof of the existence of the good ultrafilter is complicated.

Elementary proof of the conservation theorem for WNST. In the proof for NST, we replace statements from (*) to (**) by the followings: Let $\lambda = 1$. Let $\mathcal{W}_1 = [W_1, \varepsilon_1]$ be the ultrapower of $\mathcal{W}_0 = [W_0, \varepsilon_0]$ over $\mathfrak{F}$, and let $k_0^1 : W_0 \longrightarrow W_1$ be the canonical embedding. Furthermore, we replace the verification of [A.12] by the followings: To prove that $\mathcal{U}$ satisfies [A.12W], let $(A,S) \in E$, $(B,I) \in E$, and $F : {}^{o}A \rightarrow B$ (map) in $\mathcal{U}$. We write $A = h_0(a)$ $(a \in R)$ and $B = q_0(b/\mathfrak{F})$ $(b \in R^J)$, where $b(j) \neq 0$ for all $j \in J$. There is a mapping $f : \{Y \in U : (Y,B) \in E\} \longrightarrow R^J$ such that for all Y, $q_0(f(Y)/\mathfrak{F}) = Y$ and $f(Y)(j) \in b(j)$ $(j \in J)$. For each $j \in J$, we define the mapping $g(j) : a \rightarrow b(j)$ by $g(j)(x) = f(F(h_0(x)))(j)$ $(x \in a)$. Since κ (in $R = R(\kappa)$) is the limit ordinal, we have $g(j) \in R$ for all j, and so $g \in R^J$. This implies that $G = q_0(g/\mathfrak{F}) \in U$ and $(G,I) \in E$. By Łoś's theorem, we have $G : A \rightarrow B$ (map) and $G(h_0(x)) = F(h_0(x))$ $(x \in a)$ in $\mathcal{U}$. This shows that $\mathcal{U}$ satisfies [A.12W]. We proceed to [A.12E]. Let $x_1, \cdots, x_n \in U^0$. Then there exists $b \in R^J$ such that

$$\mathcal{U}^0 \models \bar{\Phi}[h_0(r), h_0(b(j)), x_1, \cdots, x_n]$$

for all $j \in J$ and $r \in j$. Let $(A,S) \in E$ and $A = h_0(r)$ $(r \in R)$. Since $t(\{r\})$ is in the filter $\mathfrak{F}$, the larger set $\{j \in J : \mathcal{U}^0 \models \bar{\Phi}[A, h_0(b(j)), x_1, \cdots, x_n]\}$ is in $\mathfrak{F}$.

Łoś's theorem shows that

$$\mathcal{U}_0 \models \Phi[A, h_1(b/\mathcal{F}), x_1, \cdots, x_n].$$

It follows that $\mathcal{U}$ satisfies [A.12E]. This completes the elementary proof for WNST.

The above simple proof that $\mathcal{U}$ satisfies [A.12E] is due to A. Robinson and T. Kamae.

§4. The development of NST (WNST).

The axiom schema [A.1] means that the standard sets can be identified with the sets in the theory ZFC, in which conventional mathematics is developed. In particular, the specific objects, for example, the set N of all natural numbers, are constructed as standard sets by ordinary set-theoretical methods. It follows from [A.12E] that every standard infinite set x has non-standard elements. In fact, this is proved by taking $a \neq b \wedge b \in x$ as $\phi(a, b, x)$. Thus non-standard elements already have been added to standard sets in our systems.

Absoluteness.

An S-formula $\phi(x_1, \cdots, x_n)$ whose free variables are all among $x_1, \cdots, x_n$ is said to be absolute if $(\forall \alpha_1, \cdots, \alpha_n)[{}^I\phi(\alpha_1, \cdots, \alpha_n) \equiv {}^E\phi(\alpha_1, \cdots, \alpha_n)]$.

Introduction of new predicate symbols.

Let $\phi(x_1, \cdots, x_n)$ be an S-formula. We introduce predicate symbols Q, ${}^IQ$, and ${}^EQ$ by

$$(\forall x_1, \cdots, x_n)\ [\ Q(x_1, \cdots, x_n) \equiv \phi(x_1, \cdots, x_n)];$$

$$(\forall \alpha_1, \cdots, \alpha_n)\ [{}^IQ(\alpha_1, \cdots, \alpha_n) \equiv {}^I\phi(\alpha_1, \cdots, \alpha_n)];$$

$$(\forall X_1, \cdots, X_n)\ [{}^EQ(X_1, \cdots, X_n) \equiv {}^E\phi(X_1, \cdots, X_n)].$$

Since [A.10] shows that $(\forall x_1, \cdots, x_n)[Q(x_1, \cdots, x_n) \equiv {}^IQ(x_1, \cdots, x_n)]$, the symbol ${}^IQ$ can be abbreviated to Q. If ϕ is absolute; that is, $(\forall \alpha_1, \cdots, \alpha_n)[{}^IQ(\alpha_1, \cdots, \alpha_n) \equiv {}^EQ(\alpha_1, \cdots, \alpha_n)]$, then Q is said to be absolute and ${}^EQ$ also can be abbreviated to Q. For example, if $\phi(x_1, x_2)$ is $(\forall u)[u \in x_1 \rightarrow u \in x_2]$ (x_1 is a subset of x_2), predicate symbols $\subset$, ${}^I\!\subset$, and ${}^E\!\subset$ are introduced. Since [A.9] implies that $\phi(x_1, x_2)$ is absolute, ${}^I\!\subset$ and ${}^E\!\subset$ are abbreviated to $\subset$.

Introduction of new operation symbols.

Let $\psi(x_1, \cdots, x_n, y)$ be an S-formula such that

(1) $$(\forall x_1, \cdots, x_n)(\exists! y)\ \psi(x_1, \cdots, x_n, y),$$

(2) $$(\forall X_1, \cdots, X_n)(\exists! Y)\ {}^E\psi(X_1, \cdots, X_n, Y),$$

where symbols $\exists!$ are used to stand for "there exists a unique". Applying [A.10] to (1), we have

(3) $$(\forall \alpha_1,\cdots,\alpha_n)(\exists!\beta)\ {}^{I}\psi(\alpha_1,\cdots,\alpha_n,\beta).$$

By (1)-(3), operation symbols T, ${}^{I}T$, and ${}^{E}T$ are introduced so that

$$(\forall x_1,\cdots,x_n)\ \psi(x_1,\cdots,x_n,\ T(x_1,\cdots,x_n));$$
$$(\forall \alpha_1,\cdots,\alpha_n)\ {}^{I}\psi(\alpha_1,\cdots,\alpha_n,\ {}^{I}T(\alpha_1,\cdots,\alpha_n));$$
$$(\forall X_1,\cdots,X_n)\ {}^{E}\psi(X_1,\cdots,X_n,\ {}^{E}T(X_1,\cdots,X_n)).$$

Since $(\forall x_1,\cdots,x_n)[T(x_1,\cdots,x_n) = {}^{I}T(x_1,\cdots,x_n)]$, the symbol ${}^{I}T$ is abbreviated to T. The operation T is said to be absolute if ψ is absolute. This means that $(\forall \alpha_1,\cdots,\alpha_n)[{}^{I}T(\alpha_1,\cdots,\alpha_n) = {}^{E}T(\alpha_1,\cdots,\alpha_n)]$, and so ${}^{E}T$ also can be abbreviated to T. For example, if $\psi(x_1,x_2,y)$ is $(\forall u)[u\in y \equiv u=x_1 \vee u=x_2]$, then (1) follows from [A.1]. Moreover, (2) follows from [A.2] and [A.3]. Thus we introduce symbols $\{\ ,\ \}$, ${}^{I}\{\ ,\ \}$ and ${}^{E}\{\ ,\ \}$. Since [A.9] shows that $\psi(x_1,x_2,y)$ is absolute, symbols ${}^{I}\{\ ,\ \}$ and ${}^{E}\{\ ,\ \}$ are abbreviated to $\{\ ,\ \}$. An operation symbol (,) (ordered pair) is introduced by $(x_1,x_2) = \{\{x_1\},\{x_1,x_2\}\}$ and is absolute. Operation symbols $\bigcup$ (union) and P (power set) are introduced by [A.4] and [A.5], respectively. Since the operation P is not absolute, the abbreviation of ${}^{E}P$ causes confusion. We note that new constant symbols are introduced in case $n=0$.

Functions.

An absolute predicate M(f,u,v) (f is a function on u to v) is defined by an S-formula $f \subset u\times v \wedge (\forall x\in u)(\exists! y)\ (x,y)\in f$. An absolute operation V(f,x) (the value of f at x) is defined by an S-formula

$$(\forall f)(\forall x)(\exists! z)(\forall u)\ [u\in z \equiv (\exists y)[u\in y \wedge (x,y)\in f] \wedge (\exists! y)\ (x,y)\in f].$$

Suppose that M(f,u,v). Then ${}^{I}M(f,u,v)$; that is, f is a function on u to v in the internal universe. For an internal α in u, ${}^{I}V(f,\alpha)$ denotes the value of f at α. In practice, both M(f,u,v) and ${}^{I}M(f,u,v)$ are written as $f: u\to v$ (map) and ${}^{I}V(f,\alpha)$ is written as $f(\alpha)$. Thus standard functions are extended naturally in our systems.

Internal definition principle.

If Φ is an I-formula and $\beta_1,\cdots,\beta_n,\delta \in I$, then $\{\alpha\in\delta : \Phi(\alpha,\beta_1,\cdots,\beta_n)\}$ is an internal set. This follows from [A.2], [A.6], [A.9], [A.10], and [A.11].

Finiteness.

A predicate F(a) is defined by an S-formula

$$(\exists n;\ \text{natural number})(\exists f)[f: n\to a\ (1{:}1,\ \text{onto})].$$

F(a) means that a is finite. For an internal set α, ${}^{I}F(\alpha)$ is usually read as "α is star-finite". Since ${}^{I}F(\alpha)$ can be abbreviated to $F(\alpha)$ in NST (WNST), we rather say, "α is finite" or "α is an internal finite set". This finiteness should be distinguished from A-finiteness in [A.12] in case of non-standard set. Applying [A.12E] to an S-formula $a \in b \wedge F(b)$, we obtain an internal finite set β such that $S \subset \beta$.

Countable saturation principle.

If a sequence $X = \{X(n)\}_{n \in N}$ of internal subsets of an internal set δ has the finite intersection property, then there exists an internal element $\alpha \in \delta$ such that $\alpha \in X(n)$ for all $n \in N$. Indeed, since $X : {}^{\circ}N \rightarrow P(\delta)(\text{map})$, the weak extension principle [A.12W] shows that there exists an internal extension γ of X such that $\gamma : N \rightarrow P(\delta)$ (map). Let $\beta(\zeta) = \bigcap\{\gamma(\eta) : \eta \in N \wedge \eta \leq \zeta\}$ $(\zeta \in N)$. Then $\beta \in I$ and $\beta : N \rightarrow P(\delta)(\text{map})$. It follows from the finite intersection property that $\beta(n) \neq 0$ for every standard $n \in N$. Since the internal set $\{\zeta \in N : \beta(\zeta) \neq 0\}$ contains the non-internal set ${}^{\circ}N$, we have $\beta(\zeta) \neq 0$ for some infinite ζ. An element α of $\beta(\zeta)$ has desired properties.

Classes.

We state how to deal with classes as external sets. Let $\Phi(\alpha, \beta_1, \cdots, \beta_n)$ be an I-formula with free variables $\alpha, \beta_1, \cdots, \beta_n$ and no other free variables. For $x_1, \cdots, x_n \in S$, an external set $\{\alpha \in I : \Phi(\alpha, x_1, \cdots, x_n)\}$ is called a class. All standard sets are classes. If a class is an internal set, then it is a standard set. From the above definition of classes, it is easily shown that classes and standard sets satisfy natural requirements, for example, axioms of Bernays-Gödel set theory.

References

[1] Nelson, E., Internal set theory: A new approach to nonstandard analysis, Bull. Amer. Math. Soc., 83 (1977), 1165-1198.

[2] Hrbacek, K., Axiomatic foundations for nonstandard analysis, Fund. Math., 98 (1978), 1-19.

[3] Kawai, T., An axiom system for nonstandard set theory, Rep. Fac. Sci. Kagoshima Univ., (Math., Phys. & Chem.), 12 (1979), 37-42.

SEMI-FORMAL FINITIST PROOF OF THE TRANSFINITE INDUCTION IN AN INITIAL SEGMENT OF CANTOR'S SECOND NUMBER CLASS *)

Shôji MAEHARA

Tokyo Institute of Technology, Tokyo

In his famous paper [2] Gentzen says as follows:

1. The task of the consistency proof of elementary number theory is to justify the disputable forms of inference on the basis of indisputable inferences (opening of III. Abschnitt);

2. The only disputable forms of inference in elementary number theory are essentially those of implication and negation (§ 10 and § 11).

In this paper we shall give such a formal system that contains no disputable forms of inference (Section I), and by help of this formal system we shall prove semi-formally but finitistically the transfinite induction on the ordinal numbers which are constructed by Ackermann [1] and whose parentheses operation is restricted to binary (Section II). This is a trial for formalizing a part of the finitistic inferences. Especially, in our formal system which admits some rules of inference related to the concept "accessibility of an ordinal number", the transfinite induction up to the first epsilon number can be proved without use of rules of inference of implication nor negation.

---

*) This research was partially supported by Grant-in-Aid for Co-operative Research Proj. No.434007, Ministry of Education, Japan.

SECTION I. Formal System

1. We shall use Ackermann's parentheses operation (α,β) in the set-theoretical meaning modified as follows:

1) $(0,\beta)=\omega^{\beta}$;

2) When $\alpha>0$, (α,β) means the β-th ordinal number ξ such that

$$(\lambda,\xi)=\xi \quad \text{for all } \xi<\alpha.$$

EXAMPLE. $(1,\beta)=\varepsilon_{\beta}$.

2. Terms are those defined by use of only one individual constant 0, free variables, and two operations $\alpha+\beta$ and (α,β). Prime formulas are those of the form s=t, s<t or J(t), where $J(\alpha)$ means informally that the ordinal number α is accessible (i.e. the transfinite induction up to α is permissible). More formulas are formed from the above prime formulas in usual way except the following: if A and B are formulas and t is a term, then $A\supset_t B$ is a formula. Any formula of the form $A\supset B$ or $\neg A$ is not used.

When Γ and Θ are finite sequences of zero or more formulas and t is a term, $\Gamma\rightarrow_t\Theta$ is called a sequent of t-type.

Abbreviation: $\Gamma\rightarrow\Theta$ means the 0-type sequent $\Gamma\rightarrow_0\Theta$.

3. The truth-value of any closed prime formula of the form s=t or s<t can be determined. The truth-values of some closed prime formulas of the form J(t) can be recognized —— the fact is all of them are true ——, but we do not presuppose the decidability of J.

4. Ordinal numbers can be defined as special closed terms expressing irredundant representations, in almost the same way as in [1].

For any closed term t, there exists only one ordinal number α such that $t=\alpha$ is true, and this α is called the value of t. The truth-values of closed formulas $s=t$ and $s<t$ can be defined independently on their set-theoretical interpretations.

Accessibility of ordinal numbers:

1) The first ordinal number 0 is considered "accessible";

2) If all ordinal numbers smaller than an ordinal number α have already been recognized as "accessible", then α is also considered "accessible".

When the value of a closed term t is accessible, the prime formula J(t) is said to be true.

5. If a closed formula contains no J, no $\supset$ nor quantifiers, we can decide in usual way whether it is true or false. As truth-values of more closed formulas we shall consider only "true" in the following sense:

1) $A \wedge B$ is true if A and B have already been recognized as true formulas;

2) $A \vee B$ is true if one of the formulas A or B has been recognized as a true formula;

3) $\forall x F(x)$ is true if we have been able to recognize the fact that $F(\alpha)$ is a true formula for an arbitrary ordinal number α;

4) $\exists x F(x)$ is true if $F(\alpha)$ has been recognized as a true formula for a definite ordinal number α;

5) The definition of the true of closed formulas of the form $A \supset_t B$ is postponed for the time being.

6. A closed sequent is said to be true, when there exists a true formula in the succedent or there exists a false formula in

the antecedent. A sequent is said to be <u>valid</u>, when any closed sequent obtained from it by such a substitution of ordinal numbers for free variables that makes all formulas of the antecedent true becomes true.

A true sequent is valid. <u>A valid closed sequent is not always true finitistically</u>.

The concept "a formula is true" has been used in the definition of the validity of sequent. <u>The concept of validity of sequent has not been and will not be used in the definition of the truth of formula</u>. If the concept of validity were used in the definition of true, it would be possible to be a source of disputability.

7. Basic sequent.

7.1. Basic mathematical sequents are chosen from among those valid sequents which contain no J nor $\supset$.

In Section II we shall use the following basic mathematical sequents:

M1. $\rightarrow t \geqq 0$

M2. $\rightarrow s < t,\ s \geqq t$

M3. $s < t,\ s \geqq t \rightarrow$

M4. $s=t \rightarrow t=s$

M5. $s=t \rightarrow (s,r)=(t,r)$

M6. $t \geqq r \rightarrow s+t \geqq r$

M7. $s_1 \geqq r \rightarrow (s_1,s_2)+t \geqq r$

M8. $s_2 \geqq r \rightarrow (s_1,s_2)+t \geqq r$

M9. $r < s+t \rightarrow r < s,\ \exists x(r=s+x \wedge x < t)$

M10. $(s_1,s_2)+t < (r_1,r_2)$

$$\rightarrow s_1 < r_1,\ s_1=r_1 \wedge s_2 < r_2,\ (s_1,s_2) < r_2$$

7.2. Basic logical sequents:

L1. $A \to A$

L2. $A, B \to A \wedge B$ L3. $A \wedge B \to A$ L4. $A \wedge B \to B$

L5. $A \vee B \to A, B$ L6. $A \to A \vee B$ L7. $B \to A \vee B$

L8. $\forall x F(x) \to F(t)$ L9. $F(t) \to \exists x F(x)$

L10. $s < t,\ A \supset_s B,\ A \to_t B$

7.3. Basic equality sequents for J:

$$s=t,\ J(s) \to J(t)$$

7.4. Basic J-sequents:

$$\forall x(x \geq t \vee J(x)) \to J(t)$$

THEOREM 1.1. Any basic sequent of 0-type is valid.

8. Rules of inference.

8.1. Structural rules are same as those of LK (Verdünnung, Zusammenziehung, Vertauschung, Schnitt), provided that one or two upper sequents and the lower sequent of each inference have one and the same type.

8.2. Rules for types:

$$\frac{\Gamma \to \Theta}{\Gamma \to_t \Theta} \qquad \frac{\Gamma \to_s \Theta}{s \leq t,\ \Gamma \to_t \Theta}$$

8.3. $\supset$-introduction:

$$\frac{A, \Gamma \to_s B}{\Gamma \to A \supset_s B}$$

8.4. $\forall$-introduction: 8.5. $\exists$-elimination:

$$\frac{\Gamma \to_s F(a)}{\Gamma \to_s \forall x F(x)} \qquad \frac{F(a), \Gamma \to_s \Theta}{\exists x F(x), \Gamma \to_s \Theta}$$

8.6. Mathematical induction on the number of steps of construction of ordinal number:

$$\frac{F(a),\ F(b),\ F(c),\ \Gamma \rightarrow_s F((a,b)+c)}{F(0),\ \Gamma \rightarrow_s F(t)}$$

8.7. Transfinite induction:

$$\frac{\forall x(x \geqq a \vee F(x)),\ \Gamma \rightarrow_s F(a)}{J(t),\ \Gamma \rightarrow_s F(t)}$$

RESTRICTION ON VARIABLES. In each application of rules 8.4 - 8.7, free variables designated by a, b, c must not occur in the lower sequent.

THEOREM 1.2. If the upper sequent of an inference which is not an application of $\supset$-introduction is valid, then the lower sequent is valid. In the case of a Schnitt, if both of the upper sequents are valid, the lower sequent is valid.

9. Truth of a formula of the form $A \supset_t B$.

Let $A \supset_t B$ be a closed formula and α be the value of t. When we have been able to find a finite sequence Γ of true formulas and a proof of the sequent $A, \Gamma \rightarrow_\alpha B$, the formula $A \supset_t B$ is said to be <u>true</u>.

THEOREM 1.3. If there exists a proof of a sequent $A, \Gamma \rightarrow_t B$, then the sequent $\Gamma \rightarrow A \supset_t B$ is valid.

PROOF. Let a proof P of the sequent $A, \Gamma \rightarrow_t B$ be given. We assume without loss of generality that $A, \Gamma \rightarrow_t B$ is closed and all formulas occurring in Γ are true, and we prove the fact that the formula $A \supset_t B$ is true.

Let α be the value of t. Adjoin

$$\dfrac{\dfrac{A,\ \Gamma \rightarrow_t B}{t \leqq \alpha,\ A,\ \Gamma \rightarrow_\alpha B}\ \text{Rule for types}}{A,\ t \leqq \alpha,\ \Gamma \rightarrow_\alpha B}$$

to the end of the proof P, then we obtain a proof of the sequent $A, t \leqq \alpha, \Gamma \rightarrow_{\alpha} B$ whose antecedent-formulas except A are true. By definition the formula $A \supset_t B$ is true

LEMMA 1. Let t be a closed term and the value of t be 0. A provable sequent of t-type is valid.

PROOF. Let S be a provable sequent of t-type and P be a proof of S. We prove by a mathematical induction on the number of the sequents occurring in P.

1) Let S be a basic sequent.

1.1) When S is a basic sequent of 0-type, S is valid (see Theorem 1.1).

1.2) When S has the form

$$s < t, A \supset_s B, A \rightarrow_t B,$$

S is valid, because the formula $s < t$ in the antecedent can not become true by any substitution of ordinal numbers for free variables.

2) Let P contain at least one inference.

2.1) When the last inference of P is an $\supset$-introduction, S is valid (see Theorem 1.3).

2.2) When the last inference of P has the form

$$\frac{\Gamma \rightarrow_s \Theta}{s \leqq t, \Gamma \rightarrow_t \Theta}.$$

We assume that the endsequent $s \leqq t, \Gamma \rightarrow_t \Theta$ is closed and the antecedent-formulas are true, and we prove the fact that there exists a true formula in Θ.

The value of s is 0, because $s \leqq t$ is true and the value of t is 0. By the hypothesis of the mathematical induction the sequent

$\Gamma \to_s \Theta$ is valid, and on the other hand the formulas of Γ are true, accordingly there exists a true formula in Θ.

2.3) When the last inference of P is one of the other inferences, we can prove the validity of S by use of the hypothesis of the mathematical induction.

LEMMA 2. Let t be a closed term and the value α of t be accessible. A provable sequent of t-type is valid.

PROOF. Let S be a provable sequent of t-type and P be a proof of S. We prove by a transfinite induction on α and by a mathematical induction on the number of the sequents occurring in P.

Hypothesis of the transfinite induction: Any provable sequent whose type has a value smaller than α is valid.

Hypothesis of the mathematical induction: If the number of the sequents occurring in a proof P* is smaller than that of P and the type of the endsequent of P* has the value α, the endsequent of P* is valid.

In the following, we assume $\alpha > 0$ is true (see Lemma 1).

1) Let S be a basic sequent. S has the form

$$s < t,\ A \supset_s B,\ A \to_t B,$$

because the value α of the type t is larger than 0. We assume that S is closed and the antecedent-formulas $s < t$, $A \supset_s B$ and A are true, and we prove the fact that B is true.

There exist a finite sequence Γ of true formulas and a proof of the sequent

$$A,\ \Gamma \to_s B,$$

because the formula $A \supset_s B$ is true. The value of the type s of the sequent is smaller than α, because the formula $s < t$ is true. By the hypothesis of the transfinite induction and the fact that the formula A and the formulas of Γ are true, we can see the fact that the formula B is true.

2) Let P contain at least one inference.

2.1) When the last inference of P has the form

$$\frac{\Gamma \rightarrow_s \Theta}{s \leqq t, \Gamma \rightarrow_t \Theta}.$$

We assume that the endsequent $s \leqq t, \Gamma \rightarrow_t \Theta$ is closed and the antecedent-formulas are true, and we prove the fact that there exists a true formula in Θ.

Let β be the value of s, then $\beta \leqq \alpha$ because the formula $s \leqq t$ is true.

2.11) When $\beta < \alpha$. By the hypothesis of the transfinite induction, we can see the above fact.

2.12) When $\beta = \alpha$. By the hypothesis of the mathematical induction, we can see the above fact.

2.2) When the last inference of P is one of the other inferences, we can prove the validity of S by use of the hypothesis of the mathematical induction. (The last inference of P is not an $\supset$-introduction, because $\alpha > 0$.)

THEOREM 1.4. If an ordinal number α is accessible, then any provable sequent of α-type is valid.

SECTION II. Proof of the Accessibility of Ordinal Numbers

THEOREM 2.1. If a sequent of the form

$$J(a), \forall x(x \geqq a \vee F(x)), \Gamma \rightarrow_s F(a)$$

is provable, then the sequent

$$J(t),\ \Gamma \rightarrow_s F(t)$$

is provable for any term t, where the free variable a must not occur in the second sequent.

PROOF is carried through by use of the basic J-sequent

$$\forall x(x \geqq a \vee J(x)) \rightarrow J(a)$$

and the transfinite induction with respect to $J(x) \wedge F(x)$.

THEOREM 2.2. If both sequents

$$\Gamma \rightarrow_s F(0)$$

and

$$(a,b)+c < r,\ F(a),\ F(b),\ F(c),\ \Delta \rightarrow_s F((a,b)+c)$$

are provable, then the sequent

$$t < r,\ \Gamma,\ \Delta \rightarrow_s F(t)$$

is provable for any term t, where free variables a, b, c must not occur in the last sequent.

PROOF is carried through by use of the mathematical induction on the number of steps of construction of ordinal number with respect to $x \geqq r \vee F(x)$ and by helps of the basic mathematical sequents M2, M3, M6, M7, M8.

THEOREM 2.3. The following sequents are provable:

1) $\rightarrow J(0)$

2) $a < b,\ J(b) \rightarrow J(a)$

3) $J(a),\ J(b) \rightarrow J(a+b)$

4) $J(a) \rightarrow J(b) \supset_a J((a,b))$

PROOF. 1) By use of the basic J-sequent

$$\forall x(x \geqq 0 \vee J(x)) \rightarrow J(0)$$

and the basic mathematical sequent M1, we can obtain a required

proof.

2) By use of the basic J-sequent

$$\forall y(y \geq c \vee J(y)) \to J(c),$$

we obtain

$$\forall x[x \geq d \vee \forall y(y \geq x \vee J(y))] \to \forall y(y \geq d \vee J(y));$$

by use of the transfinite induction with respect to $\forall y(y \geq x \vee J(y))$,

$$J(b) \to \forall y(y \geq b \vee J(y));$$

by use of the basic mathematical sequent M3,

$$a < b,\ J(b) \to J(a).$$

3) By use of M3, M4 and the basic equality sequent for J, we obtain

$$\exists x(c=a+x \wedge x < d),\ \forall x(x \geq d \vee J(a+x)) \to J(c);$$

by use of M9 and the above result of 2),

$$c < a+d,\ J(a),\ \forall x(x \geq d \vee J(a+x)) \to J(c);$$

by use of M2,

$$J(a),\ \forall x(x \geq d \vee J(a+x)) \to \forall x(x \geq a+d \vee J(x));$$

by use of the basic J-sequent,

$$J(a),\ \forall x(x \geq d \vee J(a+x)) \to J(a+d);$$

by use of the transfinite induction,

$$J(a),\ J(b) \to J(a+b).$$

4) By use of the basic logical sequent

$$u < a,\ J(v) \supset_u J((u,v)),\ J(v) \to_a J((u,v))$$

and M3, we have

(1) $u < a,\ \forall x\{x \geq a \vee \forall y[J(y) \supset_x J((x,y))]\},\ J(v) \to_a J((u,v)).$

On the other hand, by use of M3, M4, M5 and the basic equality sequent for J, we have

(2) $u=a,\ v < b,\ \forall y[y \geq b \vee J((a,y))] \to J((u,v)),$

and we have, by use of 2),

(3) $(u,v) < b,\ J(b) \rightarrow J((u,v))$.

By use of (1), (2), (3) and M10 (moreover, by help of the first rule of inference for types), we obtain

$$(u,v)+w < (a,b),\ J(v),\ J(b),\ A,\ B \rightarrow_a J((u,v)),$$

where A and B stand for the formulas

$$\forall x\{x \geqq a \vee \forall y[J(y) \supset_x J((x,y))]\} \quad \text{and} \quad \forall y[y \geqq b \vee J((a,y))],$$

respectively; by use of 3),

$$(u,v)+w < (a,b),\ J(v),\ J(w),\ J(b),\ A,\ B \rightarrow_a J((u,v)+w);$$

by use of Theorem 2.2 and 1),

$$c < (a,b),\ J(b),\ A,\ B \rightarrow_a J(c);$$

by use of M2 and the basic J-sequent,

$$J(b),\ A,\ B \rightarrow_a J((a,b)),$$

i.e. $J(b),\ A,\ \forall y[y \geqq b \vee J((a,y))] \rightarrow_a J((a,b))$;

by use of Theorem 2.1,

$$J(b),\ A \rightarrow_a J((a,b));$$

by use of the $\supset$-introduction,

$$A \rightarrow \forall y[J(y) \supset_a J((a,y))],$$

i.e. $\forall x\{x \geqq a \vee \forall y[J(y) \supset_x J((x,y))]\} \rightarrow \forall y[J(y) \supset_a J((a,y))]$;

by use of the transfinite induction,

$$J(a) \rightarrow \forall y[J(y) \supset_a J((a,y))].$$

Hence we have

$$J(a) \rightarrow J(b) \supset_a J((a,b)).$$

THEOREM 2.4. If ordinal numbers α and β are accessible, then the ordinal numbers $\alpha+\beta$ and (α,β) are accessible.

PROOF is obtained by use of the provabilities of 3) and 4) of Theorem 2.3, and by Theorem 1.4.

The ACCESSIBILITY of each of those ordinal numbers which are given in 4. of Section I results from Theorem 2.4.

REFERENCES

[1] W. Ackermann, Konstruktiver Aufbau eines Abschnitt der zweiten Cantorschen Zahlenklasse. Math. Z., 53 (1951), 403 - 413.

[2] G. Gentzen, Die Widerspruchsfreiheit der reinen Zahlentheorie. Math. Ann., 112 (1936), 493 - 565.

ON THE LENGTH OF PROOFS IN A FORMAL SYSTEM OF RECURSIVE ARITHMETIC

BY

TOHRU MIYATAKE

§0. Introduction.

In [6], R. J. Parikh proved the following result:

Theorem. (Parikh [6; Theorem 3]). For any formula A(a), the formula xA(x) is provable in PA* if and only if there is a natural number k such that all the instances $A(\bar{n})$ $(n \in \omega)$ of A(a) are provable in PA* within k steps of inferences.

, where PA* is a system for Peano arithmetic formalized with only one function symbol for successor and which represents addition and multiplication by relations, and $\bar{n}$ is a formal expression denoting the natural number n.

In this paper, we consider a system of primitive recursive arithmetic (denoted by PRA) in a formulation of logic-free equation calculus along the line of Curry [2]. PRA has S (for successor), Z (for constant 0 function), I_i^n (for projection) as initial functions and C (for composition), R (for recursion) as function connectives.

Our purpose is to show that the following (*) holds for PRA, which can be considered as an analogue of Parikh's result mentioned above, for PRA.

(*) For any equation t(a)=u(a), t(a)=u(a) is provable in PRA if and only if there is a natural number k such that all the instances $t(\bar{n})=u(\bar{n})$ of t(a)=u(a) are provable in PRA with using at most k occurrences of equations.

It seems that our result is not included in Parikh's result, and our method of proof is something different from that of Parikh's proof, which will be discussed in §5.

In §1, we describe the system PRA and, in §2 we introduce a system PRK which is a conservative first order extension of PRA and prove some preliminary facts and state our main assertion. §3 is devoted to the proof of the main assertion. The method of proof in this paper also works for other similar systems, e.g. $\mathcal{E}^n$-arithmetic (subsystem of PRA) and m-recursive arithmetic (extension of PRA) which are explained as in Cleave & Rose [3] and Rose [7]. These will be mentioned in §4.

<u>§1.</u> System PRA (logic-free equation calculus for primitive recursive arithmetic).

Let $\mathcal{L}$ be a language which consists of following symbols.

numerical constant: 0.

numerical variables: a,b,c,...

function constants: S, Z, I_i^n ($1 \leqslant i \leqslant n$).

n-function variables for each $n \geqslant 1$: f^n, g^n, h^n,...

functor connectives: C, R.

predicate symbol: =.

<u>Defnition.</u> Functors are defined as follows.

(i) S, Z, and 1-function variables are 1-functors.

(ii) I_i^n and n-function variables are n-functors.

(iii) If F is an m-functor and $G_1,\ldots,G_m$ are n-functors, then $CFG_1\ldots G_m$ is an n-functor. We write this $CF\bar{G}$ for short.

(iv) If F is an n+2-functor and G is an n-functor, then RFG is an n+1-functor.

Definition. Terms are defined as follows.

(i) 0 and numerical variables are terms.

(ii) If F is an n-functor and $t_1,\ldots,t_n$ are terms, then $F(t_1,\ldots,t_n)$ is a term. We write this $F(\bar{t})$.

Definition. Equations are expressions of the form t=u, where t and u are terms.

Axioms of PRA

(Z) $Z(a)=0$

(I) $I_i^n(\bar{a})=a_i$

(C) $CF\bar{G}(\bar{a})=F(G_1(\bar{a}),\ldots,G_m(\bar{a}))$

(R1) $RFG(\bar{a},0)=G(\bar{a})$

(R2) $RFG(\bar{a},S(b))=F(\bar{a},b,RFG(\bar{a},b))$

Inference rules of PRA

(E1) $$\frac{t=u \quad t=v}{u=v}$$

(E2) $$\frac{t_1=u_1 \ \ldots \ t_n=u_n}{F(\bar{t})=F(\bar{u})}$$

(Sb1) $\dfrac{\theta}{\theta|_t^a}$, where $\theta|_t^a$ is an equation obtained from θ by replacing all occurrences of a free variable a in θ by a term t.

(Sb2) $\dfrac{\theta}{\theta|_F^f}$, where $\theta|_F^f$ is an equation obtained from θ by replacing all occurrences of a (n-)function variable f by a (n-)functor F.

(U) $$\frac{F(\bar{a},0)=G(\bar{a},0) \quad F(\bar{a},S(b))=H|_F^f(\bar{a},b) \quad G(\bar{a},S(b))=H|_G^f(\bar{a},b)}{F(\bar{a},b)=G(\bar{a},b)}$$

In (U), F, G, H are n+1-functors, f is an n+1-funcion variable, and H is an n+1-functor of the form $CH_0H_1\ldots H_m$ where

(i) H_0 contains no occurrences of f,

(ii) H_i $(1 \leq i)$ is f or contains no occurrences of f.

Moreover, $H|^f_F$ and $H|^f_G$ are functors obtained from H by replacing all occurrences of f in H by F and G respectively. (U) expresses the fact that functions with the same defining equation are the same function.

Remark. PRA is equivalent to the system described in Goodstein [5] with function variables.

Proof figures in PRA are defined as usual (in a tree form), and $PRA \vdash \theta$ means that there is a proof figure with θ at its bottom, and $PRA \overset{k}{\vdash} \theta$ means that there is a proof of θ which has at most k equations within it.

§2. System PRK (first order extension of PRK).

Language of PRK is the first order language obtained from $\mathcal{L}$ by adding bound variables and logical symbols. Basic sequents of PRK are those sequents of the form $A \rightarrow A$, where A is a formula defined as usual.

Inference rules of PRK are all the rules of (Gentzen's) LK and the following additional ones:

(Z) $$\frac{Z(t)=0,\ \Gamma \rightarrow \Delta}{\Gamma \rightarrow \Delta}$$

(I) $$\frac{I^n_i(\bar{t})=t_i,\ \Gamma \rightarrow \Delta}{\Gamma \rightarrow \Delta}$$

(E1) $$\frac{t=u \wedge t=v \supset u=v,\ \Gamma \rightarrow \Delta}{\Gamma \rightarrow \Delta}$$

(E2) $$\frac{t_1=u_1 \wedge \cdots \wedge t_n=u_n \supset F(\bar{t})=F(\bar{u}),\ \Gamma \rightarrow \Delta}{\Gamma \rightarrow \Delta}$$

(C) $$\frac{\mathrm{CF}\bar{G}(\mathfrak{t})=F(G_1(\mathfrak{t}),\ldots,G_m(\mathfrak{t})),\Gamma\longrightarrow\Delta}{\Gamma\longrightarrow\Delta}$$

(R1) $$\frac{\mathrm{RFG}(\mathfrak{t},0)=G(\mathfrak{t}),\Gamma\longrightarrow\Delta}{\Gamma\longrightarrow\Delta}$$

(R2) $$\frac{\mathrm{RFG}(\mathfrak{t},S(u))=F(\mathfrak{t},u,\mathrm{RFG}(\mathfrak{t},u)),\Gamma\longrightarrow\Delta}{\Gamma\longrightarrow\Delta}$$

(U) $\dfrac{A,\Gamma\longrightarrow\Delta}{\Gamma\longrightarrow\Delta}$, where A is a formula of the following form:

$$F(\mathfrak{t},0)=G(\mathfrak{t},0)\wedge\forall x(F(\mathfrak{t},S(x))=H\left|{}^{f}_{F}(\mathfrak{t},x)\wedge\forall x(G(\mathfrak{t},S(x))=H\right|{}^{f}_{G}(\mathfrak{t},x)).\supset.$$
$$F(\mathfrak{t},u)=G(\mathfrak{t},u).$$

We call the formulae explicitly written in each additional schemata, the principal formulae of the inferences.

Proof figures in PRK are defined as usual, and $\mathrm{PRK}\vdash\Gamma\longrightarrow\Delta$ means that there is a proof of $\Gamma\longrightarrow\Delta$, and $\mathrm{PRK}\overset{k}{\vdash}\Gamma\longrightarrow\Delta$ means that there is a proof of $\Gamma\longrightarrow\Delta$ with at most k sequents in it.

By the same way as Gentzen's [4], we can prove.

<u>Proposition-1.</u> Cut-elimination theorem hold for PRK, i.e. every provable sequent can be proved without cut-rule.

<u>Remark.</u> By the cut-elimination procedure, the resulting cut-free proof has larger length than the original proof. But reflections on the procedure tells us that the length of the resulting cut-free proof depends only on the structure of the original proof and the numberof logical symbols which are introduced by the inference rules in the original proof. So we can conclude from the assertion "there is a proof of $\Gamma\longrightarrow\Delta$ with at most length k", that "there is a proof of $\Gamma\longrightarrow\Delta$ without cut rule at most length k_1, where k_1 is effectively calculated from k".

Now we translate quantifier-free formulae and quantifier-free sequents of PRK to equations of PRA as follows. For formulae, we make the following translation; $t=u$ to $(t \dot{-} u)+(u \dot{-} t)=0$, $t \neq 0$ to $1 \dot{-} t=0$, $t=0 \wedge u=0$ to $t+u=0$, e.t.c., where $t+u$ and $t \dot{-} u$ are abbreviations of $R(CSI_3^3)I_1^1(t,u)$ and $R(CPI_3^3)I_1^1(t,u)$ respectively, and P in the latter is $C(RI_2^3Z)I_1^1I_1^1$ which stands for predecessor function. For sequents $A_1,\ldots,A_m \rightarrow B_1,\ldots,B_n$, we take the above mentioned translation of $A_1 \wedge \ldots \wedge A_m \supset B_1 \vee \ldots \vee B_n$. If $m=0$, we take the translation of $0=0 \supset B_1 \vee \ldots \vee B_n$, and if $n=0$, we take the translation of $A_1 \wedge \ldots \wedge A_m \supset S(0)=0$ respectively.

Then the following propositions are checked easily.

Proposition-2. PRK is a conservative extension of PRA, i.e. for any equation θ, if $\rightarrow \theta$ is provable in PRK, then θ is provable in PRA.

Proposition-3. There is an effective function φ such that, for any equation θ, if $PRA \vdash^{k} \theta$, then $PRK \vdash^{\varphi(k)} \rightarrow \theta$.

Formulae of Presburger arithmetic are formulae that are constructed from numeral 0, succcesor function S, addition function +, =, and logical symbols only. We also call those formulae, naturally translted into PRK from Presburger formulae, Presburger formulae.

Proposition-4. Let A be any valid formula of Presburger arithmetic. Then $PRK \vdash \rightarrow A$.

It is easy to check that the usual method of quantifier elimination works in PRK (cf. Chang & Keisler [1]). In doing this, we use the facts that $t>u$ and $1 \dot{-} (t \dot{-} u)=0$, $t \equiv u (\mathrm{mod}\ \bar{n})$ and $rem(t \dot{-} u,\bar{n})+rem(u \dot{-} t,\bar{n})=0$ are equivalent in PRK, where rem is the usual remainder function. We remark that these functions are in $\mathcal{E}^2$, and the properties needed are provable in $\mathcal{E}^2$-arithmetic.

Now we state our assertions.

Theorem-1.

PRK $\vdash \; \rightarrow \forall x A(x)$ iff there is a k such that $(\forall n)\text{PRK} \overset{k}{\vdash} \; \rightarrow A(\bar{n})$.

(Where $\bar{n}$ is an abbreviation of the term $\underbrace{S(S(\ldots S}_{n\text{-times}}(0)\ldots)$.)

Theorem-2. Let a be a free variable. Then for any equation θ, PRA $\vdash \theta$ iff there is a k such that $(\forall n)\text{PRA} \overset{k}{\vdash} \theta\Big|^{a}_{\bar{n}}$.

Theorem-2 follows immediately from Proposition-3, Theorem-1, and Proposition-2. So we only have to prove Theorem-1.

We prove the following Main Lemma which implies Theorem-1 as its special case.

Main Lemma. Let $\Gamma(a) \longrightarrow \Delta(a)$ be a sequent of PRK. Then PRK $\vdash \Gamma(a) \longrightarrow \Delta(a)$ iff there is a k such that $(\forall n)\text{PRK} \overset{k}{\vdash} \Gamma(\bar{n}) \longrightarrow \Delta(\bar{n})$.

§3. Proof of the Main Lemma.

Let P_n's be proof figures of $\overset{k}{\vdash} \Gamma(\bar{n}) \longrightarrow \Delta(\bar{n})$. We can assume without loss of generality that P_n's are cut-free and all basic sequents in P_n's are atomic formulae.

Let a P_n be fixed.

Definition. For each term occurrence t in P_n, we define,

(i) Each maximal term occurrence t is a normal occurrence,

(ii) If t is a normal occurrence and t is of the form $F(\bar{u})$, where F is a functor different from S, then each u_i is a normal occurrence,

(iii) If t is a normal occurrence and is of the form $S(S..S(u)..)$, we write $S^i(u)$ for short, and u is not of the form $S(v)$, then u is a normal occurrence.

Now we make a blocking of P_n as follows.

(1) For the end-sequent $\Gamma(\bar{n}) \longrightarrow \Delta(\bar{n})$, we mark each normal occurrence in it with # in the following way:

$S^i(0) \Longrightarrow \#S^i\#(0)$

$S^i(a) \Longrightarrow \#S^i\#(a)$ where a is a free variable

$S^i(x) \Longrightarrow \#S^i\#(x)$ where x is a bound variable

$F(u_1,\dots,u_n) \Longrightarrow F(\tilde{u}_1,\dots,\tilde{u}_n)$ where F is a functor different from S.

$S^i(u) \Longrightarrow \#S^i\#(\tilde{u})$

In the above u, u_i's are normal occurrences and $\tilde{u}$, $\tilde{u}_i$'s are their marked occurrences.

Finally we add # to enclose those occurrences of S^n of $\bar{n}$ which are substituted for a in $\Gamma(a) \longrightarrow \Delta(a)$.

We write $\tilde{A}$ for the formula obtained from A by marking each term occurrences in A in this way.

(2) For each inference rules of LK which are different from quantifier rules: $\dfrac{\Gamma \longrightarrow \Delta}{\Pi \longrightarrow \Lambda}$ or $\dfrac{\Gamma' \longrightarrow \Delta' \quad \Gamma'' \longrightarrow \Delta''}{\Pi \longrightarrow \Lambda}$

We have a marked sequent $\tilde{\Pi} \longrightarrow \tilde{\Lambda}$ for the lower sequent. Each formula occurrence in the upper sequents have the formula occurrence corresponding naturally to it. So we transfer their marks to the formula occurrences in the upper sequent.

(3) $\forall$- or $\exists$-rules: $\dfrac{\Gamma \longrightarrow \Delta,\ A(t)}{\Gamma \longrightarrow \Delta, QxA(x)}$ or its dual inferences.

where Q is $\forall$ or $\exists$.

We have a marked sequent $\tilde{\Gamma} \longrightarrow \tilde{\Delta}, \widetilde{QxA(x)}$ for the lower sequent. Let $\tilde{t}$ be a term marked according to (1). Then we take $\tilde{\Gamma} \longrightarrow \tilde{\Delta}, \tilde{A}(\tilde{t})$ for the upper sequent, where $\tilde{A}(\tilde{t})$ is the formula obtained from $\widetilde{A(x)}$ by substituting $\tilde{t}$ for all occurrences of x in $\widetilde{A(x)}$.

(4) For the inferences not of LK:

(4.1) Rules (Z) and (I): $\dfrac{Z(t)=0, \Gamma \longrightarrow \Delta}{\Gamma \longrightarrow \Delta}$ and $\dfrac{I_i^n(\bar{t})=t_i, \Gamma \longrightarrow \Delta}{\Gamma \longrightarrow \Delta}$

Let $\widetilde{\Gamma} \longrightarrow \widetilde{\Delta}$ be for the lower sequent, then we take for the upper sequent $Z(\tilde{t})=0, \widetilde{\Gamma} \longrightarrow \widetilde{\Delta}$ and $I_i^n(\tilde{t}_1, \ldots, \tilde{t}_n)=\tilde{t}_i, \widetilde{\Gamma} \longrightarrow \widetilde{\Delta}$ respectively.

(4.2) Rule (C): $\dfrac{CF\bar{G}(\bar{t})=F(G_1(\bar{t}), \ldots, G_m(\bar{t})), \Gamma \longrightarrow \Delta}{\Gamma \longrightarrow \Delta}$

If all F and G_i's are not S, then we take for the upper sequent, $CF\bar{G}(\tilde{t}_1, \ldots, \tilde{t}_n)=F(G_1(\tilde{t}_1, \ldots, \tilde{t}_n), \ldots, G_m(\tilde{t}_1, \ldots, \tilde{t}_n)), \widetilde{\Gamma} \longrightarrow \widetilde{\Delta}$.

If some of F, G_i's are S, then we make additional marks to enclose such F or G_i's in the right hand side of the equation.
For instance,

(i) $CF\bar{G}(\tilde{t})=F(G_1(\tilde{t}), \ldots, \#G_{i_1}\#(\tilde{t}), \ldots), \widetilde{\Gamma} \longrightarrow \widetilde{\Delta}$

, if $G_{i_1}, \ldots, G_{i_r}$ are S and F and $G_j (j \neq i_1, \ldots, i_r)$ are not S.

(ii) $CFG(\tilde{t}_1, \ldots, \tilde{t}_n)=\#F\#(G(\tilde{t}_1, \ldots, \tilde{t}_n)), \widetilde{\Gamma} \longrightarrow \widetilde{\Delta}$, if F is S and G is not S.

(iii) $CFG(\tilde{t})=\#F\#(G\#(\tilde{t})), \widetilde{\Gamma} \longrightarrow \widetilde{\Delta}$, if F and G are S.

(4.3) All the remaining rules are treated similarly. We make additional marks if there is an occurrence of S explicitly denoted in the principal formula as in (4.2)(i)~(iii).

We write $\widetilde{P_n}$ for the marked proof figure. In $\widetilde{P_n}$ we define,

(i) In the end-sequent $\Gamma(\bar{n}) \longrightarrow \Delta(\bar{n})$, we call the blocks $\#S^n\#$ in $\#S^n\#(0)$, where $S^n(0)$ is an occurrence of $\bar{n}$ which is substituted for a in $\Gamma(a) \longrightarrow \Delta(a)$, designated blocks (d-blocks). All the other blocks are called invariant blocks (i-blocks).

(ii) At the stage of each inference rule, additional blocks produced (e.g. $\#F\#$, $\#G_i\#$'s in (4.2), etc.) are called i-blocks.

(iii) D-blocks and i-blocks are transfered from a lower sequent to upper sequents at each stage.

(iv) The blocks which are neither d-blocks nor i-blocks are called neutral blocks (n-blocks).

Let φ be defined inductively as follows:

$\varphi(0)=0$

$\varphi(a)=a$

$\varphi(\#S^i\#(t))=\varphi(t)+\tau_i$

$\varphi(F(\bar{t}))=F(\varphi(t_1),\dots,\varphi(t_n))$ if F is not S

, where

$$\tau_i=\begin{cases} a_0 & \text{if } \#S^i\# \text{ is a d-block} \\ \bar{1} & \text{if } \#S^i\# \text{ is an i-block} \\ b_i & \text{if } \#S^i\# \text{ is an n-block} \end{cases}$$

and $\{a_0,b_1,b_2,\dots,b_j,\dots\}$ is a set of new free variables.

(Remark) In the above a+b is an abbreciation of +(a,b) where + is the functor $R(CSI^3_3)I^1_1$ which stands for addition.

We write $\varphi(A)$ for the formula which is obtained from a blocked formula A by replacing each term occurrence t in A by $\varphi(t)$, and $\varphi(\Gamma \longrightarrow \Delta)$ for the sequent which is obtained from a blocked sequent $\Gamma \longrightarrow \Delta$ by replacing each formula occurrence A in by $\varphi(A)$.

Let $P(t_1,\dots,t_n) \rightarrow P(t_1,\dots,t_n)$ be a basic sequent in $\widetilde{P}_n$. Observe that two occurrences of t_i in this sequent may have different blocks, so we distinguish these two occurrences by denoting $P(t^1_1,\dots,t^1_n) \rightarrow P(t^2_1,\dots,t^2_n)$.

We construct a finite set of equations for each pair (t^1_i,t^2_i)

as follows. Let u_i^1 and u_i^2 are two corresponding normal occurrences in t_i^1 and t_i^2, then we construct $\Omega(u_i^1,u_i^2)$ such that $\Omega(u_i^1,u_i^2)=\Omega(v_i^1,v_i^2)\cup E(u_i^1,u_i^2)$, where $E(u_i^1,u_i^2)$ is

ϕ — If u_i^1 is $\boxed{1}\,v_i^1$, u_i^2 is $\boxed{2}\,v_i^2$, v_i^1 and v_i^2 are normal, $\boxed{1}$ and $\boxed{2}$ consist of the same(*) blocks. (*)with regard to also their kinds (d-blocks, i-block or n-block)

$h(a_0,\bar{b})=g(a_0,\bar{b})$ — Otherwise. Where h and g are corresponding terms for $\boxed{1}$ and $\boxed{2}$. For instance, if $\boxed{1}$ is $\#S^{i_1}\#S^{i_2}\#S^{i_3}\#$ and $\boxed{2}$ is $\#S^{j_1}\#S^{j_2}\#$, then $h(a_0,\bar{b})$ is $\tau_{i_3}+\tau_{i_2}+\tau_{i_1}$ and $g(a_0,\bar{b})$ is $\tau_{j_2}+\tau_{j_1}$.

$$\Omega_{P(t_1^1,\ldots,t_r^1)\to P(t_1^2,\ldots,t_r^2)}=\bigcup_{1\leq i\leq r}\Omega(t_i^1,t_i^2)$$

$\Omega_n=\bigcup\Omega_{P(t_1^1,\ldots,t_r^1)\to P(t_1^2,\ldots,t_r^2)}$, where $P(t_1^1,\ldots,t_r^1)\to P(t_1^2,\ldots,t_r^2)$ ranges over all basic sequents in $\widetilde{P_n}$.

(Remark) Reflections on the blocking procedure tell us that each normal occurrence has at most three blocks in its outermost part.

By $\varphi(\widetilde{P_n})$ we denote the figure obtained from $\widetilde{P_n}$ by replacing each sequent $\Gamma \longrightarrow \Delta$ in $\widetilde{P_n}$ by $\varphi(\Gamma \longrightarrow \Delta)$. Although inferences in $\varphi(\widetilde{P_n})$ are correct derived rules in PRK, top sequents in $\varphi(\widetilde{P_n})$ are not basic sequents. But for each such sequent $P(\varphi(t_1^1)\ldots\varphi(t_r^1)) \longrightarrow P(\varphi(t_1^2)\ldots\varphi(t_r^2))$, we can construct a proof figure in PRK of the sequent

$$\Omega_n, P(\varphi(t_1^1)\ldots\varphi(t_r^1)) \to P(\varphi(t_1^2)\ldots\varphi(t_r^2)).$$

From these proof figures and $\varphi(\widetilde{P_n})$, we obtain a proof figure of the sequent $\Omega_n, \varphi(\Gamma(\bar{n})) \rightarrow \varphi(\Delta(\bar{n}))$. It is easy to see that $\Omega_n, \varphi(\Gamma(\bar{n})) \quad \varphi(\Delta(\bar{n}))$ is equivalent in PRK to the sequent $\Omega_n, \Gamma(a_0) \longrightarrow \Delta(a_0)$. So we get the proof figure of $\Omega_n, \Gamma(a_0) \longrightarrow \Delta(a_0)$ in PRK.

Ω_n is a set of equations $h(a_0,\bar{b})=g(a_0,\bar{b})$, and h and g are of the form $\alpha_1+\ldots+\alpha_j$ (1 j 3) where α_j is one of the followings:

(i) free variables $a_0, b_1, b_2, \ldots$

(ii) numerals (bounded by some number which depends only on $\Gamma(a) \longrightarrow \Delta(a)$)

Now we define,

$$C_n(a_0) = \exists\bar{x}\left\{\bigwedge_{h=g\in\Omega_n} h(a_0,\bar{x})=g(a_0,\bar{x})\right\} .$$

(In the above, we write $\bar{b}$ for some finite sequence $b_{j_1},\ldots,b_{j_m}$ which are elements of $\{b_1,b_2,\ldots\}$ and $\exists\bar{x}$ for $\exists x_{j_1}\exists x_{j_2}\ldots\exists x_{j_m}$.)

Then we get the proof figure P_n^* of $C_n(a_0), \Gamma(a_0) \longrightarrow \Delta(a_0)$. (Note that free variables in $\bar{b}$ don't appear in $\Gamma(a_0) \longrightarrow \Delta(a_0)$.)

We claim that the number of the equations in Ω_n is bounded uniformly in n. If this is the case, then $\{C_n(a_0)\}$ can be divided into finite classes by their logical equivalence in PRK. So we pick up their representatives $C_{r_1}(a_0),\ldots,C_{r_s}(a_0)$ such that for each n there is a $C_{r_j}(a_0)$ which is equivalent to $C_n(a_0)$. Since $\text{PRK}\vdash C_{r_j}(a_0), \Gamma(a_0) \longrightarrow \Delta(a_0)$ for all $1\leq j\leq s$, we get $\text{PRK}\vdash \bigvee_{1\leq j\leq s} C_{r_j}(a_0), \Gamma(a_0) \longrightarrow \Delta(a_0)$. Here $\bigvee_{1\leq j\leq s} C_{r_j}(a_0)$ is a valid formula in Presburger arithmetic. In fact for each n, we can read off suitable numbers $m_1,\ldots$ from P_n such that $h(\bar{n},\bar{m}_1,\ldots)=g(\bar{n},\bar{m}_1,\ldots)$ is true for each $h(a_0,\bar{b})=g(a_0,\bar{b})$ in Ω_n. So $C_n(\bar{n})$ is valid. There is a $C_{r_j}(a_0)$ which is equivalent to $C_n(a_0)$, hence $C_{r_j}(\bar{n})$ is valid. Therefore $\bigvee_{1\leq j\leq s} C_{r_j}(a_0)$ is valid. By this and Proposition-4 we can get PRK $\vdash \Gamma(a_0) \longrightarrow \Delta(a_0)$.

Now we show the following claim.

<u>Claim</u>: The number of equations in Ω_n is bounded uniformly in n.

Let us ignore term occurrences in P_n and look at their logical structure of the sequents and kinds of inference rules in P_n. Let's call this a skeleton of P_n. Since length of P_n is bounded by k and all basic sequents in P_n's are atomic, skeletons arising from P_n's are finite. We call $\Box$ of blocked normal term occurrence $\Box t$ a building (blg.), where $\Box$ consists of blocks and outermost symbol of t is not S, and a blg. is regular if it consists of only one n-block. The equation h=g in Ω_n arises when two occurrences of a normal term t in a basic sequent, say $\boxed{1}\,v_1$ and $\boxed{2}\,v_2$, have different blg.'s $\boxed{1}$ and $\boxed{2}$. So it is sufficient to show that the number of the equations arising from these such pairs ($\boxed{1}$, $\boxed{2}$) is bounded uniformly in n. If $\boxed{1}$ and $\boxed{2}$ are different, then $\boxed{1}$ or $\boxed{2}$ is not regular. Observe that all the inference rules except (C), (R), (E2), (U), $\forall$-left and $\exists$-right rules always produce blg.'s which are regular. So examine these rules carefully. First we remark that the number of occurrences of bound variables in each P_n is bounded uniformly in n. For, the number of occurrences of bound variables in the end sequent depends only on the sequent $\Gamma(a) \longrightarrow \Delta(a)$, and, in tracing from the end sequent up to basic sequents, only (U)-rules produce exactly 6 new occurrences of bound variables. Since lengths of P_n's are bounded by k, the number of occurrences of bound variables in P_n is bounded uniformly in n by some k_1.

Now we consider the above mentioned rules one by one.

(1) $\forall$-left and $\exists$-right rules.

In $\dfrac{\Gamma \rightarrow \Delta \ , A(t)}{\Gamma \rightarrow \Delta, \exists x A(x)}$ and $\dfrac{A(t), \Gamma \rightarrow \Delta}{\forall x A(x), \Gamma \rightarrow \Delta}$, let $S^{i_1}x, \ldots$ $\ldots, S^{i_r}x$ be all the minimal normal term occurrences containing x in A(x). If the bounded term t is of the form $S^j u$(u is not of the form Sw), then new non-regular blg.'s $\#S^{i_1}\#S^j\#, \ldots, \#S^{i_r}\#S^j\#$ arise in the upper sequent. But, since r is bounded by k_1, the number of non-regular blg's which are produced by $\forall$-left and $\exists$-right rules is bounded uniformly in n.

(2) (E2)-rule.

In (E2), if F is S, then this rule is $\dfrac{t{=}u \supset S(t){=}S(u), \Gamma \rightarrow \Delta}{\Gamma \rightarrow \Delta}$

(if F is not S, then no new non-regular blg.'s arise in the upper sequent). So there arise exactly two non-regular blg.'s $\#S\#S^i\#$ and $\#S\#S^j\#$ in the upper sequent, where t is $S^i t'$ and u is $S^j u'$ (t' and u' are not of the form Sw). Hence the number of non-regular blg.'s which are produced by (E2) is bounded uniformly in n.

(3) (R1)-rule.

New non-regular blg.'s arise only when G is S and only one new non-regular blg. #S# is produced in the upper sequent, $RFS(0){=}\#S\#(0), \Gamma \rightarrow \Delta$.

(4) (R2)-rule.

In this case there also arise only one new non-regular blg. $\#S\#S^i\#$ in the upper sequent, $RFG(t, \#S\#S^i\#(u')){=}F(t,u,RFG(t,u)), \Gamma \rightarrow \Delta$, where u is $S^i u'$ (u' is not of the form Sw).

(5) (U)-rule

$$\frac{A,\Gamma \longrightarrow \Delta}{\Gamma \longrightarrow \Delta}$$, where A is of the form,

$$F(\mathfrak{t},0)=G(\mathfrak{t},0) \wedge \forall x(F(\mathfrak{t},S(x))=H|^f_F(\mathfrak{t},x)) \wedge \forall x(G(\mathfrak{t},S(x))=H|^f_G(\mathfrak{t},x)).\supset. F(\mathfrak{t},u)=G(\mathfrak{t},u).$$

(5.1) If F and G are not S, then exactly two occurrences of new non-regular blg. #S# (in $F(\tilde{\mathfrak{t}},\#S\#(x))$ and $G(\tilde{\mathfrak{t}},\#S\#(x))$) arise.

(5.2) If F is S and G is not S, then exactly four occurrences of new non-regular blg.'s #S# (in #F#(0), G(#S#(x)), and $\#F\#S^i\#(\tilde{u}')$, where S^iu' is u and u' is not of the form Sw), #F#S# (in #F#(S#(x))) arise.

(5.3) If F is not S and G is S, or F and G are S, there arise exactly four, or six, new non-regular blg.'s respectively.

(6) (C)-rule

$$\frac{CFG_1 \ldots G_m(\mathfrak{t})=F(G_1(\mathfrak{t}),\ldots,G_m(\mathfrak{t})),\Gamma \longrightarrow \Delta}{\Gamma \longrightarrow \Delta}$$

(6.1) If F is S and G is not S, then there arise exactly one new non-regular blg. #F# (in $CFG(\tilde{\mathfrak{t}})=\#F\#(G(\tilde{\mathfrak{t}}))$) .

(6.2) If F and G are S, then only one new non-regular blg. $\#F\#G\#S^i\#$ (in $CFG(\tilde{t})=\#F\#(G\#(S^i\#\tilde{t}'))$, where t is S^it' and t' is not of the form Sw) is produced in the upper sequent.

(6.3) If F is not S and some of $G_1,\ldots,G_m$ are S, then the upper sequent becomes

$CF\bar{G}(\hat{t})=F(G_1(\tilde{t}),\ldots,\#G_{r_1}\#(\tilde{t}),\ldots),\tilde{\Gamma} \longrightarrow \tilde{\Delta}$, where $G_{r_1},\ldots,G_{r_s}$ is S and $G_j(j\neq r_1,\ldots,r_s)$ is not S. So in this case the number of new non-regular blg.'s may not be bounded uniformly in n, but these non-regular blg.'s are all the same form $\#S\#S^i\#$ (where t is S^it' and t' is not of the form Sw).

Now we estimate the number of equations arising from the pairs $\{(\boxed{1},\boxed{2}): \boxed{1}$ or $\boxed{2}$ is non-regular$\}$.

If $\boxed{1}$ or $\boxed{2}$ comes from one of (1)~(5),(6.1),(6.2), then we can conclude, from the above considerations, that the number of these such pairs is bounded uniformly in n.

For remaining pairs, there are two cases:

(i) Both $\boxed{1}$ and $\boxed{2}$ come from (6.3).

In this case, $\boxed{1}$ and $\boxed{2}$ are the same blg.. So there arise no equations from this pair.

(ii) One comes from (6.3) and the other is a regular blg..

There are at most k applications of (C)-rules, hence there are at most k such blg.'s which are mutually distinct. Let they be $\#S\#S^{i_1}\#,\ldots,\#S\#S^{i_r}\#$ $(r\leq k)$. Then equations arising from these pairs are one of the followings,

$$b_{i_j}+\bar{1}=b_{i_j+1}\quad,\quad b_{i_j+1}=b_{i_j}+\bar{1}\qquad(1\leq j\leq r).$$

Hence we can conslude that the number of the equations in Ω_n is bounded uniformly in n. This completes the proof of the claim.

Q.E.D.

§4. Applications to similar systems.

The method of a proof, in this paper, can be used to obtain similar results for other similar systems. For instance, we mention $\mathcal{E}^n$-arithmetic and m-recursive arithmetic. Here $\mathcal{E}^n$-arithmetic is a recursive arithmetic, described as in Cleave & Rose [3] and Rose [7], for the class of $\mathcal{E}^n$-functions. $\mathcal{E}^n$-functions are defined as follows: Let $g_n(x,y)$ be given by

$$g_0(x,y)=y+1, \quad g_1(x,y)=x+y, \quad g_2(x,y)=x\cdot y$$

$$g_{n+3}(0,y)=1, \quad g_{n+3}(x+1,y)=g_{n+2}(g_{n+3}(x,y),y)$$

Then the class $\mathcal{E}^n$ is the class of functions with initial functions

$$U_0(x)=x+1, \quad U_1(x,y)=x, \quad U_2(x,y)=y, \quad \text{and } g_n(x,y)$$

and which is closed under the operations of substitution (replacing a free variables by a function, another variable or a constant) and limited recursion given by

$$h(0,x_1,\dots,x_m)=p(x_1,\dots,x_m)$$

$$h(y+1,x_1,\dots,x_m)=q(y,x_1,\dots,x_m,h(y,x_1,\dots,x_m))$$

$$h(y,x_1,\dots,x_m)\leq r(y,x_1,\dots,x_m),$$

where the functions p, q and r have already been defined.

And the system of m-recursive arithmetic is the system obtained from PRA by adding the schema for m-fold recursion.

In this section we suppose that the reader is familar with Cleave & Rose [3] and Rose [7] , for exact formulations of these systems, see [3] and [7] .

We write $\mathcal{E}^n$-AR for the system of $\mathcal{E}^n$-arithmetic, and m-RA for the system of m-recursive arithmetic. $\mathcal{E}^n$-AR is a recursive arithmetic with those functions in $\mathcal{E}^n$. We also write $\mathcal{E}^\omega$-AR for the system with those functions in $\bigcup_{n<\omega}\mathcal{E}^n$. $\mathcal{E}^\omega$-AR is equivalent to PRA.

m-RA is a recursive arithmetic with functions defined by m-fold recursion and 1-RA is equivalent to PRA. In [3] and [7] , the following (1) and (2) are proved. Let Con($\mathcal{E}^n$) and Con(m) be equations of $\mathcal{E}^n$-AR and of m-RA expressing the consistency of $\mathcal{E}^n$-AR and of m-RA respectively.

(1) For $n \geq 2$, $\mathcal{E}^{n+1}$-AR $\vdash$ Con($\mathcal{E}^n$) [3]

(2) m+1-RA $\vdash$ Con(m) [7]

Now by essentially the same argument in §1 ~ 3, we can prove,

Theorem-3. Let $n \geq 2$. For any equation θ,

$\mathcal{E}^n$-AR $\vdash$ θ iff there is a k such that $(\forall n)$ $\mathcal{E}^n\text{-AR} \overset{k}{\vdash} \theta\Big|^{a}_{\bar{n}}$.

Theorem-3'. For any equation θ,

m-RA $\vdash$ θ iff there is a k such that $(\forall n)\text{m-RA} \overset{k}{\vdash} \theta\Big|^{a}_{\bar{n}}$.

Finally,

Theorem-4.(cf. Parikh [6; Theorem 4]) Let $n \geq 2$. For any function φ from natural numbers to natural numbers, there exists a provable equation θ of $\mathcal{E}^n$-AR (resp. m-RA) and a number k such that $\mathcal{E}^{n+1}$-AR (resp. m+1-RA) $\overset{k}{\vdash}$ θ holds but $\mathcal{E}^n$-AR(resp.m-RA) $\overset{\varphi(k)}{\vdash}$ θ does not hold.

(proof: essentially the same as in [6])

Let θ(a) be an equation which expresses the consistency of $\mathcal{E}^n$ $\mathcal{E}^n$-AR (resp. m-RA). This equation is provable in n $\mathcal{E}^{n+1}$-AR (resp. m+1-RA) with, say, k equations. Take $k_1=k+1$, then for every i, $\mathcal{E}^{n+1}$-AR (resp. m+1-RA) $\overset{k_1}{\vdash} \theta\Big|^{a}_{\bar{i}}$. However, $(\forall i)$ $\mathcal{E}^n$-AR(resp. m-RA) $\overset{\varphi(k)}{\vdash} \theta\,{}^{a}_{\bar{i}}$ would give $\mathcal{E}^n$-AR(resp. m-RA) $\vdash$ θ(a). Hence there is some i such that $\mathcal{E}^n$-AR(resp. m-RA) $\overset{\varphi(k)}{\vdash}$ θ($\bar{i}$) does not hold. Take this θ($\bar{i}$) to be the desired equation. Q.E.D.

§5. Appendix.

In this paper we did not follow the argument in Parikh [6] . We discuss, in this section, the point which causes us to depart from Parikh's method.

First, there may be a suitable translation φ of equations of PRA into a formula of PA* such that,

(i) $PRA \overset{k}{\vdash} A$ implies $PA^* \overset{k'}{\vdash} \varphi(A)$ for some k' depending on k and A.

(ii) $PA^* \vdash \varphi(A)$ implies $PRA \vdash A$.

If such a φ exists, then we can prove Theorem-2 via Parikh's result as follows.

$$(\forall n) PRA \overset{k}{\vdash} A(\bar{n}) \implies (\forall n) PA^* \overset{k'}{\vdash} \varphi(A(\bar{n})) \qquad \text{(by (i))}$$
$$\implies PA^* \vdash \varphi(A(a)) \qquad \text{(by Parikh's result)}$$
$$\implies PRA \vdash A(a) \qquad \text{(by (ii))}$$

The condition (ii) seems to be necessary, for we must return to PRA from PA*. The natural translation φ via graphs of p.r. functions does not satisfy (ii), e.g. let A(a) be an equation which expresses the consistency of PRA, then clearly $PA^* \vdash \varphi(A(a))$ but $PRA \not\vdash A(a)$. Unfortunately, we don't know whether such a translation exists or not.

Now we try to follow Parikh's proof (suitably modified for PRA) precisely. Parikh's proof consists of the following (1)~(3).

(1) For any schematic system T, $T \overset{k}{\vdash} A$ implies $T \vdash^{k}_{l} A$, where $\vdash_{l}$ means that there is a proof of A with formulae with at most l logical symbols, and l is effectively calculated from k and the number of logical sbmbols in A.

(2) If, further, T has only one unary function symbol in its language, then we can construct a formula B(a) of Presburger arithmetic such that $PA^* \vdash \forall x(B(x) \leftrightarrow "T \vdash^{k}_{l} A(\bar{x})")$. Of course $"T \vdash^{k}_{l} A(\bar{x})"$

is an arithmetization of the corresponding fact.)

(3) If T is PA*, then $PA^* \vdash \forall x(\text{"}PA^* \vdash^{k}_{l} A(\bar{x})\text{"}) \rightarrow \forall x A(x)$.

From the fact that $\forall x B(x)$ is a valid sentence of Presburger arithmetic, we conclude $PA^* \vdash \forall x A(x)$.

So we want to get the following (4) ~ (6) parallel to (1) ~ (3).

(4) If $PRA \vdash^{k} A$, then $PRA \vdash^{k}_{l} A$ for some l depending on k and A.

(5) Construct a formula B(a) of Presburger arithmetic such that, $PRK \vdash \forall x(B(x) \leftrightarrow \text{"}PRK \vdash^{k}_{l} A(\bar{x})\text{"})$.

(6) Prove the partial reflection principle with respect to "$PRA \vdash^{k}_{l}$" in PRK.

In (4) ~ (6), $\vdash_{l}$ means that there is a proof with terms with at most l functor connectives.

We can consider PRK instead of PRA, in (4) ~ (6), but the essential point lies in the structure of terms. There will be no essential difficulties in (6). And if Parikh's method can be completed for (4), then (5) will be automatically established. So we consider (4) in details.

In the following we consider the system obtained by omitting (Sb1), (Sb2) and , instead of (U), we take

$$\frac{F(\mathfrak{t},0)=G(\mathfrak{t},0) \qquad F(\mathfrak{t},S(a))=H\Big|^{f}_{F}(\mathfrak{t},a) \qquad G(\mathfrak{t},S(a))=H\Big|^{f}_{G}(\mathfrak{t},a)}{F(\mathfrak{t},u)=G(\mathfrak{t},u)}$$

, hereafter we mean this rule by (U). It suffices to establish (4) ~ (6) to get Theorem-2. We also call this system PRA.

To simplify the argument, for a term t of PRA, we define t* as follws:

(i) (0)* is 0, (a)* is a, where a is a free variable.

(ii) $(F(t_1,\ldots,t_n))^*$ is $F(t_1^*,\ldots,t_n^*)$, where F is not S.

(iii) $(S(t))^*$ is t^*.

We introduce two kinds of meta-variables and extend the language of PRA.

(i) term-variables (t-var.) : σ, τ, ...

(ii) functor-variables (f-var.) : φ, ψ, ...

<u>Definition.</u> Analysis is a tree, in which a node is one of the followings; (Z), (I), (C), (R1), (R2), (E1), (E2) and (U;f,a), and such that

(i) top node is one of (Z), (I), (C), (R1), and (R2),

(ii) all the other nodes are one of (E1), (E2), and (U;f,a),

(iii) a node (E1) has exactly two nodes immeadiately above it,

(iv) a node (U;f,a) has exactly three nodes immeadiately above it.

The intended meaning is clear. So we give no explanations.

<u>Definition.</u> Let be given an analysis. A diagram with respect to (w.r.t.) this analysis is obtained by assigning an equation (of the extended language) to each node of the analysis.

Let be given an analysis and a diagram w.r.t. this analysis. If there is a substitution of terms (of the original language) to t-var.'s and a substitution of functors (of the original language) to f-var.'s, then we can construct a formula B of Presburger arithmetic w.r.t. this substitution such that,

PRK$\vdash$ B $\leftrightarrow$ "there is a proof with the given analysis" .

So we try to find a diagram w.r.t. the given analysis and an equation, and to find a suitable substitution for this diagram.

<u>1st step.</u> To the bottom node of the analysis, we assign the given equation (of PRA).

<u>2nd step.</u> We construct a diagram by assigning equations to nodes of the analysis from the bottom to the top nodes. And, at the same time, we construct finite sets Ω_I, Ω_C, Ω_{R1}, Ω_{R2}, Ω_U, $\Omega_{U'}$,

and $\Omega_=$. Suppose we have already assigned equations as follows.

$$\begin{array}{c} t_1=t_2 \\ \downarrow \\ \theta \end{array}$$, where θ is the given equation corresponding to the bottom,

(2.1) Case (E1), i.e. the node corresponding to $t_1=t_2$ is (E1).

In this case we extend as follows.

$$\begin{array}{ccc} \sigma=t_1 & & \sigma=t_2 \\ & \diagdown \diagup & \\ & t_1=t_2 & \end{array}$$, where σ is a new t-var.

(2.2) Case (E2).

(2.2.1) The case that t_1 (or t_2) has a form $F(\bar{u})$.

(2.2.1.1) If t_2 (or t_1) is a t-var. σ, then we replace all the occurrences of σ, in the so far constructed diagram and in $\Omega_I \sim \Omega_=$, by $F\bar{\delta}$, where $\delta_1, \ldots, \delta_n$ are new t-var.'s. And extend thus,

$$\begin{array}{ccc} u_1=\delta_1 & \cdots & u_n=\delta_n \\ & \diagdown \mid \diagup & \\ & F\bar{u}=F\bar{\delta} & \end{array}$$.

(2.2.1.2) If t_2 (or t_1) has a form $G\bar{v}$, then we add a pair (F,G) to $\Omega_=$, and extend as follows.

$$\begin{array}{ccc} u_1=v_1 & \cdots & u_n=v_n \\ & \diagdown \mid \diagup & \\ & F\bar{u}=G\bar{v} & \end{array}$$.

(2.2.1.3) If t_2 (or t_1) has a form $\varphi\bar{v}$ (φ:f-var.), then we replace all occurrences of φ by F as in (2.2.1.1) and extend as follows.

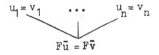

$$\begin{array}{ccc} u_1=v_1 & \cdots & u_n=v_n \\ & \diagdown \mid \diagup & \\ & F\bar{u}=F\bar{v} & \end{array}$$.

(2.2.2) The case that both of t and t are t-var.'s σ and τ.

We replace σ and τ by $\varphi\bar{\delta}$ and $\varphi\bar{\nu}$, where φ, $\delta_1, \ldots, \delta_n$, $\nu_1, \ldots, \nu_n$ are new f-var. and new t-var.'s, respectively as in

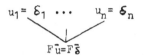

(2.2.1.1), and extend thus,

$$\frac{\delta_1 = \nu_1 \quad \cdots \quad \delta_n = \nu_n}{\varphi\bar{\delta} = \varphi\bar{\nu}}.$$

(2.3) Case (U;f,a).

(2.3.1) The case that t_1 (or t_2) has a form $F(\bar{u}, u_0)$.

(2.3.1.1) If t_2 (or t_1) is a t-var. σ, then we replace σ by $\varphi_0(\bar{u}, u_0)$ as in (2.2.1.1) and extend as follows.

$$\frac{F(\bar{u},0) = \varphi_0(\bar{u},0) \qquad F(\bar{u},a) = \varphi_1(\bar{u},a) \qquad \varphi_0(\bar{u},a) = \varphi_2(\bar{u},a)}{F(\bar{u},u_0) = \varphi_0(\bar{u},u_0)}$$

, where φ_0, φ_1, φ_2 are new f-var.'s. And we add (φ_1, φ_2) to Ω_U.

(2.3.1.2) If t_2 (or t_1) has a form $G(\bar{v}, v_0)$, then we extend as follows,

$$\frac{F(\bar{u},0) = G(\bar{v},0) \qquad F(\bar{u},a) = \varphi_1(\bar{u},a) \qquad G(\bar{v},a) = \varphi_2(\bar{v},a)}{F(\bar{u},u_0) = G(\bar{v},v_0)}$$

, where φ_1 and φ_2 are new f-var.'s, and we add (φ_1, φ_2) to Ω_U, and (u_i, v_i) $(i=0,\ldots,n)$ to $\Omega_=$.

(2.3.2) The case that both of t_1 and t_2 are t-var.'s σ and τ.

We extend as follows,

$$\frac{\sigma_0 = \tau_0 \qquad \sigma_1 = \gamma_1 \qquad \tau_1 = \gamma_2}{\sigma = \tau}$$

, where σ_i, τ_i, γ_j $(i=0,1 ; j=1,2)$ are new t-var.'s, and we add (σ_0, σ_1), (τ_0, τ_1) to Ω_U, and (γ_1, γ_2) to Ω_U.

(2.4) Case (Z).

(2.4.1) If t_1 is a t-var. σ, then we replace σ by $Z(\tau)$ and t_2 by 0, where τ is a new t-var. (in this case, t_2 must be an atomic expression if not, there is no proof with this analysis).

(2.4.2) If t_1 has a form $F(u)$, replace F by Z and t_2 by 0 (in this case, F must be Z or a f-var. and t_2 must be as in (2.4.1)).

(2.5) Case (C).

(2.5.1) The case that t_1 has a form $CF_0F_1\ldots F_m(\bar{u})$.

(2.5.1.1) If t_2 has a form $F(\bar{v})$, then we add ($v_i, F_i(\bar{u})$)'s and (F_0, F) to $\Omega_=$.

(2.5.1.2) If t_2 is a t-var. σ, then we replace σ by $F_0(F_1(\bar{u})\ldots F_m(\bar{u}))$.

(2.5.2) The case that t_2 has a form $F(\bar{v})$ and t_1 is a t-var. σ.

We replace σ by $CF\varphi_1\ldots\varphi_m(\bar{\tau})$, where $\varphi_1,\ldots,\varphi_m$ are new f-var.'s, $\tau_1,\ldots,\tau_n$ are new t-var.'s and add ($\varphi_i(\bar{\tau})$, v_i)'s to $\Omega_=$.

(2.5.3) For remaining cases, we add (t_1, t_2) to Ω_C.

All the remaining cases (I), (R1), and (R2) are treated similarly to the case of (C).

3rd step. By 1st and 2nd steps, we have sets of pairs Ω_C, Ω_I, Ω_{R1}, Ω_{R2}, Ω_U, $\Omega_{U'}$ and $\Omega_=$. We replace each pair in $\Omega_=$ by suitable pairs, according to the structures of t_1, t_2 such that for each pair (t_1, t_2) in $\Omega_=$, t_1 or t_2 consists of a single symbol (i.e. atomic expression). We suppose that this can be done, for otherwise there is no proof with this analysis. Let $\Omega_=$ be $\{(t_i^1, t_i^2)\}$ and and t_i^1 be atomic. Then we replace one of a t-var.'s or f-var.'s which is maximal in a partial ordering defined by;

(i) if (σ, t) is in $\Omega_=$ and τ occurs in t, then $\sigma \gtrsim \tau$ (and $\sigma \sim \tau$, if t is τ),

(ii) if (φ, F) is in $\Omega_=$ and ψ occurs in F, then $\varphi \gtrsim \psi$ (and $\varphi \sim \psi$, if F is ψ).

If φ is maximal and (φ, RFG) is in $\Omega_=$, then we first replace all occurrences of φ and ψ (for $\varphi \sim \psi$) in $\Omega_I, \ldots, \Omega_=$ by $R\varphi_1\varphi_2$ where φ_1 and φ_2 are new f-var.'s, and replace the pairs

(φ, RFG) and $(\psi, RF'G')$'s by (φ_1, F), (φ_2, G), (φ_1, F')'s and (φ_2, G')'s. If we define $d(\Omega_=)$ by $\prod 3^{c_i}$, where c_i+1=the number of symbols in t_i^2, then by this replacement $d(\Omega_=)$ decreases.

All the other cases are treated similarly.

4th step.

(4.1) The case that there is a pair (F_1, F_2) or (γ_1, γ_2) in Ω_U such that one of F_i's (or γ_i's) has a form $CH_0 \ldots H_m$ (or $CH_0 \ldots H_m(\bar{t}, t_0)$).

Suppose this is F_1 (or γ_1), then we select $H_{i_1}, \ldots, H_{i_r}$ from $H_1, \ldots, H_m$ and make H by replacing H_{i_j}'s by a function variable f according to the node (U;f,a). Now we omitt the pair from Ω_U and add (H_{i_j}, H_{i_k}) $(1 \leq j < k \leq r)$ to $\Omega_=$ and add $(F_2, H|^f_\varphi)$ (or $(\gamma_2, H|^f_\varphi(\bar{t}, t_0))$) to $\Omega_=$, where φ is a new f-var.. Further we replace t_0 by a and, for corresponding pairs (u_0, u_1), (v_0, v_1) in $\Omega_{U'}$, we omitt these pairs and add $(u_0, H_{i_1}(\bar{t}, 0))$, $(u_1, H_{i_1}(\bar{t}, a))$, $(v_0, \varphi(\bar{t}, 0))$ and $(v_1, \varphi(\bar{t}, a))$ to $\Omega_=$.

(4.2) The case that there are pairs (t_0, t_1), (u_0, u_1) in $\Omega_{U'}$ such that one of them has a form $F(\bar{v}, v_0)$.

In this case, similar operations as in (4.1) are carried out so that $\Omega_=$ represents necessary informations for a proof.

(4.2) The case that there is a pair (t ,u) in Ω_C such that t has a form $CF_0 \ldots F_m(\bar{v})$ or u has a form $F_0(F_1(\bar{v}), \ldots, F_m(\bar{v}))$.

In this case, similar operations as in (4.1) are carried out, i.e. omitt this pair from Ω_C and add $(u, F_0(F_1(\bar{v}), \ldots, F_m(\bar{v}))$ (or $(t, CF\varphi_1 \ldots \varphi_m(\bar{v}))$) to $\Omega_=$.

All the other cases corresponding to Ω_I, Ω_{R1}, Ω_{R2} are treated similarly.

Hereafter we repeat 3rd and 4th steps over and over again. Although, in 3rd step, $d(\Omega_=)$ may increase, but the number of pairs in $\tilde{\Omega} = \Omega_I \cup \Omega_C \cup \Omega_{R1} \cup \Omega_{R2} \cup \Omega_{U'} \cup \Omega_U$ decreases. So in a finitely many steps we have $d(\Omega_=)=1$, i.e. all expressions in $\Omega_=$ are atomic.

If $\tilde{\Omega}$ is empty, we may substitute certain simple terms and functors to t-var.'s and f-var.'s in $\Omega_=$ to get a substitution of the diagram. And from this we can construct a formula B(a) of Presburger arithmetic, expressing the condition of suitable adjunctions of consecutive S's in front of each subterm occurrences in a so far constructed diagram to make this diagram into a correct proof with the given analysis. This is exactly the case in Parikh's proof where only the set $\Omega_=$ is constructed. But in our situation there is the set $\tilde{\Omega}$ which may be non-empty. Thus obtained $\tilde{\Omega}$ and $\Omega_=$ express the minimum conditions which must be satisfied by a proof with the given analysis. So we must give a method by which, if there is a proof with this analysis, a proof can be transformed into another correct proof with terms of simpler structures. This is the point that we meet difficulties.

For instance, suppose Ω_U is not empty and consider the corresponding (U)-rule.

$$\frac{F(\mathfrak{t},0)=G(\mathfrak{t},0) \qquad F(\mathfrak{t},a)=\varphi_1(\mathfrak{t},a) \qquad G(\mathfrak{t},a)=\varphi_2(\mathfrak{t},a)}{F(\mathfrak{t},t_0)=G(\mathfrak{t},t_0)}, \quad (\varphi_1,\varphi_2)\in\Omega_U.$$

φ_1 and φ_2 are supposed to be $H|^f_F$ and $H|^f_G$ for some functor H of the form $CH_0 \ldots H_m$. The simplest way is to suppose that H contains no occurrences of f and transform a proof with this analysis into the one which satisfy this supposition. With this supposition, we add all pairs (u,v)'s to $\Omega_=$, where $\varphi_1\mathfrak{t}=u$ and $\varphi_2\mathfrak{t}=v$ appear at the top of the diagram, and carry out 3rd and 4th steps again to get new $\Omega_=$ with $d(\Omega_=)=1$. This $\Omega_=$ and new $\tilde{\Omega}$ represent minimum conditions which must be satisfied by a proof enjoying the supposition. But in doing this, we may come to a contraditory point as mentioned in (2.4.1) or (2.4.2) or 3rd step (i.e. we cannot transform this $\Omega_=$ to the one such that for all pairs (t_1, t_2) in it, t_1 or t_2 is atomic).

If this occurs, we cannot assume that H contains no occurrences of f in it.

There also may be a pair $(\psi(\bar{t}), \varphi(\bar{u}))$ or (t,u) in Ω_I or Ω_C. In this case we can only know that leftmost symbol of ψ or t is I or C, and we cannot make a suitable choice of ψ or t a priori.

If we take the strategy that starting from the simplest choice, e.g., I_1^1 for ψ or $C\psi_1\psi_2$ for ψ etc., where ψ_1 and ψ_2 are unary f-var.'s, and determinig the remaining expressions as far as possible. In the course of doing this, we may confront a difficulty, for instance, by a pair in Ω_U, in other words, there may be a case that supposed form of H cannot be determined because of the so far determined form of φ_1 and φ_2, i.e. φ_1 is $CH_0\ldots$ and φ_2 is $CH_0'\ldots$ and H_0 and H_0' have different argument numbers or occurrences of f in φ_1 and φ_2 for which so far determined F and G are substituted do not coincide, hence we must change the structures of φ_1 and φ_2 so far determined. At this point we select, so to speak, some simple compatible structures for φ_1 and φ_2, then back to the starting point. But repeating this procedure with this new ad hoc structures may come to another difficulties, and this procedure gives us no assurance of termination.

It seems to us that this is the point corresponding to the last sentence in Parikh's paper "The problem, of course, is ..., since atomic formula of PA can be much more complex than those of PA*.".

References

[1] Chang, C.C. & Keisler, H.J. : Model Theory, North-Holland, Amsterdam, 1973.

[2] Curry, H.B. : A formalization of recursive arithmetic, Amer. J. Math. 63 (1941), 263-282.

[3] Cleave, J.P. & Rose, H.E. : ^{n}-arithmetic, in "Sets, models and recursion theory" ed. by Crossley, North-Holland, Amsterdam, 1967.

[4] Gentzen, G. : Investigations into logical deduction, in "The collected papers of G. Gentzen", ed. by M. E. Szabo, North-Holland, 1969.

[5] Goodstein, R.J. : Recusive number theory, Amsterdam, 1957.

[6] Parikh, R.J. : Some results on the length of proofs, Trans. Amer. Math. Soc. 177 (1973), 29-36.

[7] Rose, H.E. : On the consistency and undecidability of recursive arithmetic, Zeitschr. f. math. Logik u. Grundlagen d. Math., Bd. 7. S.124-135, (1961).

[8] Miyatake, T. : On the length of proofs in formal systems, Tsukuba J. Math. 4 (1980), 115-125.

[9] Yukami, T. : A theorem on the formalized arithmetic with function symbol ' and +, Tsukuba J. Math. 1 (1977), 195-211.

Homogeneous formulas and definability theorems

Nobuyoshi MOTOHASHI
Institute of Mathematics, University of Tsukuba
Sakura-mura, Ibaraki, Japan

Let L be a first order classical predicate calculus with equality LK , or a first order intuitionistic predicate calculus with equality LJ . For the sake of simplicity, we assume that L has neither function symbols nor individual constant symbols. By n-ary formulas in L , we mean formulas $F(\bar{a})$ in L with a sequence $\bar{a}$ of distinct free variables of length n such that every free variable in F occurs in $\bar{a}$. Let $\mathbf{R}$ be a finite set of predicate symbols such that the equality symbol $=$ belongs to $\mathbf{R}$. For each non-negative integer k and disjoint two sequences $\bar{a} = \langle a_1, a_2, \ldots, a_n \rangle$, $\bar{b} = \langle b_1, b_2, \ldots, b_n \rangle$ of distinct free variables of the same length n , the 2n-ary homogeneous formula over $\mathbf{R}$ of degree k , denoted by $\mathrm{Hom}^k_{\mathbf{R}}(\bar{a};\bar{b})$, is the 2n-ary formula defined by :

$$\mathrm{Hom}^0_{\mathbf{R}}(\bar{a};\bar{b}) = \bigwedge_{R \in \mathbf{R}} \bigwedge_{1 \leq i_1, i_2, \ldots, i_r \leq n} (R(a_{i_1}, a_{i_2}, \ldots, a_{i_r}) \equiv R(b_{i_1}, b_{i_2}, \ldots, b_{i_r})),$$

$$\mathrm{Hom}^{k+1}_{\mathbf{R}}(\bar{a};\bar{b}) = \begin{cases} \forall x_{n+1} \exists y_{n+1} \mathrm{Hom}^k_{\mathbf{R}}(\bar{a}^\frown x_{n+1}; \bar{b}^\frown y_{n+1}) & \text{if } k \text{ is even,} \\ \forall y_{n+1} \exists x_{n+1} \mathrm{Hom}^k_{\mathbf{R}}(\bar{a}^\frown x_{n+1}; \bar{b}^\frown y_{n+1}) & \text{if } k \text{ is odd.} \end{cases}$$

By (2n-ary) homogeneous formulas over $\mathbf{R}$, we mean (2n-ary) homogeneous formulas over $\mathbf{R}$ of degree k for some k . Note the sentences;

$\forall \bar{x} \mathrm{Hom}^k_{\mathbb{R}}(\bar{x};\bar{x})$,

$\forall \bar{x} \forall \bar{u} \forall \bar{y} \forall \bar{v}(\mathrm{Hom}^k_{\mathbb{R}}(\bar{x}^\frown \bar{u};\bar{y}^\frown \bar{v}) \supset \mathrm{Hom}^k_{\mathbb{R}}(\bar{x};\bar{y}))$,

$\forall \bar{x} \forall \bar{y}(\mathrm{Hom}^{k+1}_{\mathbb{R}}(\bar{x};\bar{y}) \supset \mathrm{Hom}^k_{\mathbb{R}}(\bar{x};\bar{y}))$, $k=0,1,2,\ldots$

are all provable in L , but

$\forall \bar{x} \forall \bar{y}(\mathrm{Hom}^k_{\mathbb{R}}(\bar{x};\bar{y}) \supset \mathrm{Hom}^k_{\mathbb{R}}(\bar{y};\bar{x}))$, $k=0,1,\ldots$

are not, generally, provable in L . (Counter-example : Let L = LK , k=1 , and $R = \{ < , = \}$, where $<$ is a binary predicate symbol. Then, $\mathrm{Hom}^1_{\mathbb{R}}(a;b)$ is equivalent to the formula

$\forall x \exists y(x < a \equiv y < b .\wedge. a < x \equiv b < y .\wedge. x=a \equiv y=b)$

in all ordered structures. Let M be the ordered structure figured by the tree 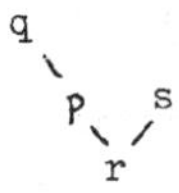

.

Then, $\mathrm{Hom}^1_{\mathbb{R}}(s;p)$ is true in M , but $\mathrm{Hom}^1_{\mathbb{R}}(p,s)$ is not.) suppose that I is a binary predicate symbol in L , which does not belong to $\mathbb{R}$. Let $\mathrm{Iso}_{\mathbb{R}}(I)$ be the set of the following sentences in L ;

$\forall x \exists y I(x,y)$, $\forall y \exists x I(x,y)$,

$\forall \bar{x} \forall \bar{y}(I(\bar{x};\bar{y}) \supset . R(\bar{x}) \equiv R(\bar{y}))$, $R \in \mathbb{R}$,

where $I(\bar{a};\bar{b})$ is the formula

$I(a_1,b_1) \wedge I(a_2,b_2) \wedge \ldots\ldots \wedge I(a_r,b_r)$

if $\bar{a} = \langle a_1,\ldots,a_r \rangle$, $\bar{b} = \langle b_1,\ldots,b_r \rangle$.

$\mathrm{Iso}_{\mathbb{R}}(I)$ is a theorey which means that I is an automorphism which preserves all R in $\mathbb{R}$, i.e. $\mathbb{R}$- automorphism. For each predicate symbol R , R-free formulas are formulas which have no occurrences of R . Let $A(\bar{a},\bar{b})$ be an I-free, 2n-ary formula in L , and T an I-free theory in L , i.e.. T is a set of I-free sentences in L . Then, $A(\bar{a},\bar{b})$ is said to be weakly preserved under $\mathbb{R}$-automorphism with respect to T in L, if the sentence $\forall\bar{x}\forall\bar{y}(I(\bar{x};\bar{y}) \supset A(\bar{x},\bar{y}))$ is provable from T and $Iso_{\mathbf{R}}(I)$ in L, i.e. $T, Iso_{\mathbf{R}}(I) \vdash_L \forall\bar{x}\forall\bar{y}(I(\bar{x};\bar{y}) \supset A(\bar{x},\bar{y}))$ 2n-ary homogeneous formulas over $\mathbf{R}$ are examples of 2n-ary formulas which are weakly preserved under $\mathbf{R}$-automorphism with respect to any theory T in L. A k-ary formula $F(\bar{a})$ has a k-ary formula $G(\bar{a})$ as a sub-relation with respect to T in L, if the sentence $\forall\bar{x}(G(\bar{x}) \supset F(\bar{x}))$ is provable from T in L.

In this paper, we shall show that every 2n-ary formula which is preserved under $\mathbf{R}$-automorphism with respect to T in L, has a 2n-ary homogeneous formula over $\mathbf{R}$ as a sub-relation with respect to T in L.

THEOREM A. An I-free, 2n-ary formula is weakly preserved under $\mathbf{R}$-automorphism with respect to T in L if and only if it has a 2n-ary homogeneous formula over $\mathbf{R}$, as a sub-relation with respect to T in L.

We give, here, a proof of Theorem A by using the simple approximation theorem of uniqueness conditions by existence conditions in Motohashi [4]. Since "if-part" of Theorem A is obvious, it is sufficient to prove "only-if-part" of Theorem A. Assume that $A(\bar{a},\bar{b})$ is an I-free, 2n-ary formula in L, which is weakly preserved under $\mathbf{R}$-automorphism with respect to T in L. Let $\mathbf{U} = \{ \forall\bar{u}\forall\bar{v}(I(\bar{u};\bar{v}) \supset .R(\bar{u}) \equiv R(\bar{v})) \mid R \in \mathbf{R} \}$, $\mathbf{E} = \{ \forall x\exists yI(x,y), \forall y\exists xI(x,y) \}$, and $\mathbf{X} = \{ I(a_1,b_1),\ldots,I(a_n,b_n) \}$, where $\bar{a} = \langle a_1,\ldots,a_n \rangle$, $\bar{b} = \langle b_1,\ldots,b_n \rangle$. Then, $A(\bar{a},\bar{b})$ is provable from $T,\mathbf{U},\mathbf{E},\mathbf{X}$ in L. Note that $\mathbf{U}$ is a set of normal uniqueness conditions of I, $\mathbf{E}$ is a set of simple existence conditions of I, and $\mathbf{X}$ is a set of I-atomic formulas (see [4] for notions). So, by the simple approximation theorem in [4], we have a simple approximation

$C(\bar{a},\bar{b})$ of $\mathfrak{U}$ by $\mathfrak{E}$ over $\mathfrak{X}$ such that the sentence $\forall\bar{x}\forall\bar{y}(C(\bar{x},\bar{y})\supset A(\bar{x},\bar{y}))$ is provable from T in L. Since A and every sentence in T have no occurrences of I, the sentence $\forall\bar{x}\forall\bar{y}(C^*(\bar{x},\bar{y})\supset A(\bar{x},\bar{y}))$ is provable from T in L, where $C^*(\bar{a},\bar{b})$ is the formula obtained from $C(\bar{a},\bar{b})$ by replacing every occurrence of I of the form $I(u,v)$ by the true sentence $\top$. On the other hand, simple approximations of $\mathfrak{U}$ by $\mathfrak{E}$ over $\mathfrak{X}$ are, essentially, formulas obtained from the definition of homogeneous formulas over $\mathfrak{R}$ by replacing $Hom^0_{\mathfrak{R}}(\bar{a},\bar{b})$ by $Hom^0_{\mathfrak{R}}(\bar{a},\bar{b})\wedge I(\bar{a};\bar{b})$. Therefore, every formula obtained from a simple approximation of $\mathfrak{U}$ by $\mathfrak{E}$ over $\mathfrak{X}$ by replacing every occurrence of I of the form $I(u,v)$ by the true sentence $\top$, is equivalent to a homogeneous formulas over $\mathfrak{R}$ in L. Hence, $C^*(\bar{a},\bar{b})$ is equivalent to a 2n-ary homogeneous formula in L. Therefore, $A(\bar{a},\bar{b})$ has a 2n-ary homogeneous formulas over $\mathfrak{R}$, as a sub-relation with respect to T in L. This completes our proof of Theorem A. An I-free, n-ary formula $B(\bar{a})$ in L, is said to be preserved under $\mathfrak{R}$-automorphism with respect to T in L, if the sentence $\forall\bar{x}\forall\bar{y}(I(\bar{x};\bar{y})\supset.B(\bar{x})\equiv B(\bar{y}))$ is provable from T, $Iso_{\mathfrak{R}}(I)$ in L.

COROLLARY B. An I-free, n-ary formula $B(\bar{a})$ is preserved under $\mathfrak{R}$-automorphism with respect to T in L, if and only if the sentence

$$\forall\bar{x}(B(\bar{x})\equiv \exists\bar{y}(B(\bar{y})\wedge Hom^k_{\mathfrak{R}}(\bar{y};\bar{x})))$$

is provable from T in L, for some k.

PROOF.

Since if-part of this corollary is obvious, it is sufficient to prove only-if-part.

$$T\ ,\ Iso_{\mathbf{R}}(I) \quad \vdash_L \forall\bar{x}\forall\bar{y}(I(\bar{x};\bar{y}) \supset (B(\bar{x}) \equiv B(\bar{y})))$$

$\Longrightarrow$

$$T \quad \vdash_L \forall\bar{y}\forall\bar{x}(Hom^k_{\mathbf{R}}(\bar{y};\bar{x}) \supset (B(\bar{y}) \equiv B(\bar{x}))) \text{ for some } k$$

(By Theorem A)

$\Longrightarrow$

$$T \quad \vdash_L \forall\bar{x}\forall\bar{y}(B(\bar{y}) \wedge Hom^k_{\mathbf{R}}(\bar{y};\bar{x}) \ .\supset B(\bar{x})) \text{ for some } k$$

$\Longrightarrow$

$$T \quad \vdash_L \forall\bar{x}(\exists\bar{y}(B(\bar{y}) \wedge Hom^k_{\mathbf{R}}(\bar{y};\bar{x})) \supset B(\bar{x})) \text{ for some } k$$

$\Longrightarrow$

$$T \quad \vdash_L \forall\bar{x}(B(\bar{x}) \equiv \exists\bar{y}(B(\bar{y}) \wedge Hom^k_{\mathbf{R}}(\bar{y};\bar{x}))) \text{ for some } k\ ,$$

because $T \quad \vdash_L \forall\bar{x}(B(\bar{x}) \supset \exists\bar{y}(B(\bar{y}) \wedge Hom^k_{\mathbf{R}}(\bar{y};\bar{x})))$ for all k.

(q.e.d.)

From Corollary B, we have the following definability theorem.

COROLLARY C. Suppose that $P \notin \mathbf{R} \cup \{I\}$, $Q \notin \mathbf{R} \cup \{I\}$ are two distinct n-ary predicate symbols, $T(P)$ is a theory in L such that every predicate symbol which occurs in a formula in $T(P)$, belongs to $\mathbf{R} \cup \{P\}$, and $T(Q)$ is the theorey obtained from $T(P)$ by replacing every occurrence of P by Q . Then, the following three conditions (i), (ii), (iii) are all equivalent :

(i) The sentence $\forall\bar{x}(P(\bar{x}) \equiv Q(\bar{x}))$ is provable from $T(P)$, $T(Q)$, $Iso_{\mathbf{R}}(I)$, $\forall\bar{x}\forall\bar{y}(I(\bar{x};\bar{y}) \supset .\ P(\bar{x}) \equiv Q(\bar{y}))$ in L .

(ii) The n-ary formula $P(\bar{a})$ is preserved under $\mathbf{R}$-automorphism with respect to $T(P)$ in L .

(iii) The sentence $\forall\bar{x}(P(\bar{x}) \equiv \exists\bar{y}(P(\bar{y}) \wedge Hom^k_{\mathbf{R}}(\bar{y};\bar{x})))$ is provable from $T(P)$ in L .

PROOF. By Corollary B, (ii) and (iii) are equivalent. Also, obviously (ii) implies (i). Assume (i). By replacing every occurrence $Q(\bar{u})$ of Q by $P^*(\bar{u})$ in a derivation of $\forall\bar{x}(P(\bar{x})\equiv Q(\bar{x}))$ from $T(P)$, $T(Q)$, $Iso_{\mathbb{R}}(I)$, $\forall\bar{x}\forall\bar{y}(I(\bar{x};\bar{y})\supset .P(\bar{x})\equiv Q(\bar{y}))$ in L, we obtain a derivation of $\forall\bar{x}(P(\bar{x})\equiv P^*(\bar{x}))$ from $T(P), T(P^*), Iso_{\mathbb{R}}(I)$, $\forall\bar{x}\forall\bar{y}(I(\bar{x};\bar{y})\supset . P(\bar{x})\equiv P^*(\bar{y}))$ in L, where $P^*(\bar{a})$ is the formula $\forall\bar{z}(I(\bar{z};\bar{a})\supset P(\bar{z}))$. Since $Iso_{\mathbb{R}}(I) \vdash_L \forall\bar{x}\forall\bar{y}(I(\bar{x};\bar{y})\supset .P(\bar{x})\equiv P^*(\bar{y}))$, we have that $Iso_{\mathbb{R}}(I) \vdash_L F(P)\equiv F(P^*)$ for any sentence $F(P)$ in $T(P)$. Therefore, $T(P), Iso_{\mathbb{R}}(I) \vdash_L \forall\bar{x}(P(\bar{x})\equiv P^*(\bar{x}))$. Hence, $T(P), Iso_{\mathbb{R}}(I) \vdash_L \forall\bar{x} (P(\bar{x})\equiv \forall\bar{z}(I(\bar{z};\bar{x})\supset P(\bar{z})))$. This clearly implies (ii). (q.e.d.)

Next, we consider the first order classical predicate calculus with equality LK. By $\mathbb{R}$-formulas, we shall mean formulas which have no occurrences of predicate symbols not in $\mathbb{R}$. By the definition of homogeneous formulas over $\mathbb{R}$, every 2n-ary homogeneous formula over $\mathbb{R}$, $Hom^k_{\mathbb{R}}(\bar{a};\bar{b})$, is equivalent to formula of the form $\bigvee_{i=1}^{m}(A_i(\bar{a})\wedge B_i(\bar{b}))$, for some n-ary $\mathbb{R}$-formulas $A_1(\bar{a}),\dots,A_m(\bar{a}),B_1(\bar{b}),\dots,B_m(\bar{b})$. Hence, we have :

COROLLARY D. An I-free, 2n-ary formula $A(\bar{a},\bar{b})$ is weakly preserved under $\mathbb{R}$-automorphism with respect to T in LK if and only if there are n-ary $\mathbb{R}$-formulas $A_1(\bar{a}),\dots,A_m(\bar{a}),B_1(\bar{b}),\dots,B_m(\bar{b})$ such that the 2n-ary formula $\bigvee_{i=1}^{m}(A_i(\bar{a})\wedge B_i(\bar{b}))$ is weakly preserved under $\mathbb{R}$-automorphism with respect to T in LK, and $A(\bar{a},\bar{b})$ has all the 2n-ary formulas $A_1(\bar{a})\wedge B_1(\bar{a}),\dots,A_m(\bar{a})\wedge B_m(\bar{b})$ as sub-relations with respect to T in LK.

Also, from Corollary C, we have :

COROLLARY E. Assume all the hypotheses in Corollary C. Then, the following three conditions are all equivalent :

(i) The sentence $\forall\bar{x}(P(\bar{x}) \equiv Q(\bar{x}))$ is provable from $T(P)$, $T(Q)$, $Iso_{\mathbb{R}}(I)$, $\forall\bar{x}\,\forall\bar{y}(I(\bar{x};\bar{y}) \supset .P(\bar{x}) \equiv Q(\bar{y}))$ in LK .

(ii) The n-ary formula $P(\bar{a})$ is preserved under $\mathbb{R}$-automorphism with respect to $T(P)$ in LK .

(iii) The sentence $\bigvee_{s=1}^{N} \forall\bar{x}(P(\bar{x}) \equiv C_s(\bar{x}))$ is provable from $T(P)$ in LK , for some n-ary $\mathbb{R}$-formulas $C_1(\bar{a}),\ldots,C_N(\bar{a})$.

PROOF. By Corollary C, it is sufficient to prove that (iii) above follows from the following $(iv)_k$, for each k .

$(iv)_k$ $\quad T(P) \vdash_{LK} \forall\bar{x}(P(\bar{x}) \equiv \exists\bar{y}(P(\bar{y}) \wedge Hom^k_{\mathbb{R}}(\bar{y};\bar{x})))$

Assume $(iv)_k$. Then, there are n-ary $\mathbb{R}$-formulas $A_1(\bar{a}),\ldots,A_m(\bar{a})$, $B_1(\bar{b}),\ldots,B_m(\bar{b})$ such that

$$T(P) \vdash_{LK} \forall\bar{x}(P(\bar{x}) \equiv \exists\bar{y}(P(\bar{y}) \wedge (\bigvee_{i=1}^{m} (A_i(\bar{y}) \wedge B_i(\bar{x}))))).$$

Hence,

$$T(P) \vdash_{LK} \forall\bar{x}(P(\bar{x}) \equiv \bigvee_{i=1}^{m}(B_i(\bar{x}) \wedge \exists\bar{y}(P(\bar{y}) \wedge A_i(\bar{y})))).$$

For each set $s \subsetneq \{1,2,\ldots,m\}$, let C_s be the sentence

$$\bigwedge_{i\in s} \exists\bar{y}(P(\bar{y}) \wedge A_i(\bar{y})) \wedge \bigwedge_{i\notin s} \neg\exists\bar{y}(P(\bar{y}) \wedge A_i(\bar{y})) \quad .$$

Then, clearly , $\vdash_{LK} \bigvee_{s \subsetneq \{1,2,\ldots,m\}} C_s$ and $T(P), C_s \vdash_{LK} \forall\bar{x}(P(\bar{x}) \equiv \bigvee_{i\in s} B_i(\bar{x}))$ for each $s \subsetneq \{1,2,\ldots,m\}$. Hence, $T(P) \vdash_{LK} \bigvee_{s \subsetneq \{1,2,\ldots,m\}} \forall\bar{x}(P(\bar{x}) \equiv \bigvee_{i\in s} B_i(\bar{x}))$. This means that (iii) above holds. (q.e.d.)

The equivalence between (i) and (iii) in Corollary E, is a syntactical form of Svenonius' definability theorem (cf. Motohashi [3]). Therefore, Corollary C can be considered as a Svenonius' type definability theorem which holds both in LK and LJ . Note that C. Mizutani gave, in[2], a similar definability theorem in LJ by using Motohashi P-formulas (cf. [1], [2]). Our Corollary C gives a Svenonius' type definability theorem in LJ by using special Motohashi P-formulas in Harnik-Makkai [1], which is a refinement of Mizutani's theorem. Moreover, we can easily extend our results in the first order infinitary logic $L_{\omega_1 \omega}$, and obtain Theorem 2.1 in [1], by using the simple approximation theorem in $L_{\omega_1 \omega}$ (cf. [4]). Also, by replacing the set $Iso_{\mathbf{R}}(I)$, the axioms of $\mathbf{R}$-automorphism, by a set of the axioms of another morphism, e.g. $\mathbf{R}$-homomorphism, $\mathbf{R}$-embedding, etc, we obtain "definability theorems" with respect to this morphism in the style of Corollary C (cf. [5]).

References

[1] V.Harnik and M.Makkai, Applications of Vaught sentences and the covering theorem, J.S.L., vol. 41 (1976), pp.171-187.

[2] C.Mizutani, Definability theorem for the intuitionistic predicate logic with equality, to appear.

[3] N.Motohashi, A new theorem on definability in a positive second order logic with countable conjunctions and disjunctions, Proc. Japan Acad., vol. 48 (1970), pp.153-156.

[4] N.Motohashi, Approximation theorems of uniqueness conditions by existence conditions, to appear.

[5] N.Motohashi, A theorem in the theory of definition, J. Japan Math. Soc., vol. 22 (1970), pp.490-494.

Boolean valued combinatorics

by Kanji Namba

In this paper, we state some elementary properties and problems of infinite combinatrics in Boolean valued set theory introduced by D. Scott and R. M. Solovay. This notion is an algebraic description of concepts introduced by P. J. Cohen. He originally introduced the notion in order to prove some independence proof in set theory and to construct many variety of models of set theory. The notions are

"forcing condition" and "generic filter"

and the modified notions by D. Scott and R. M. Solovay are

"Boolean algebra" and "dual space"

and they added many things to the original concepts.

1. Boolean algebra

1.1 Boolean algebra and its dual space

Boolean algebra is an algebraic structure with two binary operations + and . corresponding to "or" and "and" and a unary operation - corresponding to "not". They satisfy the following conditions

$$a+b = b+a \qquad ab = ba$$

$$a+(bc) = (a+b)(a+c) \qquad a(b+c) = ab+ac$$

$$a+(b.-b) = a \qquad a(b.-b) = a$$

These properties imply the following conditions

$$a+(b+c) = (a+b)+c \qquad a(bc) = (ab)c$$

$$a+ab = a \qquad a(a+b) = a$$

A Boolean algebra is considered as a partial order structure $(B,\leq)$ with the order relation introduced by the equivalence

$$a \leq b \equiv ab = a \equiv a+b = b$$

The elements $0 = a.-a = b.-b$ and $1 = a+-a = b+-b$ are the smallest and the largest elements of B, which corresponds to the truth value "false" and "true" respectively. The Boolean algebra

$$2 = \{0,1\}$$

is the smallest non-trivial complete Boolean algebra.

There is also natural and important binary operation $\rightarrow$ corresponding to "imply" defined by

$$b^a = a\rightarrow b = -a+b$$

And they satisfy the following law of exponentiation

$$a^b.a^c = a^{b+c} \qquad a^c.b^c = (ab)^c$$

$$(a^b)^c = a^{bc} = a^b + a^c$$

So we may consider "or", "and" and "imply" as "sum", "product" and "power" respectively. The fundamental relation between $\leq$ and $\rightarrow$ is

$$a \leq b \equiv a\rightarrow b = 1$$

A Boolean algebra B is called complete if every subset A of B has the least upper bound, namely if it has an element p such that

$$\forall x\in A(x \leq p) \qquad \forall y\in B(\forall x\in A(x \leq y) \rightarrow p \leq y)$$

and of course such p is unique and it is denoted by

$$\sum_{x\in A} x \qquad \text{or} \qquad \sup_{x\in A} x$$

which corresponds to the quantifier $\exists$ meaning "exist". Since the automorphism $-:B \rightarrow B$ transposes the order and exchange + and .

$$-(a+b) = -a.-b \qquad -(ab) = -a+-b$$

$$a \leq b \equiv -b \leq -a$$

we have that if every subset of B has the least upper bound then it has the greatest lower bound and it is denoted by

$$\prod_{x\in A} x \qquad \text{or} \qquad \inf_{x\in A} x$$

which corresponds to the quantifier $\forall$ meaning "for all" and satisfies de Morgan's law

$$-\sum_{x\in A} x = \prod_{x\in A} -x \qquad -\prod_{x\in A} = \sum_{x\in A} -x$$

Let B and B' be Boolean algebras and let h be a homomorphism $h:B \to B'$, then the inverse image

$$h^{-1}(1) = \{a \in B \mid h(a) = 1\}$$

is a filter and dually $h^{-1}(0)$ is an ideal. The filter $F = h^{-1}(1)$ and the ideal $I = h^{-1}(0)$ is related by

$$a \in F \equiv -a \in I$$

in such a case F is called co-filter of I and I is called co-ideal of F. Since the condition being a filter or an ideal is finitary property, every filter or ideal can be extended to a maximal filter or ideal by using the axiom of choice or equivalently Zorn's lemma. This means that there is a homomorphism $k:B' \to 2$ such that following diagram commutes

$$\begin{array}{ccc} B & \overset{kh}{\to} & 2 \\ h\searrow & & \nearrow k \\ & B' & \end{array}$$

which means that $2 = \{0,1\}$ is a terminal object in the category of Boolean algebra homomorphisms. Conversely if $h:B \to 2$ is a homomorphism then $F = h^{-1}(1)$ is a maximal filter and $I = h^{-1}(0)$ is a maximal ideal. Maximal filter is also called ultra-filter and it is characterized by $a \in F$ or $-a \in F$ for every $a \in B$. Usually the notion corresponding to $h:B \to 2$ is called prime filter.

Now we consider the topological space B*, called the dual space of B, defined by the set of all homomorphisms

$$B^* = \{h:B \to 2\}$$

with the topology induced by basic open sets

$$a^* = \{h \in B^* \mid h(a) = 1\}$$

The topological space B* thus topologized satisfies the property:

B* is compact, totally disconnected Baire space

Since the equation $\sum_\nu a_\nu = 1$ is equivalent to the subset $\bigcup_\nu a_\nu^*$ is dense open in the dual space B*, so the Baire property of B* is equivalent to Rasiowa-Sikorski theorem on the existence of $h:B \to 2$ such that

$$h(\sum_{\nu} a_{n\nu}) = \sum_{\nu} h(a_{n\nu})$$

for every $n \in \omega$.

1.2 Order and topology

Let $(P,\leq)$ be a partial order structure. Then P is considered as a topological space, namely every point p has unique neighbourhood

$$V(p) = \{x \in P \mid p \leq x\}$$

Asymmetric law means that the space P is a Kolmogorov space. This topological space is characterized by the property that

open sets are closed under arbitrary intersection.

Conversely, Kolmogorov space with this property determines a partial order introduced by

$$p \leq q \equiv \forall U{:}\mathrm{open}(p \in U \to q \in U)$$

and the topological space introduced by this order coincides with the original topology.

Let X be a topological space and let $\Diamond A$ and $\Box A$ be the closure $\bar{A}$ and the interior int(A) of A respectively. Then two modal operations are related by the relation

$$-\Diamond A = \Box - A \qquad\qquad -\Box A = \Diamond - A$$

and they correspond to "may" and "must" respectively.

The operations $\Diamond$, $\Box$ and their products $\Diamond\Box$, $\Box\Diamond$ are idempotent, that is

$$\Diamond\Diamond A = \Diamond A \qquad\qquad \Box\Box A = \Box A$$

$$\Diamond\Box\Diamond\Box A = \Diamond\Box A \qquad\qquad \Box\Diamond\Box\Diamond A = \Box\Diamond A$$

Let N be the ideal of all nowhere dense subsets of X, namely the kernel of the map $\Box\Diamond : P(X) \to P(X)$. Then the boundary

$$A - \Box A$$

of closed or open set is nowhere dense. Let # be the boundary map

$\#(A) = \Diamond A - \Box A$, then the kernel C of

$$P(X) \overset{\#}{\to} P(X) \overset{\Box\Diamond}{\to} P(X)$$

forms a Boolean algebra and includes every closed set and open set.

Now we consider following exact sequence of mappings

$$1 \to N \subset C \to B \to 1$$

then the quotient object B is a complete Boolean algebra. The idempotency of the operators $\Diamond\Box$ and $\Box\Diamond$ means that the image of the mappings is just the set of all fixed points. Such a set is called regular open if $\Box\Diamond A = A$ and regular closed if $\Diamond\Box A = A$. By the relation

$$\Diamond A - \Box A = (\Diamond A - A) \cup (A - \Box A)$$

the Boolean algebra B just introduced is isomorphic to the complete Boolean algebra of regular open sets and also isomorphic to regular closed sets of X. For regular open sets, we have

$$A+B = \Box\Diamond(A \cup B) \qquad A.B = A \cap B$$

$$-A = \Box(X-A)$$

$$\sum_{x\in\Lambda} A_x = \Box\Diamond \bigcup_{x\in\Lambda} A_x \qquad \prod_{x\in\Lambda} A_x = \Box\Diamond \bigcap_{x\in\Lambda} A_x$$

Especially if X is the topological space introduced by a partial ordering, then we have

$$\prod_{x\in\Lambda} A_x = \bigcap_{x\in\Lambda} A_x$$

and the notion of regular open set

$$p \in \Box\Diamond A \equiv \forall q \geq p \, \exists r \geq q (r \in A)$$

corresponds to the notion of "weak forcing".

Let F be a directed set in $(P,\leq)$, namely compatibility

$$\forall x\in F \, \forall y\in F \, \exists z\in F (x,y \leq z)$$

then the set F* of regular open sets A such that

$$A \in F^* \equiv \exists x\in F (V(x) \subset A)$$

is a filter on the Boolean algebra B of regular open sets of P.

So it can be extended to a maximal filter. Let $P^{\#}$ be the set of all maximal directed sets and let $W(x)$ be the set defined by

$$F \in W(x) \equiv x \in F$$

and consider the topological space generated by open basis $W(x)$. Then the space $P^{\#}$ become a Hausdorff space, but in general, $P^{\#}$ and the dual space B^* is not isomorphic. However B is isomorphic to the complete Boolean algebra of regular open sets of $P^{\#}$.

Let B be a Boolean algebra, Then Wallman-Stone representation theorem means that B is represented as sets of all closed and open, shortly clopen, sets of the dual space B^*, which means that every continuous function

$$f:B^* \to 2$$

represents an element a of B by the relation

$$\forall h \in B^*(f(h) = h(a))$$

Hence we can introduce a function $a^*:B^* \to 2$ defined by

$$a^*(h) = h(a)$$

for every h in B^*. Since $2 = \{0,1\}$ is the complete Boolean algebra of sets of singleton space $1 = \{0\}$, we may write $2 = 1^*$. And identify an element a of B with a function $a:1 \to B$, then

$$a:1 \to B \qquad\qquad a^*:B^* \to 1^*$$

are considered as contravariant relation under the transformation $*$.

This kind of view points will lead to the fundamental description of Boolean valued set theory and sheaf theoretical description of them. Though the dual space B^* of B is very special space, the above relation naturally extends to Heyting algebra valued set theory and sheaf theoretical description of them and it is very interesting subject to extend them to more general operator valued set theory, especially for non-commutative cases.

Let $B_1 \times B_2$ be the product of two Boolean algebras. We consider the dual space $(B_1 \times B_2)^*$ of the product algebra. Then by

$$h(1,1) = h(0,1)+h(1,0) = 1$$

and dually $h(0,1).h(1,0) = 0$, we have either

$$h(0,1) = 1 \qquad\qquad h(1,0) = 1$$

We consider $h_1 : B_1 \to 2$ defined by $h_1(a) = h(a,0)$ and similarly for $h_2(a) = h(0,a)$. This means that

$$(B_1 \times B_2)^* = B_1^* + B_2^*$$

where + in the left side means the disjoint union of topological spaces.

Next we consider the product space of two dual spaces $B_1^* \times B_2^*$. What would be the corresponding notion in the sense of algebra. An open set A of the product space $B_1^* \times B_2^*$ is determined by a set of pairs $(a,b) \in B_1 \times B_2$ such that

$$a^\# \times b^\# \subset A$$

where $a^\# = \{h \in B^* | h(a) = 1\}$. So the function $f_A : B_1 \times B_2 \to 2$ defined by

$$f_A(a,b) = 1 \equiv a^\# \times b^\# \subset A$$

is monotone decreasing. The notion of monotone decreasing function corresponds to open set and increasing function to closed set. A function satisfying

$$\inf_{q \le p} \sup_{r \le q} f(r) = f(p)$$

corresponds to the notion of regular open set.

What would be natural topology on the space of all continuous functions between two dual spaces $B_1^* \to B_2^*$, including algebraic correspondant to this seems to be interesting.

Any how for an ordered structure $(P, \le)$ the notions

$$\inf_{y \le x} \qquad\qquad \sup_{y \le x}$$

correspond to the operation of "interior" and "closure" respectively, and they are idempotent and also for their products.

1.3 Quotient algebras

Bet B be a Boolean algebra, by additive number of B, we mean the least ordinal number, if there exist, such that there is a function $f:\kappa \to B$ for which

$$\sum_{\nu<\kappa} f(\nu)$$

does not exist in B. The additive number of B is denoted by

$$\mathrm{add}(B)$$

if there is no such function, then we put $\mathrm{add}(B) = \infty$. $\mathrm{add}(B)$ is always a regular infinite cardinal or ∞.

By saturation number of B, we mean the least ordinal number κ such that there is no function $f:\kappa \to B-\{0\}$ for which

$$\forall \nu < \kappa\ \forall \mu < \nu (f(\nu).f(\mu) = 0)$$

that is, there is no pairwise disjoint family of κ positive elements in B. The saturation number of B is denoted by $\mathrm{sat}(B)$. $\mathrm{sat}(B)$ is always a regular cardinal, and uncountable if it is not finite, but it can be weakly inaccessible as in the case of regular open sets of weak product space

$$X = \prod_{\nu<\kappa} \nu$$

where ν is considered as discrete space and κ is weakly inaccessible.

The following property states a fundamental relation of additive number, saturation number and the completeness of the quotient algebra.

Let B be a Boolean algebra, I be its ideal and $C = B/I$ be the quotient algebra, that is

$$1 \to I \to B \to B/I \to 1$$

be an exact sequence. And suppose

$$\mathrm{add}(I) \geq \kappa \qquad \mathrm{add}(B) \geq \kappa^+ \qquad \mathrm{sat}(C) \leq \kappa^+$$

then the quotient algebra C is complete.

Typical example of this is the case of quotient algebra

$$B = \mathcal{B}/I_{\mu}$$

where $\mathcal{B}$ is a Borel family and I_{μ} is the measure ideal

$$I_{\mu} = \{A \in \mathcal{B} \mid \mu(A) = 0\}$$

determined by a σ-finite, finitely additive measure μ defined on $\mathcal{B}$.

2. Boolean valued structure

2.1 Structure and interpretation

By a type we mean a category consisting of functions whose domain and co-domain are singletons. For the first order logic calculus, the category T of types consists of morphisms

0	object
1	truth value
$0\ldots0 \to 0$	function
$0\ldots0 \to 1$	predicate
$1\ldots1 \to 1$	logical connective
$01 \to 1$	quantifier

By a language of type T, we mean the following diagram with inclusions and natural transformations

$$\begin{array}{ccc} L & \subset & F \\ \cup & & \downarrow\tau \\ C & \to & T \end{array}$$

F is called formal expressions, the generator L of F is called a language, C is called constant symbols and L-C is called variables.

By a generator, we mean that for any category M of functions such that

$$\begin{array}{ccc} L & \overset{\psi}{\to} & M \\ & {}_{\tau}\searrow & \downarrow \\ & & T \end{array}$$

can be uniquely factored through F, namely there is unique natural transformation $\to$ so that the following diagram commutes

$$\begin{array}{ccc} L & \overset{\psi}{\rightarrow} & M \\ \cap & \nearrow^{\phi} & \downarrow \\ F & \underset{\tau}{\rightarrow} & T \end{array}$$

In this case M is called a model or a structure and the morphism ψ is called a interpretation.

For the constant symbols in C, we use, for example, the followings

0	0 1 e π ϕ ω
1	0 1
00 → 0	+ × ∪ ∩
00 → 1	= ≤ ε ⊂
11 → 1	∧ ∨ → ≡
01 → 1	∃ ∀

and for the variable symbols, for example

0	a b c ... x y z ...
0...0 → 0	f g h ...
0...0 → 1	P Q R ...

The inclusion map L ⊂ F is called the formation rule and usually we use free product expression $f(t_1,\ldots,t_n)$ from symbols $f,t_1,\ldots,t_n$, so that the following diagram commutes

$$\begin{array}{cccccc} t_1 & \cdots & t_n & \overset{f}{\rightarrow} & f(t_1,\ldots,t_n) \\ \downarrow & & \downarrow & \downarrow & \downarrow \\ 0 & \cdots & 0 & \underset{0\ldots0\rightarrow0}{\rightarrow} & 0 \end{array}$$

and for quantifier we form $\exists xA(x,t_1,\ldots,t_n)$ from symbols $\exists$, x and expression $A(x,t_1,\ldots,t_n)$ so that the following diagram commutes

$$\begin{array}{cccc} x & A(t_1,\ldots,t_n) & \overset{\exists}{\rightarrow} & \exists xA(x,t_1,\ldots,t_n) \\ \downarrow & \downarrow & \downarrow & \downarrow \\ 0 & 1 & 01\rightarrow1 & 1 \end{array}$$

The constant symbols

$$\neg \quad \vee \quad \wedge \quad \rightarrow \quad \equiv \quad \exists \quad \forall$$

are interpreted in Boolean algebra as functions

$$- \quad + \quad . \quad \rightarrow \quad \leftrightarrow \quad \Sigma \quad \Pi$$

where the inverse image of type mapping τ of type 1 is a Boolean algebra, namely $B = \tau^{-1}(1)$. Such a model is called a Boolean valued model or atructure.

There is a natural partial order defined on the formal expressions F, which is called sub-expression relation and it is well-founded. By this ordering, F can be considered as a topological space, and with this topology the isolated points are just the elements of L and it is dence open in F, namely $F = \Diamond L$ and $L = \Box F$. Therefore every function on L, which is always continuous, can be uniquely extended to a continuous function on F.

Let $\phi,\psi : L \to M$ be two interpretations, then we write

$$\phi \underset{x}{\equiv} \psi$$

for that the value of ϕ and ψ are the same except x, namely

$$\forall y \neq x(\phi(y) = \psi(y))$$

Note that for any x of L and u of M of type 0, there is ψ such that

$$\phi \underset{x}{\equiv} \psi \qquad \psi(x) = u$$

such ψ is sometimes denoted by

$$\phi\binom{x}{u}$$

For a expression of the form $f(t_1,\ldots,t_n)$, the interpretation ϕ uniquely extends by the commutative diagram

$$\begin{array}{ccc} t_1 \;\cdots\; t_n & \xrightarrow{f} & f(t_1,\ldots,t_n) \\ \downarrow \qquad \downarrow & & \downarrow \\ \phi(t_1)\ldots\phi(t_n) & \xrightarrow[\phi f]{} & \phi(f(t_1,\ldots,t_n)) \end{array}$$

or simply by

$$\phi(f(t_1,\ldots,t_n)) = \phi f(\phi(t_1),\ldots,\phi(t_n))$$

And for quantifier, we have for example

$$\begin{array}{ccc} x \quad A(t_1,\ldots,t_n) & \xrightarrow{\exists} & \exists x A(x,t_1,\ldots,t_n) \\ \downarrow \qquad\qquad \downarrow & & \downarrow\phi \\ \phi(x) \quad \phi(A(t_1,\ldots,t_n)) & \xrightarrow[\Sigma]{} & \sum_{\phi \underset{x}{\equiv} \psi} \psi(A(x,t_1,\ldots,t_n)) \end{array}$$

This means that it is introduced by the simultaneous induction on on all the interpretations, and it is also denoted by

$$\phi(\exists x A(x,t_1,\dots,t_n)) = \sum_{\phi \underset{x}{=} \psi} \psi(A(x,t_1,\dots,t_n))$$

Free variables V(e) of an expression e, is defined as unique extension satisfying

$$V(e) = \begin{cases} \phi & \text{if } e \in C \\ \{e\} & \text{if } e \in L-C = V \end{cases}$$

and for example

$$V(f(t_1,\dots,t_n)) = V(t_1) \cup \dots \cup V(t_n)$$

$$V(\exists x A(x,t_1,\dots,t_n)) = V(A(x,t_1,\dots,t_n)) - \{x\}$$

The set of all free variables V(e) in a formal expression is always a finite set, and it is a support of the expression, in the sense that the value of interpretation ϕ at e, namely $\phi(e)$ is uniquely determined by the values of ϕ on V(e). If V(e) is a subset of $\{x_1,\dots,x_n\}$, then e is usually written as

$$e(x_1,\dots,x_n)$$

For an interpretation ϕ such that

$$\phi(x_1) = u_1 \ \dots \ \phi(x_n) = u_n$$

uniquely determined value $\phi(e(x_1,\dots,x_n))$ is usually denoted as

$$[e(u_1,\dots,u_n)]$$

or even simply $e(u_1,\dots,u_n)$. Especially if the structure M is formal expressions F, then an interpretation is called a substitution. Interpretation is sometimes called an evaluation in M.

If e is of type 0, then [e] is an element of the set $U = \tau^{-1}(0)$ which is called underlying set or universe of M, and if e is of type 1, then [e] is an element of $B = \tau^{-1}(1)$ which is a complete Boolean algebra.

We consider Gentzen type sequent

$$A_1,\dots,A_m \to B_1,\dots,B_n$$

then the value of it is defined by

$$-[A_1] + \dots + -[A_m] + [B_1] + \dots + [B_n]$$

If the above value is 1, the largest element of B, then it is called true under the interpretation, and it is equivalent to the equality

$$[A_1] \cdot \dots \cdot [A_m] \le [B_1] + \dots + [B_n]$$

A sequent true under all interpretations is called valid. Concerning these notations, well-known fundamental relation is the equivalence of

"validity and provability"

of sequences. Non-trivial derection is called completeness theorem.

3. Boolean valued model for set theory

3.1 D. Scott and R. M. Solovay introduced Boolean valued model for set theory $V^{(B)}$, which is a Boolean version of von Neumann's construction of the universe of sets V, namely

$$R(\alpha+1) = P(R(\alpha)) \qquad \alpha \text{ successor}$$

$$R(\alpha) = \bigcup_{\beta<\alpha} R(\beta) \qquad \alpha \text{ limit}$$

The equality R(0) = 0 follows from the convention on the operation. Fundamental principle here is the identification of "set" and its "representing function". Hence the intended notion is

$$v(u) = [u \varepsilon v]$$

But if we accept this, the domain of v would be a proper class and not a set. So we consider only functions with "set" support, that is we understand the values out side its domain is always 0, the least element of B.

By this principle, Boolean valued power set $P^{(B)}(u)$ will correspond to the set of all B-valued functions

$$P^{(B)}(u) = \{x \mid x;u \to B\}$$

where $x;u \to B$ is an abbreviation of $x:\mathrm{dom}(x) \to B$ and $\mathrm{dom}(x) \subset u$.

Now the Boolean valued universe $V^{(B)}$ is introduced by

$$V^{(B)}(\alpha+1) = P^{(B)}(V^{(B)}(\alpha)) \qquad \alpha \text{ successor}$$

$$V^{(B)}(\alpha) = \bigcup_{\beta<\alpha} V^{(B)}(\beta) \qquad \alpha \text{ limit}$$

and put $V^{(B)}$ be their union, namely

$$V^{(B)} = \bigcup_{\alpha} V^{(B)}(\alpha)$$

This is an explicit definition of $V^{(B)}$. Corresponding implicit definition of this is the smallest class W satisfying

$$\forall u(u \varepsilon W \equiv u;W \to B)$$

Here we consider some steps of definition of functions such as sum, product, power and etc. such that

$$\begin{cases} n + 0 = n \\ n + m' = (n + m)' \end{cases} \qquad \begin{cases} n \times 0 = 0 \\ n \times m' = (n \times m) + n \end{cases}$$

$$\begin{cases} n^0 = 1 \\ n^{m'} = n^m \times n \end{cases} \qquad \begin{cases} n^{(0)} = 1 \\ n^{(m')} = n^{n^{(m)}} \end{cases}$$

Then the "fourth" function satisfies the relation

$$n^{(m)} = \left.n^{n^{\cdot^{\cdot^{\cdot^{n}}}}}\right\} m$$

Consider this to slightly general structure of sets, then the notions would be "direct sum", "direct product", "exponential" and etc. Any how R(n) may be considered as

$$R(n) = 2^{(n)}$$

Extending this definition to ordinal numbers and replace $2 = \{0,1\}$ by Boolean algebra B, we have

$$R^{(B)}(\alpha) = B^{(\alpha)}$$

or when we consider pertial B-valued functions, replace B by

$$B' = B \cup \{B\}$$

In this respect, set theory can be considered as a theory dealing with up to "fourth" functions.

3.2 Dual notions

Let B be a complete Boolean algebra. Then the dual space B* is the space of all homomorphisms defined on B and values in 2, namely

$$u \in B^* \equiv u:B \to 2$$

and let B** be the dual algebra of continuous functions defined on B* and values in 2, namely

$$v \in B^{**} \equiv v:B^* \to 2$$

According to Stone's representation theorem, we have that B** is isomorphic to the algebra B. This means that we have the following duality relation

algebra		space
B	——	B^*
|		|
2^{B^*}	——	2^B
function		morphism

Now, we shall state some initial steps of construction of $V^{(B)}$ and its dual $V^{(B)*}$

$$V_0^{(B)} = 0 \qquad V_0^{(B)*} = 0$$
$$V_1^{(B)} = 0 \qquad V_1^{(B)*} = 0^*$$
$$V_2^{(B)} = B \qquad V_2^{(B)*} = 2^{B^*}$$

where we have made the following identifications

$$0 \to a = a \quad \text{for} \quad a \in B$$

$$a^*(h) = h(a) \quad \text{for} \quad h \in B^*$$

and we have the following equivalence

$$0 \to a \in V_2^{(B)} \equiv a \in B \equiv a^* \in V_2^{(B)*}$$

For a successor ordinal, the dual notions are

$$u \in V_{\alpha+1}^{(B)} \equiv u:V_\alpha^{(B)} \to B$$
$$u^* \in V_{\alpha+1}^{(B)*} \equiv u^*:B^* \to 2^{V_\alpha^{(B)*}}$$

We have the following fundamental relation

$$(v^*)u^*(h) = h(u(v)) = (u(v))^*(h)$$

in the above, we write the argument of the function $u^*(h)$ left side, namely as $(v^*)u^*(h)$, so that we have symbolically

$$v^*u^* = (uv)^*$$

For a limit ordinal, the dual notions are

$$V_\alpha^{(B)} = \bigcup_{\beta<\alpha} V_\beta^{(B)} \qquad V_\alpha^{(B)*} = \bigcup_{\beta<\alpha} V_\beta^{(B)*}$$

According to the exponential law, we have

$${}^{V_\alpha^{(B)}}B = (2^{B^*})^{V_\alpha^{(B)}} \overset{*}{=} 2^{B^* \times V_\alpha^{(B)*}} = (2^{V_\alpha^{(B)*}})^{B^*}$$

We consider the following space

$$V_\alpha^{(B)} \overset{*}{=} V_\alpha^{(B)*}$$

as a discrete topological space and form the product space $B^* \times V_\alpha^{(B)*}$ and consider the set of all continuous functions

$$u : B^* \times V_\alpha^{(B)*} \to 2$$

Shortly speaking $V^{(B)}$ is the class of all hereditary B valued functions and $V^{(B)*}$ is the class of all hereditary B^* domained functions.

In this case $2 = \{0,1\}$ is considered simultaneously as a topological space and as a Boolean algebra. Important point here is the notion of duals

$$\begin{array}{ccc} \text{algebra} & & \text{space} \\ A & \text{———} & A^* \\ | & & | \\ C^{A^*} & \text{———} & C^{*A} \\ \text{function} & & \text{morphism} \end{array}$$

where A, C and A^*, C^* are dual spaces and algebras. Sometimes they are self-adjoint as in the case of 2, Euclidean space or Hilbert space. And to form cumulative hierarchy, we use the notion of coupling

$$C^{(A^* \times V^{(A)*})}$$

in this case C is called coupling space or dually coupling algebra.

3.3 Elementary properties

We summerize here some well-known properties of $V^{(B)}$. The class $V^{(B)}$ is defined by the induction so the corresponding principle is the rule of induction

Induction: Let P(u) be a property. Then

$$\forall u{:}V^{(B)} \to B(\forall x \in \mathrm{dom}(u)P(x) \to P(u))$$

implies

$$\forall u \in V^{(B)}\ P(u).$$

We restate here the notion introduced by D. Scott and R. M. Solovay concerning the truth values of the formulas $u \in v$ and $u = v$. They are defined by simultaneous induction as

$$[u \in v] = \sum_{x\ \mathrm{dom}(v)} v(x).[x = u]$$

$$[u = v] = \prod_{x \in \mathrm{dom}(u)} (u(x) \to [x \in v]). \prod_{x \in \mathrm{dom}(v)} (v(x) \to [x \in u])$$

By this definition, we have the equality axiom

$$u_1 = v_1, \ldots, u_n = v_n, A(u_1, \ldots, u_n) \to A(v_1, \ldots, v_n)$$

is true in $V^{(B)}$. By this we have the relation

$$[A(x)] = \sum_{x \in V^{(B)}} [A(y)][y = x]$$

corresponding property in analysis is the equality

$$f(x) = \int f(y)\delta(y,x)dx$$

This means that $[x = y]$ corresponds to so called delta function, namely we have

$$[x = y] \qquad\qquad \delta(x,y)$$

For the restricted quantifiers $\exists x \in u$ and $\forall x \in u$, we have

$$[\exists x \in uA(x)] = \sum_{x \in \mathrm{dom}(u)} u(x).[A(x)]$$

$$[\forall x \in uA(x)] = \prod_{x \in \mathrm{dom}(u)} u(x) \to [A(x)]$$

Concerning the above equality the corresponding notion to $\exists x \varepsilon u$ is integration, and the corresponding notion to $\forall x \varepsilon u$ and $\rightarrow$ would be the most important and interesting concepts in general space or algebra valued set theory.

We remark here about the relation of dual notions. The operations Σ^* and Π^* for continuous functions B^* to 2 are defined to be the least upper bound and greatest lower bound for continuous functions. These operations satisfy the following relations

$$\sum_\nu{}^* a_\nu^* = (\sum_\nu a_\nu)^* \qquad \prod_\nu{}^* a_\nu^* = (\prod_\nu a_\nu)^*$$

The continuous functions $[u^* \varepsilon v^*]$ and $[u^* = v^*]$ are defined similarly as follows

$$[u^* \varepsilon v^*] = \sum_{x^* \varepsilon cod(v^*)}{}^* (x^*)v^*.[x^* = u^*]$$

$$[u^* = v^*] = \prod_{x^* \varepsilon cod(u^*)}{}^* ((x^*)u^* \rightarrow [x^* \varepsilon v^*]). \prod_{x^* \varepsilon cod(v^*)}{}^* ((x^*)u^* \rightarrow [x^* \varepsilon u^*])$$

Then we have the following commutative relations

$$[u^* \varepsilon v^*] = [u \varepsilon v]^*$$

$$[u^* = v^*] = [u = v]^*$$

We list here some elementary properties of Boolean models

1) If a se quent $A_1,\ldots,A_m \rightarrow B_1,\ldots,B_n$ is provable then its value is 1 in $V^{(B)}$.

2) All axioms of ZFC are true, namely value 1 in $V^{(B)}$.

Continuity and compactness of values of formulas.

3) (maximum principle) For any formula $A(x)$, there is an element u of $V^{(B)}$ such that

$$[\exists x A(x)] = \sum_{x \varepsilon V^{(B)}} [A(x)] = [A(u)]$$

dually there is an element v of $V^{(B)}$ such that

$$[\forall x A(x)] = \prod_{x \varepsilon V^{(B)}} [A(x)] = [A(v)]$$

4) (intermediate value theorem) If $[A(u)] \leq a \leq [A(v)]$, then there is an element w of $V^{(B)}$ such that $a = [A(w)]$.

5) (conditional maximum principle) For any formulas $A(x)$ and $B(x)$, there is an element u of $V^{(B)}$ such that

$$[\exists x A(x)] = [A(u)]$$

$$[\exists x(A(x) \wedge B(x))] = [B(u)]$$

This property is extended to the case of monotone well-ordered sequence of properties, namely if

$$[\forall \alpha < \rho \; \forall \nu < \alpha \; \forall x(A(\nu,x) \rightarrow A(\alpha,x))] = 1$$

then there is an element u of $V^{(B)}$ such that

$$[\exists x A(\nu,x)] = [A(\nu,u)]$$

for all $\nu < \alpha$.

3.4 Complete homomorphisms

Let $h:B \rightarrow C$ be a complete homomorphism. Then there is cannonical map

$$h:V^{(B)} \rightarrow V^{(C)}$$

which is determined by the relation

$$h(x) \in \mathrm{dom}(h(u)) \equiv x \in \mathrm{dom}(u)$$

$$h(u)(h(x)) = \sum_{\substack{h(y)=h(x) \\ y \in \mathrm{dom}(u)}} h(u(y))$$

Consider epi-mono factrization

$$B \overset{p}{\rightarrow} D \overset{i}{\rightarrow} C$$

where p is an epimorphism and i is a monomorphism. Then the maps

$$V^{(B)} \overset{p}{\rightarrow} V^{(D)} \overset{i}{\rightarrow} V^{(C)}$$

are also epi- and mono- morphisms.

1) (epimorphism property) If p is an epimorphism, then for any formula

$$p([A(u_1,\ldots,u_n)]) = [A(p(u_1),\ldots,p(u_n))]$$

Suppose the formulas $A(x,u_1,\ldots,u_n)$ and $B(x,u_1,\ldots,u_n)$ be restricted formulas and the formula

$$\exists x A(x,u_1,\ldots,u_n) \equiv \forall x B(x,u_1,\ldots,u_n)$$

be provable in ZFC. then the equivalent formula is called Δ_1^1-formula.

2) (Δ_1^1-property) Let $P(u_1,\ldots,u_n)$ be a Δ_1^1-formula. Then

$$h([P(u_1,\ldots,u_n)]) = [P(h(u_1),\ldots,h(u_n))]$$

for any complete homomorphism h.

4. Combinatorial properties of $V^{(B)}$

4.1 Representation of elements

Let $^-:2 \to B$ be the cannonical inclusion and let

$$^-:V^{(2)} \to V^{(B)}$$

be the induced inclusion. Since $V \simeq V^{(2)}$, V is included in $V^{(B)}$. Assume that $[u \in V] = 1$, then we have a partition of unity

$$[u \in V] = \sum_{x \in V} [u = x] = 1$$

because we have $[\bar{x} = \bar{y}] = 0$ for $x \neq y$ and

$$[u = \bar{x}].[u = \bar{y}] \leq [\bar{x} = \bar{y}]$$

Hence u is represented as

$$u = \sum_{x \in s} [u = \bar{x}].\bar{x}$$

where s is the set defined by

$$\{x | [u = \bar{x}] > 0\}$$

This is so called orthgonal divelopment of u. Conversely let $\{a_x | x \in s\}$ be a partition of unity indexed by a set s. Then

$$u = \sum_{x \in s} a_x.\bar{x}$$

is an element of V. Namely "partition of unity" is just the element of V.

Similarly a subset u of $\bar{a}$ in $V^{(B)}$ corresponds to a function $u:a \to B$ by

$$u(\bar{x}) = [\bar{x} \in u]$$

which is a B-valued "vector" of size a. And a relation between $\bar{a}$ and $\bar{b}$ corresponds to a function

$$A: a \times b \to B$$

which is a B-valued "matrix" of type (a,b). Let R(x,y) be a relation, then the corresponding matrix is defined by

$$A(x,y) = [R(\bar{x},\bar{y})]$$

Let A and C be matrices of type (a,b) and (b,c) respectively. Then the product AC of type (a,c) is defined by

$$AC(x,z) = \sum_{y \varepsilon b} A(x,y).C(y,z)$$

of course this notion corresponds to the product of relations

$$\exists y \varepsilon b(R(x,y) \wedge S(y,z))$$

Let u be a function defined on $\bar{a}$ and values in $\bar{b}$ in $V^{(B)}$. Then the properties to be a function are

$$\forall x \varepsilon a \exists y \varepsilon b((\bar{x},\bar{y}) \varepsilon u)$$

$$\forall x \varepsilon a \forall y_1 \varepsilon b \forall y_2 \varepsilon b((\bar{x},\bar{y}_1) \varepsilon u \wedge (\bar{x},\bar{y}_2) \varepsilon u \to \bar{y}_1 = \bar{y}_2)$$

So the corresponding relations in matrix are

$$\sum_{y \varepsilon b} A(x,y) = 1$$

$$\sum_{x \varepsilon a} A(x,y_1).A(x,y_2) = 0 \text{ for } y_1 \neq y_2$$

this means that the "length" of row vector is 1 and the "inner product" of different column vectors is 0, namely they are orthogonal

Additional properties such as "onto" and "one to one" are

$$\forall y \varepsilon b \exists x \varepsilon a((x,y) \varepsilon u)$$

$$\forall y \varepsilon b \forall x_1 \varepsilon a \forall x_2 \varepsilon a((x_1,y) \varepsilon u \wedge (x_2,y) \varepsilon u \to x_1 = x_2)$$

Hence the corresponding relations in matrix are the length of column vectors are 1 and row vectors are orthogonal. So onto one to one, namely isomorphism, corresponds to "orthogonal" matrix. In general, there are many variety of orthogonal matrices of type $(\omega_\alpha,\omega_\beta)$ for different different cardinal numbers ω_α and ω_β. This property corresponds to the fact that the cardinality of $\bar{\omega}_\alpha$ and $\bar{\omega}_\beta$ are equal in $V^{(B)}$. The notion of one to one onto map on a set is just the notion of "unitary" transformation.

Many properties in elementary set theory are stated using Boolean valued matrices.

1) (box argument) Let A be a (n,n) matrix where $n \in \omega$, and A^* be its transposition. Then we have

$$A^*A = E_n \rightarrow AA^* = E_n$$

where E_n is the unit matrix.

2) (axiom of choice) Let A be a matrix of type (a,b) such that

$$A^*A \geq E_a \qquad AA^* = E_b$$

Then there is a matrix B of type (b,a) such that

$$BB^* \leq E_a \qquad B^*B = E_b \qquad AB = E_a$$

3) (Cantor-Bernstein theorem) Let A and B be matrices of type (a,b) and (b,a) respectively such that

$$A^*A = E_a \qquad AA^* \leq E_b \qquad B^*B = E_b \qquad BB^* \leq E_a$$

Then there is an orthogonal matrix C of type (a,b) such that

$$C^*C = E_a \qquad CC^* = E_b$$

Such C is defined by

$$C = \sum_{n\in\omega} (BA)^n(\bar{a}-B\bar{b})A + \sum_{n\in\omega} (AB)^n(\bar{b}-A\bar{a})B^* + \prod_{n\in\omega} (BA)^n\bar{a}A$$

4) (Cantor theorem) There is a natural correspondence between a function $f:a \rightarrow P(a)$ and a square matrix A of type (a,a). And the diagonal argument is to consider the vector

$$u(x) = -A(x,x)$$

Hence for any square matrix A of type (a,a), there is a vector u on a such that for every $x \in a$

$$\sum_{y\in a} u(y)-A(x,y) + \sum_{y\in a} A(x,y)-u(y) = 1$$

Similar description is possible for König's theorem. An element of direct sum corresponds to a partition of unity indexed by the direct sum, and an element of direct product

$$\prod_{x \in c} b(x)$$

corresponds to a vector of size c of partitions of unity indexed by b(x)'s.

Suppose that B is an atomless complete Boolean algebra. We consider a partition of unity consisting of non-zero elements of B such that

$$\sum_{n,m \in \omega} a_{n,m} = 1$$

Then the cardinality of B is the cardinality of the product

$$\prod_{n,m \in \omega} B(a_{n,m})$$

where $B(a_{n,m})$ is the restriction algebra of B to $a_{n,m}$. We can chose such $a_{n,m}$'s so that the cardinality of the set

$$\sum_{m \in \omega} B(a_{n,m})$$

is always the same for all $n \in \omega$. By the following cardinal equality

$$\#(\prod_{n,m \in \omega} B(a_{n,m})) = \#(\prod_{m \in \omega} B(a_{n,m}))$$

we have that

$$\#(B^{\omega}) = \#B$$

Hence the set of all B-valued vector of size ω and the set of all partitions of unity indexed by ω have the same cardinality in V. But they correspond to the sets $P(\omega)$ and ω respectively, so they have different cardinality in $V^{(B)}$. Clearly the above argument does not extend to ω_1 but it may give some informations on the cardinality of Boolean algebras. As a special case we have that every atomless Borel family satisfies the above cardinal equation.

4.2 Measure and category

Let X be a Baire space, and let B be the complete Boolean algebra of regular open sets of X.

According to D. Scott, a real number in $V^{(B)}$ is defined by a Dedekind cut of rational numbers Q, which is a monotone function

$$f : Q \to B$$

with the property that

$$\prod_{r \in Q} f(r) = 0 \qquad \sum_{r \in Q} f(r) = 1$$

Since f(r) is a regular open set in X, there is a function f* defined on co-meager set of X such that

$$f^*(x) = \inf \{ r \in Q \mid x \in f(r) \}$$

This means that for every $r \in Q$ and co-meager x in X, we have

$$f^*(x) < r \equiv x \in f(r)$$

Since $x \in f(f^*(x)+\varepsilon)$ for every $\varepsilon > 0$ and it is open, there is a neighbourhood U(x) of x such that

$$x \in U(x) \subset f(f^*(x)+\varepsilon)$$

which means that

$$y \in U \to f^*(y) < f^*(x)+\varepsilon$$

Considering lower cut, we have that f* is a continuous function defined on a co-meager set of X. Such function is of course different from the function which is defined on X with co-meager continuous points.

The dual space B* of a Boolean algebra is a Baire space, every real number in $V^{(B)}$ is represented by a continuous real-valued function defined on a co-meager set of B*. But if a real number in $V^{(B)}$ is bounded, namely if $[-\bar{n} \le u \le \bar{n}] = 1$ for some $n \in \omega$, then since f(x) can be taken as clopen set, we have that f* is a totally defined continuous function on B*.

Let $\mathcal{B}$ be a Borel family of a topological space X, and μ be a σ-finite, namely ω_1-saturated, measure on $\mathcal{B}$. And let I_μ be the measure ideal

$$I_\mu = \{ A \in \mathcal{B} \mid \mu(A) = 0 \}$$

Consider the quotient algebra $B = \mathcal{B}/I_\mu$. As before a real number in $V^{(B)}$ is represented by a monotone function $f:Q \to R$. But in this case $f(r)$ is as a representative an element of $\mathcal{B}$. So the function f^* satisfying

$$r < f^*(x) \equiv x \in f(r)$$

is a Borel measurable function. Conversely such function determines a real number in $V^{(B)}$.

Let ν be another measure absolutely continuous over μ. Then by Radon-Nikodym, there is a measurable function f such that

$$\nu(A) = \int_A f(x)d\mu(x)$$

or equivalently

$$\frac{d\nu}{d\mu} = f$$

So real number in $V^{(B)}$ is also considered as absolutely continuous measure over μ. Characterization of reals by measurable function is due to R. M. Solovay.

4.3 Representation of subsets

Let B be the complete Boolean algebra of regular open sets of Baire space X. Let (Y,U) be another topological space. Then $(\bar{Y},\bar{U})$ is a topological space in $V^{(B)}$.

For a subset A of the product space $X \times Y$, we define A^x and A_x by the following property

$$y \in A^x \equiv x \in A_y \equiv (x,y) \in A$$

Assume that $[u \in \bar{Y}] = 1$, then we say that $u \in V^{(B)}$ is represented by A if

$$[y \in u] = [A_y]$$

where $[A_y]$ is the element of B determined by A_y.

Now we consider an open set v of $\bar{Y}$ in $V^{(B)}$. Then we have

$$[\forall y(y \in v \equiv \exists u \in \bar{U}(u \subset v))] = 1$$

This means that

$$[y \in v] = \sum_{u \in U(y)} [u \subset v]$$

Since the set $[u \subset v]$ is a regular open set in X, following set

$$\bigcup_{u \in U} [u \subset v] \times u$$

is an open set in the product space $X \times Y$ and it represents v. Conversely any open set of $X \times Y$ represents an open set of $\bar{Y}$ in $V^{(B)}$.

Suppose $\bar{\omega}_1 = \omega_1$ in $V^{(B)}$, then the notion of representations commutes with the operations of countable union and complementation. Let $u^{\#}$ be the representation of u. Then we have for example

$$(\bigcup_{n \in \omega} u_n)^{\#} = \bigcup_{n \in \omega} u_n^{\#}$$

Hence every Borel set of $X \times Y$ represent a Borel set of $\bar{Y}$ in $V^{(B)}$. Suppose that $\mathrm{sat}(B) = \omega_1$, then every Borel set of $\bar{Y}$ in $V^{(B)}$ is represented by a Borel set of $X \times Y$. But if $\mathrm{sat}(B) > \omega_1$, even in the case $\bar{\omega}_1 = \omega_1$, it is not generally true.

As a general remark, let $a_\nu (\nu < \lambda)$ be pairwise disjoint family of regular open sets of X and let u_ν be represented by $u_\nu^{\#}$, then

$$(\sum_{\nu<\lambda} a_\nu u_\nu)^{\#} = \bigcup_{\nu<\lambda} u_\nu^{\#} | a_\nu$$

where the restriction is defined by

$$u^{\#} | a = \{(x,y) \in u^{\#} | x \in a\}$$

The left hand side of this equation is not a Borel operation if $\lambda \geq \omega_1$.

Concerning denseness, we have that $\upsilon \subset \bar{Y}$ is dense open in $V^{(B)}$ if and only if its representation $u^{\#}$ is dence open in $X \times Y$. Hence if $\mathrm{sat}(B) = \omega_1$ then a subset $u \subset \bar{Y}$ is meager or co-meager if and only if $u^{\#}$ is meager or co-meager in $X \times Y$. General behavior of these properties seems to be an interesting subject.

4.4 Commutativity of ideal product

Let I_X and I_Y be ideals of subsets of X and Y respectively. Then the product of them is defined by the following equivalence

$$A \notin I_X \times I_Y \equiv \{y \in Y | A_y \notin I_X\} \notin I_Y$$

First we consider the case of measure algebra.

Let B_X and B_Y be Borel families on X and Y, μ and ν be their measures, I_μ and I_ν be corresponding measure ideals. Then we consider the product Borel family and product measure $\mu \times \nu$ on it. Let $\tilde{B}_Y$ be the Borel family generated by $\bar{B}_Y$ in $V^{(B)}$, where $B = B_X/I_\mu$ Then by $\mathrm{sat}(B) = \omega_1$, every element of $\tilde{B}_Y$ is represented by an element of the product family $B_X \times B_Y$.

For $u \in \tilde{B}_Y$, we put

$$f(x) = \nu(u^{\#x})$$

then $f: X \to R$ is a measurable function, so it represents a real number in $V^{(B)}$. This means that ν has a natural extension $\tilde{\nu}$ on $\tilde{B}_Y$ in $V^{(B)}$. To see that $\tilde{\nu}$ is defined as a function in $V^{(B)}$, we must check the extensionality condition. By Fubini theorem we have

$$\mu \times \nu(A) = \int_X \nu(A^x)d\mu(x) = \int_Y \mu(A_y)d\nu(y)$$

So $\mu \times \nu(A) = 0$ if and only if $f(x) = \nu(A^x)$ represent 0 in $V^{(B)}$. By the induction on the steps to construct an element of $\tilde{B}_Y$ from $\bar{B}_Y$ in $V^{(B)}$, we have that

$$[u = v] - \{x \in X | \nu(u^{\#x}) \neq \nu(v^{\#x})\}$$

has μ measure 0. Hence

$$[u = v] \leq \{x \in X | \nu(u^{\#x}) = \nu(v^{\#x})\}$$

which means the extensionality condition.

As before Fubini theorem implies

$$A \notin I_{\mu \times \nu} \equiv A \notin I_\mu \times I_\nu \equiv A \notin I_\nu \times I_\mu$$

This means that these ideals are equal, namely

$$I_{\mu \times \nu} = I_\mu \times I_\nu \simeq I_\nu \times I_\mu$$

The quotient algebra $\tilde{B}_Y/I_{\tilde{\nu}}$ in $V^{(B)}$ is represented by the quotient algebra $B_X \times B_Y/I_{\mu \times \nu}$.

Next we consider the case of category algebra.

Let T be a topological space, by N_T we denote the ideal of all nowhere dense sets of T. For closed set F in X × Y, if $F \notin N_{X \times Y}$, then there are open sets U and V such that $U \times V \subset F$, which means

$$\{y \in Y | F_y \notin N_X\} \notin N_Y$$

namely $F \notin N_X \times N_Y$.

Suppose that Y satisfies 2nd axiom of countability, and let U_n $(n \in \omega)$ its countable open basis. Let F be a closed set such that

$$D = \{y \in Y | F_y \notin N_X\} \notin N_Y$$

Then by the assumption on X, we have

$$D = \bigcup_{n \in \omega} \{y \in Y | U_n \subset F_y\}$$

Therefore we have n such that $\{y \in Y | U_n \subset F_y\} \notin N_Y$. Since F is closed, we have $V \times U \subset F$ where

$$V = \Box\Diamond \{y \in Y | U_n \subset F_y\}$$

which means that $F \notin N_{X \times Y}$.

It seems to be essential that for some condition on Y is nessesary to show $N_X \times N_Y = N_Y \times N_X$ or $N_{X \times Y} = N_X \times N_Y$. It is very interesting to have "good" criterion for the commutativity.

Now we consider mixed case of measure and category.

Take for example X = Y = R and let B be the set of all Borel sets, I_μ be the measure ideal generated by Lebesgue measure and I_c be the ideal of meager sets. Let us consider their products

$$I_\mu \times I_c \qquad\qquad I_c \times I_\mu$$

Let $I_{\mu \times c} = I_{c \times \mu}$ be the ideal generated by countable union of rectangles

$$\{C \times D \mid C \in I_\mu \wedge D \in I_c\}$$

Then by the above argument, we have

$$I_{c \times \mu} = I_{\mu \times c} = I_\mu \times I_c$$

and one direction inclusion

$$I_\mu \times I_c \subset I_c \times I_\mu$$

But there is a set D such that

$$D \in I_\mu \times I_c \qquad D \notin I_c \times I_\mu$$

To show this, we consider a set which is meager and its complement is measure zero. And let D be the set defined by

$$D = \{(x,y) \mid x+y \in A\}$$

Then for every x the set D^x is congruent to A and so $D^x \in I_c$, this means that

$$\{x \in R \mid D^x \notin I_c\} = \phi \in I_\mu$$

but D_y also congruent to A and so $D_y \notin I_\mu$ for every y, this means

$$\{y \in R \mid D_y \notin I_\mu\} = R \notin I_c$$

This means that the topological space B = $\mathcal{B}/I_\mu$ deteremined by its order does not satisfy 2nd axiom of countability, namely in this case not separable. To be more presice the topology of B is determined by the unique neighbourhood

$$V(a) = \{b \in B \mid 0 < b \leq a\}$$

for each $a \in B - \{0\}$.

We can give direct proof of this for every non-atomic measure algebra. Suppose $\{a_n \mid n \in \omega\}$ is a dense set, then there is a sequence $\{b_n \mid n \in \omega\}$ such that

$$0 < b_n < a_n \qquad \mu(b_n) < \frac{1}{2^{n+2}}\mu(a_0)$$

and put

$$b^* = a_0 - \sum_{n\in\omega} b_n$$

Then there is no a_n such that $0 < a_n < b^*$.

Let sep(X) be the smallest cardinality of dense set in X. Then the above remark means that

$$\mathrm{sep}(B) \geq \omega_1$$

for every non-atomic measure algebra. General behavior of sep(B) for measure algebras in the case $2^{\omega} > \omega_1$ seems to be interesting.

Let $\mathcal{B}$ be the set of all Borel subsets of R^2, and consider two complete Boolean algebras

$$B_{c\mu} = \mathcal{B}/I_c \times I_\mu \qquad\qquad B_{\ c} = \mathcal{B}/I_\mu \times I_c$$

By the inclusion

$$I_\mu \times I_c \subset I_c \times I_\mu$$

we have a projection

$$p : B_{\mu c} \to B_{c\mu}$$

If this mapping is complete, then by projection theorem

$$p([\![A(u_1,\ldots,u_n)]\!]) = [\![A(p(u_1),\ldots,p(u_n))]\!]$$

for every $u_1,\ldots,u_n$ in $V^{(B_{\mu c})}$. This means that there is no formula which is true in $V^{(B_{\mu c})}$ but not true in $V^{(B_{c\mu})}$.

Let $\tilde{\mu}$ be Lebesgue measure on R in $V^{(B_c)}$ and let $B_{\tilde{\mu}}$ be the complete Boolean algebra determined by $\tilde{\mu}$ in $V^{(B_c)}$. Then $V^{(B_{\mu c})}$ is isomorphic to $V^{(B_c)(B_{\tilde{\mu}})}$.

Let $\tilde{B}_\mu$ be the completion of Boolean algebra $\bar{B}_\mu$ in $V^{(B_c)}$. Then $V^{(B_{c\mu})}$ is isomorphic to $V^{(B_c)(\tilde{B}_\mu)}$ and it is also isomorphic to $V^{(B_\mu)(B_{\tilde{c}})}$ where $B_{\tilde{c}}$ is the complete Boolean algebra of regular open sets of R in $V^{(B_\mu)}$.

Interesting problem concerning this is

whether $B_{c\mu}$ and $B_{\mu c}$ isomorphic?

in various situation.

5. Boolean valued matrix

5.1 Decomposition by cardinality

Since the property to be a cardinal number is a Π^1_1-property we have that

$$[\mathrm{Card}(\kappa)] \neq 0 \rightarrow \mathrm{Card}(\kappa)$$

By the axiom of choice, we have a divelopment by cardinality

$$u = \sum_{\kappa} [\bar{\bar{u}} = \bar{\kappa}]u$$

This means that on $[\bar{\bar{u}} = \bar{\kappa}]$, we have a function f such that

$$f: u \overset{1-1}{\underset{\text{onto}}{\rightarrow}} \kappa$$

A relation R on u is represented by the relation R_f on κ defined by

$$[xRy] = [f(x)R_f f(y)]$$

which is a square Boolean valued matrix of size κ. Consider another isomorphism g, then

$$f^{-1}g: \kappa \rightarrow \kappa$$

is an isomorphism, so it is represented by unitary matrix. This means that the property invariant under unitary transformation corresponds to set theoretical property in $V^{(B)}$.

5.2 Eigen values and vectors

Now we consider Boolean valued matrix and its invariant sets. Let A be a square Boolean valued matrix of type (a,a), namely A is a binary relation on $\bar{a}$ in $V^{(B)}$. We consider an equation of the form

$$Au = bu$$

where b is an element of B and u is a vector of size a, namely a subset of $\bar{a}$ in $V^{(B)}$.

In the above, if $u \neq 0$ then b is called an eigen value of matrix A and u is called an eigen vector corresponding to b. The contents of the above equation is

$$b \leq [Au = u]$$

namely the set u is invariant under the relation A in $V^{(B)}$ with the possibility greater or equal to b.

We consider some properties and construction of such invariant sets. We define the power of matrix A by the induction on the ordinal numbers as follows

$$A^{\alpha+1} = AA^{\alpha} \qquad\qquad A^{\alpha} = \bigcap_{\beta<\alpha} A^{\beta}$$

Then each A^{α} is considered as a function defined on the sets of all B valued vectors of size a to itself, anmely

$$A^{\alpha}:P^{(B)}(a) \to P^{(B)}(a)$$

The operation is monotone, that is

$$u \subset v \to A^{\alpha}u \subset A^{\alpha}v$$

Therefore if $Au \subset u$ then we have by the induction on ordinals

$$\alpha < \beta \to A^{\beta}u \subset A^{\alpha}u$$

Hence there is an ordinal number α such that

$$u^* = AA^{\alpha}u = A^{\alpha}u$$

this means that u^* is an invariant set with respect to A.

The set $\bar{a}$ in $V^{(B)}$ clearly satisfies the condition $A\bar{a} \subset \bar{a}$, so we have that a^* is the largest invariant set with eigen value 1. The set u^* may be empty, namely

$$u^* = A^{\alpha}u = 0$$

in such case we say that A is nilpotent on u, especially if $a^* = 0$, then A is called nilpotent matrix.

The property of being an invariant subset is the conjunction of two formulas

$$\forall x \in u\ \exists y \in u((x,y) \in A)$$

$$\forall x \in a\ \forall y \in u((x,y) \in A \to x \in u)$$

Now we consider the transitive B-valued relation defined by

$$A^* = \sum_{n<\omega} A^n = (1-A)^{-1}$$

$$A^+ = \sum_{n<\omega} A^{n+1} = A(1-A)^{-1}$$

and the notations

$$x \geq_A y \equiv (x,y) \in A^* \qquad x >_A y \equiv (x,y) \in A^+$$

The relation $x >_A y$ means that

$$x_1,\ldots,x_n \in a((x,x_1) \in A \wedge \ldots \wedge (x_n,y) \in A)$$

meaning that there is a chain connecting x to y in A. Note that in general $x >_A y$ does not imply $x \neq y$ but

$$x \geq_A y \equiv x >_A y \vee x = y$$

The relation $>_A$ is a transitive relation on a in $V^{(B)}$, and

$$\forall x \in a\ \forall y \in u((x,y) \in A \rightarrow x \in u)$$

means that

$$x \geq_A y,\ y \in u \rightarrow x \in u$$

which also means that u is an open set in the topological space $(a,\geq_A)$ with the order topology.

The condition

$$\forall x \in u\ \exists y \in u((x,y) \in A)$$

means that every x in u has a predecessor in u under A. Namely every element x of u has a descending sequence beginning with x

$$x = x_0 >_A x_1 >_A \cdots >_A x_n >_A \cdots$$

hence every element of u is not well-founded, and the set of all not well-founded elements

$$w = \{x \in a \mid x \text{ is not well-founded}\}$$

determines the decomposition of $\bar{a}$ into well-founded part $\bar{a}$-w and not well-founded part w. The well-founded part $\bar{a}$-w may be called initial well-founded part of $>_A$. The ordinal number α of this well-founded part satisfies the condition

$$A^\alpha(\bar{a}-w) = 0$$

This means that the maximal nilpotent set with respect to A is the initial well-founded part of $>_A$. And the set w of all not well-founded elements is the largest invariant set of a.

5.3 Minimal invariant sets

Conjunction of the relation $x \geq_A y$ and $y \geq_A x$ means that there is a loop in which x and y appear, and it defines an equivalence relation on a set $\bar{a}$ in $V^{(B)}$. For every A-invariant set u, we have

$$x =_A y, x \in u \rightarrow y \in u$$

where $x =_A y$ is the equivalence relation mentioned above.

We consider $a^{\#}$ of equivalence classes devided by $=_A$. Then $a^{\#}$ is considered as a partial ordered structure by the induced order defined by

$$x \geq_A^{\#} y \equiv \exists s \in x \exists t \in y (s \geq_A t)$$

and in this case $\geq_A^{\#}$ is asymmetric ordering. An element x is called reflexible with respect to A if $x >_A x$.

We consider how independent sets depend on the partial order generated by the relation defined above. Let $(P, \geq, R)$ be a asymmetric partial order structure with the set of reflexible elements R.

Let V(x) be the open basis defined by

$$V(x) = \{y \in P | y \geq x\}$$

Then as remarked before any invariant set is an open set. If x is reflexible, namely if $x \in R$, then V(x) is an invariant set, so if u is minimal invariant and $u \cap R \neq 0$, then it must be a singleton $\{x\}$ and $u = V(x)$.

Now we consider the case $u \cap R = 0$. The condition

$$\forall x \in u \exists y \in u((x,y) \in A)$$

means that every x in u has a predecessor under A, namely every element x of u has a descending sequence begining with x

$$x = x_0 >_A x_1 >_A \cdots >_A x_n >_A \cdots$$

Conversely given any such descending sequence, the smallest open set, namely the smallest closed set of dual order topology, including the sequence

$$u = \{x \in P | \ n \in \omega (x \geq x_n)\}$$

is the smallest invariant set including the above sequence. And it satisfies the compatibility condition

$$\forall x,y \in u \ \exists z \in u(x \geq z \wedge y \geq z)$$

This means that minimal invariant set is compatible.

Let u be a minimal invariant set and $x \in u$, then $V(x)$ is well-founded, because if it is not well-founded then there is an descending sequence in $V(x)$, so the smallest open set including the sequence is invariant and does not include x, which contradicts to the minimality of u.

In general, the condition $V(x)$ is well-founded is not a sufficient condition to be an element of a minimal invariant set.

A subset u is minimal invariant if and only if its co-initiality is ω, namely there is a descending sequence of type ω, and for any descending sequence $x_1, x_2, \ldots, x_n, \ldots$ and any element x of u there is $n \in \omega$ such that $x \geq x_n$.

Now we consider some special examples. Suppose the relation A represent a function f. Then the sequence

$$x, f(x), f^2(x), \ldots, f^n(x), \ldots$$

determines a minimal invariant set u. Any element y of u satisfies the condition that there are natural numbers k and n such that

$$f^k(x) = f^n(y)$$

Since this defines an equivalence relation, so the domain of f is devided into disjoint invariant sets.

If A^{-1} represent a function f, then

$$\bigcup_{\alpha} (f^{\alpha}(\bar{a}) - f^{\alpha+1}(\bar{a}))$$

is the nilpotent part and the components is the union of minimal invariant sets, but in general they are not disjoint if f is not one to one.

Especially, if A is unitary, namely it represent one to one onto function, then the domain is the union of finite or infinite cycles and they are minimal invariant sets.

If the relation is symmetric, then the domain is devided into connected components and isolated points. If the relation is idempotent then initial points are nilpotent and each fixed points forms minimal invariant set.

5.4 Matrix and measure

Let B be a measure algebra $B = \mathcal{B}/I_\mu$. For a Boolean valued matrix $A = (c_{xy})$ we can form a real valued matrix

$$\mu(A) = (\mu(c_{xy}))$$

We consider the case that a matrix represent a function, namely

$$\forall x \in a\, \forall y_1, y_2 \in b((x,y_1) \in f \wedge (x,y_2) \in f \rightarrow y_1 = y_2)$$

which means that the Boolean valued elements

$$c_{xy} = [(x,y) \in f]$$

are disjoint as a family indexed by $y \in b$.

Another notion is "independence" of two matrices, namely $A = (c_{xy})$ and $B = (d_{yz})$ are independent if

$$\mu(c_{xy}d_{yz}) = \mu(c_{xy})\mu(d_{yz})$$

Therefore if A represent a function and A, B are independent, then by disjointness, we have

$$\mu(\sum_{y\in b} c_{xy}d_{yz}) = \sum_{y\in b} \mu(c_{xy}d_{yz}) = \sum_{y\in b} \mu(c_{xy})\mu(d_{yz})$$

which means that the measure commutes with the notion of products of Boolean valued and real valued matrices.

Any how corresponding to Scott-Solovay's notion of

Boolean algebra and its dual space

it seems to be interesting to factorize "analysis" into

category and measure.

References

[1] P. J. Cohen: The independence of the continuum hypothesis I, II, Proc. Nat. Acad. US, 50 (1963) 1143-1148; 51 (1964) 105-110.

[2] K. Gödel: The consistency of the axiom of choice and of the generalized continuum hypothesis with the axioms of set theory, Ann. Math. Studies, Princeton Univ. Press 1940.

[3] T. Jech: Set theory, Academic Press 1978.

[4] K. Kunen: Inaccessibility properties of cardinals, Stanford Univ. Ph.D. Thesis.

[5] D. A. Martin, R. M. Solovay: Internal Cohen extensions, Ann. Math. Logic 2 (1970) 143-178.

[6] K. Namba: On the closed unbounded ideal of ordinal numbers, Comm. Math. Univ. St. Pauli, Tokyo 22 (1973) 33-56.

[7] D. S. Scott, R. M. Solovay: Boolean valued models for set theory, Lecture note, Summer Institute on Axiomatic Set Theory, UCLA, 1967.

[8] J. H. Silver: Some applications of model theory in set theory, Ann. Math. Logic 3 (1971) 45-110.

[9] R. M. Solovay: A model of set theory in which every set of reals is Lebesgue measurable, Ann. Math. 92 (1970) 1-56.

[10] R. M. Solovay, S. Tennenbaum: Iterated Cohen extensions and Souslin's problem, Ann. Math. 94 (1971) 201-245.

UNDECIDABILITY OF EXTENSIONS OF THE MONADIC FIRST-ORDER THEORY OF SUCCESSOR AND TWO-DIMENSIONAL FINITE AUTOMATA

Hiroakira Ono

Faculty of Integrated Arts and Sciences
Hiroshima University, Hiroshima

Akira Nakamura

Department of Applied Mathematics
Hiroshima University, Hiroshima

Introduction

The decision problems of various monadic second-order theories have been solved affirmatively by using the theory of automata, for example, Büchi [1], [2] and Rabin [7]. In [1], Büchi showed that the decision problem of the weak monadic second-order theory of successor $Th_2^w< x+1 >$ can be reduced to the emptiness problem of finite automata and then derived the decidability of the theory $Th_2^w< x+1 >$, since the emptiness problem is recursively solvable. In order to generalize this result, he next introduced finite automata on infinite sequences. Then, he got in [2] also the decidability of the monadic second-order theory of succesor $Th_2< x+1 >$ by using the similar reduction. In [7] Rabin made an interesting and important progress in this direction. In fact, he developed the theory of automata on infinite trees, proved the decidability of the emptiness problem of these automata and got the decidability result of the monadic second-order theory of two successor functions, since the latter can be reducible to the emptiness problem of automata on infinite trees, also in this case. As the decision problem of various second-order theories can be reduced to that of the monadic second-order theory of two successors, it turns out that his method of using the theory of automata is a very powerful tool for getting the decidability.

On the other hand, we have shown in [5] the undecidability of the monadic first-order arithmetic $Th_1 < P ; 2x, x+1 >$ with a single monadic predicate symbol P and functions 2x and x+1, by proving that the meeting problem, a kind of decision problems, of finite causal ω^2-systems, whose undecidability is shown by the second author in [4], can be reduced to the satisfiability problem of $Th_1 < P ; 2x, x+1 >$. Here, a finite causal ω^2-system is a kind of two-dimensional finite automata, and hence our result can be regarded as an application of the theory of two-dimensional finite automata to the decision problem of monadic first-order theories. In this sense, our method forms a remarkable contrast to the method developed by Büchi and Rabin. From the above result, the undecidability of the monadic second-order theory $Th_2 < 2x, x+1 >$ of functions 2x and x+1 by R.M. Robinson [8] and of the monadic first-order theory $Th_1^{=} < P ; x+y >$ with the equality symbol, a monadic predicate symbol P and addition by H. Putnam [6], follows immediately. It should be noticed that almost all of methods of proving the undecidability employed so far consist of showing the definability of addition and multiplication (or divisibility) on natural numbers in a given theory.

In this note, we will give some explications of our basic idea employed in [5] and show the undecidability result of various extensions of the monadic first-order theory of successor by generalizing our method. In the next section, we will make a rough sketch of our proof of the undecidability of the theory $Th_1 < P ; 2x, x+1 >$, to make this note as self-contained as possible.

Then, we will extend the result to various monadic first-order theories of successor with a monotone increasing function, in Section 2. On the other hand, we will remark in Section 3 that almost all monadic second-order theories of a single strictly monotone increasing function are decidable, as a corollary of Rabin's result [7]. By combining this remark with the results in Section 2, we can say that for a great many of functions which increase more rapidly than x, our result is critical.

1. Finite causal ω^2-systems and the undecidability of the monadic first-order arithmetic Th_1< P ; 2x, x+1 >

First, we will mention about finite causal ω^2-systems and their meeting problem, which has been investigated in [4]. Before giving a precise definition, we will give an intuitive explanation. A finite causal ω^2-system is a kind of two-dimensional automata consisting of an infinite array of cells, all alike, on the two-dimensional plane, each of which is in a (virtual) quiescent state at time t = 0. At time t = 1, the (1,1)-cell falls into the initial state and then each (m,n)-cell changes its state one after another into another stable state according to neighboring state functions of the system. The meeting problem means the problem of deciding whether or not there exists a cell which will take a given special state eventually in a given finite causal ω^2-system.

Now let us define finite causal ω^2-systems mathematically. For our present purpose, we will define them in the following form. (As for the original form of finite causal ω^2-systems, see [4]. Finite causal ω^2-systems in this note are nothing but *modified finite causal ω^2-systems of the third type* in [5].)

<u>Definition 1.1.</u> A *finite causal ω^2-system* (or a *FC system* for short) is a quintuple ($Q, q_1, \tau_1, \tau_2, \tau$), where

1) Q is a nonempty finite set of *states*,
2) q_1 is an element of Q, called the *initial state*,
3) τ_1 and τ_2 are functions from Q to Q, and τ is a function from $Q \times Q$ to Q. Sometimes, they are called *neighboring state functions*.

With each FC system S, we will associate a function τ^+, called the *allocation function* determined by S, from the set $U = \{ (m,n) ;$ m and n are positive integers such that $1 \leq n \leq m \}$ to Q. The function τ^+ is defined inductively as follows.

1) $\tau^+(1,1) = q_1$,
2) $\tau^+(m,1) = \tau_1(\tau^+(m-1,1))$ for $m > 1$,
3) $\tau^+(m,m) = \tau_2(\tau^+(m-1,m-1))$ for $m > 1$,
4) $\tau^+(m,n) = \tau(\tau^+(m,n-1),\tau^+(m-1,n))$ for $1 < n < m$.

Intuitively, $\tau^+(m,n)$ denotes the state taken by the (m,n)-cell after the transition of states. Now, the *meeting problem* of FC systems is a problem of deciding whether or not there exists a pair (m,n) in the set U such that $\tau^+(m,n) = q_\delta$ in a given FC system S, where q_δ is a distinguished element in the set of states of S.

Let M be any Turing machine. We can show that the tape configuration at each step of computation of M can be represented by a FC system S_M, when the initial tape is blank. Thus, if M halts in a special state q then $\tau^+(m,n)$ takes a special state of S_M corresponding to q for some (m,n). But, it is well-known that the halting problem of Turing machines (more precisely, the problem of deciding for any Turing machine whether or not it halts eventually if the initial tape is blank) is recursively unsolvable. Thus, we have the following.

<u>Theorem 1.1.</u> ([4],[5]) The meeting problem of FC systems is recursively unsolvable.

Now, let $Th_1 < \{P_n\}_{n \in N}; 2x, x+1 >$ be the monadic first-order arithmetic with functions 2x, x+1 and countably many monadic predicate symbols $P_1, P_2, \ldots$. Firstly, we will show the undecidability of $Th_1 < \{P_n\}_{n \in N}; 2x, x+1 >$ and then derive the undecidability of $Th_1 < P ; 2x, x+1 >$ by *encoding* these predicate symbols by a single predicate symbol P. For a given FC system S and given state q_δ of S, we will construct such a formula $B_{S,\delta}$ of $Th_1 < \{P_n\}_{n \in N}; 2x, x+1 >$ that $B_{S,\delta}$ is satisfiable in the domain N of the set of natural numbers if and only if no cells take the state q_δ in S. If we succeed it, the undecidability result follows from Theorem 1.1. So, we will show how to construct the formula $B_{S,\delta}$, in the following.

Take an arbitrary FC system S = (Q, q_1, τ_1, τ_2, τ), where $Q = \{q_1, q_2, \ldots, q_r\}$. Corresponding to Q, we will take monadic predicate symbols $P_1, P_2, \ldots, P_{4r}$. For the sake of brevity, we sometimes write P_{4i-3} as D_i, P_{4i-2} as H_i, P_{4i-1} as R_i and P_{4i} as X_i for $1 \leq i \leq r$. We will give some intuitive explanations of the construction of the formula $B_{S,\delta}$. It will be necessary that the formula $B_{S,\delta}$ contains all the informations about the transition of states of S and it also implies that no cells take the state q_δ . Let $\alpha(x) = 2x+1$. Then clearly, $\alpha^n(1) = 2^{n+1} - 1$.

Let $p_{m,n}$ denote the number $2^{m-n}\alpha^{n-1}(1)$ ($= 2^m - 2^{m-n}$) for $1 \leq n \leq m$. It is easy to see that

$$1 = p_{1,1} < p_{2,1} < p_{2,2} < \dots < p_{m,1} < p_{m,2} < \dots$$
$$< p_{m,k} < p_{m,k+1} < \dots < p_{m,m} < p_{m+1,1} < \dots .$$

We will associate each (m,n)-cell with the number $p_{m,n}$ and will interpret predicate symbols D_i, H_i, and R_i as follows.

1) $D_i(x)$ holds if and only if $x = p_{n,n}$ and $\tau^+(n,n) = q_i$ for some $n \geq 1$,

2) $H_i(x)$ holds if and only if $x = p_{m,1}$ and $\tau^+(m,1) = q_i$ for some $m \geq 2$,

3) $R_i(x)$ holds if and only if $x = p_{m,n}$ and $\tau^+(m,n) = q_i$ for some m, n such that $1 \leq n < m$.

Let us consider functions φ_1, φ_2, φ, which correspond to neighboring state functions τ_1, τ_2, τ, respectively, satisfying the conditions that

1) $\varphi_1(p_{m-1,1}) = p_{m,1}$ for $m > 1$,

2) $\varphi_2(p_{m-1,m-1}) = p_{m,m}$ for $m > 1$,

3) $\varphi(p_{m,n-1}, p_{m-1,n}) = p_{m,n}$ for $1 < n < m$.

Then, we can define φ_1 and φ_2 by

$$\varphi_1(x) = 2x \quad \text{and} \quad \varphi_2(x) = \alpha(x) = 2x + 1.$$

Hence, both of them are definable in terms of functions 2x and x+1. So, we will be able to express the transition by neighboring state functions τ_1 and τ_2 by using them. On the other hand, we can not define such a function φ, in terms of these two functions. This causes some difficulties of expressing the transition by τ. To avoid them, we will introduce auxiliary predicate symbols X_i's, which transmit informations step by step. More precisely, predicate symbols X_i's can be interpreted as follows.

4) $X_i(x)$ holds if and only if $p_{m,n} < x < p_{m,n+1}$ and $\tau^+(m,n) = q_i$ for some m, n such that $1 \leqq n < m$.

Taking these into consideration, we come to the following definitions. Firstly, define formulas T_0, $T_1(x;i,j)$ for such i, j that $\tau_1(q_i) = q_j$, $T_2(x;i,j)$ for such i, j that $\tau_2(q_i) = q_j$, $T_3(x;i,j,k)$ and $T_4(x;i,j,k)$ for such i, j, k that $\tau(q_i,q_j) = q_k$, $T_5(x;i,j)$ for every i, j such that $1 \leqq i, j \leqq r$ and $T_6(x;i,j)$ for such i, j that $1 \leqq i, j \leqq 4r$ and $i \neq j$, as follows.

$$T_0 \leftrightarrow \exists y(D_1(y) \wedge H_{i_0}(2y)) \qquad \text{if} \quad \tau^+(2,1) = q_{i_0},$$

$$T_1(x;i,j) \leftrightarrow [H_i(x) \to (H_j(2x) \wedge X_j(2x+1))],$$

$$T_2(x;i,j) \leftrightarrow [D_i(x) \to D_j(2x+1)],$$

$$T_3(x;i,j,k) \leftrightarrow [(R_j(x+1) \wedge X_i(2x)) \to (R_k(2(x+1)) \wedge X_k(2(x+1)+1))],$$

$$T_4(x;i,j,k) \leftrightarrow [(D_j(x+1) \wedge (H_i(2x) \vee R_i(2x))) \to R_k(2(x+1))],$$

$$T_5(x;i,j) \leftrightarrow [(X_i(x+1) \wedge (H_j(2x) \vee R_j(2x) \vee X_j(2x))) \to (X_j(2(x+1)) \wedge X_j(2(x+1)+1))],$$

$$T_6(x;i,j) \leftrightarrow \neg(P_i(x) \wedge P_j(x)).$$

Now, define $B_{S,\delta}$ by

$$\begin{aligned} B_{S,\delta} \leftrightarrow T_0 &\wedge \bigwedge_{i,j} \forall x T_1(x;i,j) \wedge \bigwedge_{i,j} \forall x T_2(x;i,j) \\ &\wedge \bigwedge_{i,j,k} \forall x T_3(x;i,j,k) \wedge \bigwedge_{i,j,k} \forall x T_4(x;i,j,k) \\ &\wedge \bigwedge_{i,j} \forall x T_5(x;i,j) \wedge \bigwedge_{i,j} \forall x T_6(x;i,j) \\ &\wedge \forall x \neg(D_\delta(x) \vee H_\delta(x) \vee R_\delta(x)), \end{aligned}$$

where each i, j (or i, j, k) in $\bigwedge_{i,j}$ (or $\bigwedge_{i,j,k}$) ranges over all pairs of (i,j) (or all triples of (i,j,k)) such that $T_s(x;i,j)$ (or $T_s(x;i,j,k)$, respectively) is defined. Notice that the last conjunctive component means that no cells take the state q_δ. Clearly, $B_{S,\delta}$ is a formula of the language L of $Th_1 < \{P_n\}_{n\in N}; 2x, x+1 >$. A formula F of L is said to be *satisfiable* in the set of natural numbers N if there is a structure with the domain N giving an interpretation of each predicate symbol P_n appearing in F, in which F holds. Now we can show the following.

Lemma 1.2. The formula $B_{S,\delta}$ is satisfiable in N if and only if there exists no (m,n) in U such that $\tau^+(m,n) = q_\delta$ in S.

From this with Theorem 1.1, the following theorem follows.

Theorem 1.3. ([5]) The monadic first-order arithmetic $Th_1 < \{P_n\}_{n \in N};\ 2x,\ x+1 >$ is undecidable.

Next, we will encode predicate symbols $P_1, P_2, \ldots, P_{4r}$ by using only the single monadic predicate symbol P. Let u be the minimum number such that $r+4 \leq 2^u$. For each predicate symbol P_k for $1 \leq k \leq 4r$, define a formula $Q_k(x)$ of the language $\tilde{L}$ of $Th_1 < P\ ;\ 2x,\ x+1 >$ as follows.
For each i, t such that $0 \leq i < r$ and $1 \leq t \leq 4$,

$$\begin{aligned} Q_{4i+t}(x) \leftrightarrow\ & \neg P(2^u x) \wedge \ldots \wedge \neg P(2^u x+t-2) \wedge P(2^u x+t-1) \\ & \wedge \neg P(2^u x+t) \wedge \ldots \wedge \neg P(2^u x+i+3) \wedge P(2^u x+i+4) \\ & \wedge \neg P(2^u x+i+5) \wedge \ldots \wedge \neg P(2^u x+2^u-1). \end{aligned}$$

Our coding corresponds to the representation of a number 4i+t for $0 \leq i < r$ and $1 \leq t \leq 4$ by a sequence of binary digits 0...010...010...0 of length 2^u in which the 1's appear in t th and (i+5)th places from the left. Let $\tilde{B}_{S,\delta}$ be the formula of L obtained from $B_{S,\delta}$ by replacing each occurrence of the form $P_k(w)$ by $Q_k(w)$ for any k such that $1 \leq k \leq 4r$. Then, it is easy to see that $\tilde{B}_{S,\delta}$ is satisfiable in N if and only if $B_{S,\delta}$ is satisfiable in N. Thus, we have the following.

Corollary 1.4. ([5]) The monadic first-order arithmetic $Th_1 < P ; 2x, x+1 >$ is undecidable.

2. Undecidability of extensions of monadic first-order theory of successor

In this section, we will extend the result obtained in the preceding section. We will first give some explanations of our notations and terminology. Let L_1 be a first-order language whose non-logical symbols consist of the set of individual variables, monadic predicate symbols P_α for each $\alpha < \lambda$ and unary function symbols $\bar{f}_\beta$, each of which represents a function f_β on N, for each $\beta < \mu$. The satisfiability of a formula of L_1 in N can be defined similarly. Of course, in this case each function symbol $\bar{f}_\beta$ is interpreted as a function f_β. A formula A is *valid*, if $\neg A$ is unsatisfiable. Now $Th_1 < \{P_\alpha\}_{\alpha<\lambda}; \{\bar{f}_\beta\}_{\beta<\mu} >$ denotes the set of valid formulas of L_1. When the language L_1 contains also the equality symbol, we will describe the set of valid formulas by $Th_1^= < \{P_\alpha\}_{\alpha<\lambda}; \{\bar{f}_\beta\}_{\beta<\mu} >$. Secondly, let L_2 be a second-order language, whose non-logical symbols consist of the set of individual variables, the set of monadic predicate variables and unary function symbols $\bar{f}_\beta$ for each $\beta < \mu$. In this case, the language L_2 contains also monadic second-order quantifiers. A formula A of L_2 is said to be valid, if A holds in every (monadic second-order) structure, which assigns a function f_β to each function symbol $\bar{f}_\beta$ and in which every predicate variable ranges over the power set of N. Then $Th_2 < \{\bar{f}_\beta\}_{\beta<\mu} >$ denotes the set of valid formulas of L_2. In the following, we will use a letter f_β for a function symbol $\bar{f}_\beta$

by the abuse of symbols. Now we will prove the following theorem, which is a generalization of Theorem 1.3.

Theorem 2.1. The monadic first-order theory of successor $Th_1^= < \{P_i\}_{i \in N};\ f,\ x+1 >$ with a function f is undecidable, if f satisfies either of the following conditions I and II.

I. 1) f is strictly monotone increasing, i.e,

$$x < y \quad \text{implies} \quad f(x) < f(y),$$

2) for some integer $t \geq 1$, $f(x)+t < f(x+t)$ for every x.

II.1) f is monotone increasing, i.e,

$$x < y \quad \text{implies} \quad f(x) \leq f(y),$$

2) for some integer $t \geq 2$, $f(x+t) < f(x)+t$ for every x,

3) f is an onto-mapping from N to N.

Proof. Suppose first that f satisfies the condition I. Our proof proceeds similarly as that of Theorem 1.3. So, we will show only how to construct the formula $B_{S,\delta}$, in this case. Formulas $T_0, \ldots, T_6$ are defined for the same i, j (and k) as in Theorem 1.3. Also, $T_5'(x;i,j)$ will be defined for every i, j such that $1 \leq i,\ j \leq r$. Let r be an integer satisfying the condition I 2). Define a function g by $g(x) = f(x+t)$ for every $x \in N$. Clearly, g is definable in L_1. Now, let

$$T_0 \leftrightarrow D_1(\bar{0}) \wedge H_{i_0}(\overline{g(0)}) \qquad \text{if} \quad \tau^+(2,1) = q_{i_0},$$

$$T_1(x;i,j) \leftrightarrow [H_i(x) \rightarrow H_j(g(x))],$$

$$T_2(x;i,j) \leftrightarrow [D_i(x) \rightarrow D_j(g(x)+1)],$$

$$T_3(x;i,j,k) \leftrightarrow [R_j(x+1) \wedge X_i(g(x)) \rightarrow R_k(g(x+1))]$$

$$T_4(x;i,j,k) \leftrightarrow [(R_j(x+1) \vee D_j(x+1)) \wedge (H_i(g(x)) \vee R_i(g(x)) \rightarrow R_k(g(x+1))],$$

$$T_5(x;i,j) \leftrightarrow [X_i(x+1) \wedge (H_j(g(x)) \vee R_j(g(x)) \vee X_j(g(x))) \rightarrow X_j(g(x+1))],$$

$$T_5'(x;i,j) \leftrightarrow [(X_j(x) \vee H_j(x) \vee R_j(x)) \wedge \neg \exists y(\ g(y) = x+1\ ) \rightarrow X_j(x+1)],$$

$$T_6(x;i,j) \leftrightarrow \neg(P_i(x) \wedge P_j(x)).$$

Notice here that each numeral $\bar{n}$, which represents a natural number n, is definable in L_1, since L_1 contains the equality symbol and the successor function. The formula $B_{S,\delta}$ is defined similarly as that in Theorem 1.3. But in this case, we must add $\bigwedge_{i,j} \forall x T_5'(x;i,j)$ as another conjunctive component. We can also show the similar lemma as Lemma 1.2. In this case, $p_{m,n}$ is defined by $p_{m,n} = g^{m-n}h^{n-1}(0)$ for $1 \leq n \leq m$, where $h(x) = g(x)+1$. Thus, we have our theorem for the case I.

Next, suppose that f satisfies the condition II. Define a function f* by

$$f^*(x) = \min_z(\ f(z) = x\ ).$$

By II 3), f*(x) is defined for every $x \in N$. Furthermore, we can show that f* satisfies the condition I. By the above proof, we know that $Th_1^=\langle \{P_i\}_{i\in N}: f^*, x+1 \rangle$ is undecidable. So, we have only to show that the function f* is definable, or more precisely, the relation ' f*(x) = y ' is definable by using the function f and the successor. Indeed, it can be defined as

$$\text{' } f^*(x) = y \text{ '} \leftrightarrow (\ f(y) = x \wedge \forall z(\ z+1 = y \rightarrow \neg(\ f(z) = x\ ))).$$

Thus, our theorem holds also for the case II. Q.E.D.

As a corollary, the following strengthened form of the result by Elgot and Rabin [3] follows. Here a function f is *hypermonotonic*, if $f(x)+1 < f(x+1)$ for every x. For example, $[e^x]$ and $[qx/p]$ (for $2p \leq q$) are hypermonotonic, where [w] denotes the integer part of the real number w.

Corollary 2.2. If f is a hypermonotonic function, then $Th_1^= \langle \{P_i\}_{i\in N}; f, x+1 \rangle$ is undecidable.

As other examples, x^n (for n > 1) and [qx/p] (for p < q < 2p) satisfy the condition I, and $[\sqrt[n]{x}]$, $[log x]$ and [qx/p] (for q < p) satisfy the condition II. Thus, in particular, the following corollary holds, which is a strengthened form of Theorem 2 in Siefkes [9].

Corollary 2.3. $Th_1^= \langle \{P_i\}_{i\in N}; [qx/p], x+1 \rangle$ is undecidable if and only if $p \neq q$ and $q \neq 0$.

Moreover, we can obtain the first-order form of Theorems 6 and 8 in [9].

We notice here that the condition I 2) can be rewritten as

$$1 + \frac{1}{t} \leqq \frac{f(x+t) - f(x)}{t} .$$

So, if f satisfies the condition that df/dx > 1 but $\lim_{x\to\infty} df/dx = 1$ (*in an approximate sense*), then the condition I 2) can not be satisfied. Thus, for example, we don't know whether $Th_1^= \langle \{P_i\}_{i\in N}; x+[\sqrt{x}], x+1 \rangle$ is undecidable or not. On the other hand, $Th_2\langle x+[\sqrt{x}], x+1 \rangle$ is shown to be undecidable as an instance of Theorem 3 in Thomas [10].

It is possible to extend our coding technique used in Corollary 1.4 to some other cases. At present, we can show the following.

Corollary 2.4. If f is a hypermonotonic function, then $Th_1^= \langle P; f, x+1 \rangle$ is undecidable.

3. Comparison with decidability results

In Section 1, we have shown the undecidability of $Th_1 < P ; 2x, x+1 >$. On the other hand, monadic second-order theories $Th_2 < x+1 >$ and $Th_2 < 2x, 2x+1 >$, and hence $Th_2 < 2x >$, are decidable by results of Büchi and Rabin. So, it will be interesting to give some conditions on a function f, which imply the decidability of the monadic second-order theory $Th_2 < f >$, since this will tell us about the boundary between decidability results and undecidability results shown in the preceding section.

A pair $< A, <f_\beta>_{\beta<\mu} >$ is called an *algebra* (of *type* μ) if A is a nonempty set and f_β is a unary function from A to A for each $\beta < \mu$. Two algebras of the same type $< A, <f_\beta>_{\beta<\mu} >$ and $< B, <g_\beta>_{\beta<\mu} >$ are *isomorphic* if there exists a bijective mapping φ from A to B such that $\varphi(f_\beta(x)) = g_\beta(\varphi(x))$ for any $x \in A$ and any $\beta < \mu$. An element $x \in A$ is an *atom* of an algebra $< A, <f_\beta>_{\beta<\mu} >$ if x is not in the range of any f_β ($\beta < \mu$). An element x is a *fixed point* of a function f if $x = f(x)$ holds. Notice that if f is strictly monotone increasing, y is a fixed point of f and $x < y$ then x is also a fixed point of f. It is also clear that if φ is an isomorphism between $< A, <f_\beta>_{\beta<\mu} >$ and $< B, <g_\beta>_{\beta<\mu} >$ and x is an atom of $< A, <f_\beta>_{\beta<\mu} >$ (or a fixed point of f_β ($\beta < \mu$)), then $\varphi(x)$ is also an atom of $< B, <g_\beta>_{\beta<\mu} >$ (or a fixed point of g_β, respectively).

Now, let L_2 and L_2' be second-order languages, whose sets of function symbols consist of f_β ($\beta < \mu$) and g_β ($\beta < \mu$), respectively. For any formula A of L_2, define A* to be a formula of L_2' obtained from A by replacing every occurrence of f_β in A by g_β for each $\beta < \mu$. By a well-known theorem on model theory, we have that if two algebras $< N, \langle f_\beta \rangle_{\beta<\mu} >$ and $< N, \langle g_\beta \rangle_{\beta<\mu} >$ are isomorphic, then for any formula A of L_2, A is valid if and only if A* is valid. In other words, $A \in Th_2< \{f_\beta\}_{\beta<\mu} >$ if and only if $A^* \in Th_2< \{g_\beta\}_{\beta<\mu} >$. So, we have the following.

Lemma 3.1. If two algebras $< N, \langle f_\beta \rangle_{\beta<\mu} >$ and $< N, \langle g_\beta \rangle_{\beta<\mu} >$ are isomorphic and $Th_2< \{g_\beta\}_{\beta<\mu} >$ is decidable, then $Th_2< \{f_\beta\}_{\beta<\mu} >$ is also decidable.

In the following, we will show the decidability of $Th_2< f >$ for a large number of strictly monotone increasing function f, using results by Rabin and Büchi. Let us define the *type* of a function f on N as follows. A function f is of the type (m,n), if m is the number of atoms of $< N, f >$ and n is the number of fixed points of f, where m and n may be infinite.

Theorem 3.2. Suppose that the type (m,n) of a function f is given. Then the monadic second-order theory $Th_2< f >$ is decidable, if f is strictly monotone increasing.

Proof. We notice first that if n is infinite then $f(x) = x$ for every $x \in N$ and hence clearly, $Th_2< f >$ is decidable. So, we can assume that n is finite. Now suppose that m is infinite. If $n = 0$, then it is easy to show that $< N, f >$ and $< N^+, 2x >$ are isomorphic, where N^+ denotes the set of positive integers. Thus, $Th_2< f >$ is decidable by Lemma 3.1. If $n > 0$, then the fixed points of f are just 0, 1,..., n-1. In this case, we can show that $< N, f >$ and $< N, g_n >$ are isomorphic, where g_n is a function defined by

$$g_n(x) = \begin{cases} x & \text{if } x < n \\ 2x & \text{if } x \geq n. \end{cases}$$

Now let S2S be the monadic second-order theory of two successor functions and T_2 be the infinite binary tree (see [7]). Then, we can construct such an algebra $< A, h_n >$ that A is a definable subset (in S2S) of T_2, h_n is a definable function (in S2S) on T_2 and $< A, h_n >$ and $< N, g_n >$ are isomorphic. From the definability of $< A, h_n >$ in the decidable theory S2S, follows the decidability of $Th_2< h_n >$ (on A). Thus, $Th_2< g_n >$ and hence $Th_2< f >$ (on N) are also decidable. Next suppose that m is finite. Then, we can show that $< N, f >$ and $< N, g_{m,n} >$ are isomorphic, where $g_{m,n}$ is a function defined by

$$g_{m,n}(x) = \begin{cases} x & \text{if } x < n \\ x+m & \text{if } x \geq n. \end{cases}$$

Without difficulty, we can verify that $g_{m,n}$ is definable in $Th_2< x+1 >$. Hence, $Th_2< f >$ is decidable also in this case.

To cite an example, let us take any function f satisfying the condition I of Theorem 2.1. Then, $f(x) - x$ is unbounded, i.e, for any z there exists x such that $f(x) - x$, since for any integer k $f(kt+1) - (kt+1) \geq k$ holds, where t is the integer given by the condition I 2). On the other hand, we can verify that for every strictly monotone increasing function f, $f(x) - x$ is unbounded if and only if $< N, f >$ has infinitely many atoms. So, $Th_2< f >$ is decidable when the number of fixed points of f is explicitly given. It will be also interesting to compare our result with the undecidability result by Thomas [10]. As for functions increasing more rapidly than the identity function, his theorem says that if f is strictly monotone increasing and $f(x) - x$ is unbounded then $Th_2 < f, x+1 >$ is undecidable.

We say that a function f is *weakly hypermonotonic*, if there exists $n \geq 0$ such that $f(x) = x$ for every $x < n$ and $f(x)+1 < f(x+1)$ for every $x \geq n$. For such a function f, define a function f^+ by $f^+(x) = f(x)+1$.

Lemma 3.3. If f is a weakly hypermonotonic function whose fixed points are just 0, 1, ... , n-1, then every y greater than n-1 can be represented uniquely of the form $f_1 f_2 \ldots f_m(x)$, where each f_i is either f or f^+, $m \geq 0$ and x is either n or an atom of $< N, <f,f^+> >$.

Proof. The possibility of such a representation can be proved by using induction on y. The uniqueness follows from the fact that both f and f^+ are injective.

We can easily show the following lemma.

Lemma 3.4. If f is a weakly hypermonotonic function, then $< N, <f,f^+> >$ has infinitely many atoms if and only if the function $f(x) - 2x$ is unbounded.

Theorem 3.5. If f is a weakly hypermonotonic function with n fixed points ($n \geq 0$) such that $f(x) - 2x$ is unbounded, then the monadic second-order theory $Th_2 < f,f^+ >$ is decidable, where f^+ is a function defined by $f^+(x) = f(x)+1$.

Proof. We notice first that $f^+(i) = i+1$ for every i such that $0 \leq i \leq n$. We define a subset B of the infinite binary tree T_2 by

$$B = \{ 1^k ; 1 \leq k \leq n \} \cup \{ 1^{n+1}z ; z \in \{0,1\}^* \} \cup \{ 0^k 1z ; k \geq 1, z \in \{0,1\}^* \}$$

(using the notation in [7]). We also define a function $r_0^{(n)}$ on B by

$$r_0^{(n)}(x) = \begin{cases} x & \text{if } x = 1^k \text{ for some k such that } 1 \leq k \leq n \\ r_0(x) & \text{otherwise,} \end{cases}$$

where r_j is a function satisfying $r_j(x) = xj$ for $j = 0, 1$ and $x \in T_2$. It is clear that both B and $r_0^{(n)}$ are definable in S2S. We will show that $< N, <f,f^+> >$ and $< B, <r_0^{(n)}, r_1> >$ are isomorphic. Let $a_1, a_2, a_3, \ldots$ be an enumeration of atoms of $< N, <f,f^+> >$. Notice that $< N, <f,f^+> >$ has infinitely many atoms by Lemma 3.4. For $\varepsilon \in 0,1$, define f^ε by $f^\varepsilon = f$ if $\varepsilon = 0$ and $f^\varepsilon = f^+$ if $\varepsilon = 1$. We will define a mapping φ from B to N by

$$\varphi(x) = \begin{cases} k-1 & \text{if } x = 1^k \text{ for } 1 \leq k \leq n \\ f^{\varepsilon 1} f^{\varepsilon 2} \ldots f^{\varepsilon p}(n) & \text{if } x = 1^{n+1}\varepsilon_p \ldots \varepsilon_2 \varepsilon_1 \text{ for } \varepsilon_i \in \{0,1\} \\ f^{\varepsilon 1} f^{\varepsilon 2} \ldots f^{\varepsilon p}(a_j) & \text{if } x = 0^j 1 \varepsilon_p \ldots \varepsilon_2 \varepsilon_1 \\ & \text{for } \varepsilon_i \in \{0,1\} \text{ and } j \geq 1. \end{cases}$$

Then, $\varphi(r_0^{(n)}(x)) = f(\varphi(x))$ and $\varphi(r_1(x)) = f^+(\varphi(x))$ hold for every $x \in B$. By this with Lemma 3.3, we have that φ is an isomorphism between $< B, <r_0^{(n)}, r_1> >$ and $< N, <f,f^+> >$. Since $< B, <r_0^{(n)}, r_1> >$ is definable in the decidable theory S2S, $Th_2 < f,f^+ >$ is decidable.

Of course, we can extend the above theorem to the case where $f(x) - 2x$ is bounded and the number of atoms of $< N, <f,f^+> >$ is explicitly given. On the other hand, we can not prove analogous results on functions which increase more *slowly* than x. To clarify the difference between them, let us consider algebras $< N, 2x >$ and $< N, [x/2] >$. It is easy to see that $< N, 2x >$ and $< N^+, x^2 >$ are isomorphic. But, algebras $< N, [x/2] >$ and $< N^+, [\sqrt{x}] >$ are not isomorphic, though each of these latter functions is the 'inverse' function of the former. Moreover, we can prove that if algebras $< N, f >$ and $< N, [x/2] >$ are isomorphic, where f is a monotone increasing function such that $f(0) = 0$ and $f(x) < x$ for each $x > 0$, then f(x) is identical to [x/2]. Thus, we know only a little about the decision problem of theories of these slowly increasing functions, at present. As for a decidable example, we can take the theory $Th_2< [x/k] >$ for $k \geq 1$. For, the relation '$y = [x/k]$' is definable in the monadic second-order theory of k successor functions SkS, which is shown to be decidable by Rabin [7]. Concerning the undecidability, we can remark that the first-order theory $Th_1^{=}< f >$ and hence the monadic second-order theory $Th_2< f >$ are undecidable for uncountably many functions f satisfying the condition II of Theorem 2.1. This make a remarkable contrast with Theorem 3.2. So, we will give the proof in the following. Let S be any non-recursive subset of N, containing neither 0 nor 1. We enumerate elements of S as K_1, k_2, k_3, ... , according to size. Define a function f_S from N to N by $f_S(n) = \sum_{i=1}^{n} k_i$. Of course, f_S is a non-recursive, strictly monotone increasing function such that 0 is only one fixed point and $f_S(x) - x$ is unbounded. Next define the 'inverse' function g_S of f_S by

$$g_S(m) = n \quad \text{if and only if} \quad f_S(n) \leq m < f_S(n+1).$$

It is obvious that g_S satisfies the condition II of Theorem 2.1. We will show that $Th_1^{=}< g_S >$ is undecidable. For any $k > 0$, define a formula F_k by

$$\exists x_1 \ldots \exists x_k \exists y (\bigwedge_{1 \leq i < j \leq k} \neg (x_i = x_j) \wedge \bigwedge_{1 \leq i \leq k} (g_S(x_i) = y) \wedge \forall z (g_S(z) = y \rightarrow \bigwedge_{1 \leq i \leq k} (z = x_i))).$$

Then, $F_k \in Th_1^=\langle g_S \rangle$ if and only if for some n there exist exactly k integers m such that $g_S(m) = n$, if and only if for some n $f_S(n+1) - f_S(n) = k$, if and only if $k \in S$. Since S is non-recursive, $Th_1^=\langle g_S \rangle$ must be undecidable. Clearly, there exist uncountably many non-recursive subsets of N, containing neither 0 nor 1, and for such subsets S and S', $g_S \neq g_{S'}$ if $S \neq S'$. It should be noticed that for every function f_S in the above, $Th_2\langle f_S \rangle$ is decidable by Theorem 3.2. On the other hand, g_S satisfies also the condition that both $g_S(x)$ and $x - g_S(x)$ are unbounded. Thus, Theorem 3 in [10], which says the undecidability of $Th_2\langle f, x+1 \rangle$ for functions f satisfying a certain condition, can be strengthened to the first-order theory $Th_1^=\langle f \rangle$ for uncountably many functions f.

References

[1] J.R. Büchi, Weak second order arithmetic and finite automata, *Z. Math. Logik Grundlagen Math.*, 6 (1960) 66-72.

[2] J.R. Büchi, On a decision method in restricted second order arithmetic, *Proc. Internat. Congr. Logic, Method. and Philos. Sci. 1960*, Stanford Univ. Press, 1962, 1-11.

[3] C.C. Elgot and M.O. Rabin, Decidability and undecidability of second (first) order theories of (generalized) successor, *J. Symbolic Logic*, 31 (1966) 169-181.

[4] A. Nakamura, On causal ω^2-systems, *J. of Computer and System Sciences*, 10 (1975) 235-265.

[5] H. Ono and A. Nakamura, Undecidability of the first-order arithmetic A[P(x),2x,x+1], *J. of Computer and System Sciences*, 18 (1979) 243-253.

[6] H. Putnam, Decidability and essential undecidability, *J. Symbolic Logic*, 22 (1957) 39-54.

[7] M.O. Rabin, Decidability of second-order theories and automata on infinite trees, *Trans. Amer. Math. Soc.*, 141 (1969) 1-35.

[8] R.M. Robinson, Restricted set-theoretical definitions in arithmetic, *Proc. Amer. Math. Soc.*, 9 (1958) 238-242.

[9] D. Siefkes, Undecidable extensions of monadic second order successor arithmetic, *Z. Math. Logik Grundlagen Math.*, 17 (1971) 385-394.

[10] W. Thomas, A note on undecidable extensions of monadic second order successor arithmetic, *Arch. math. Logik*, 17 (1975) 43-44.

SECTIONS AND ENVELOPES OF TYPE 2 OBJECTS

JUICHI SHINODA
Nagoya University

In [5: 11.21], Kleene posed the following problem.

> Do there exist type 2 objects F and G such that those degrees are different but $1\text{-sc}(F) = 1\text{-sc}(G)$?

This problem was answered by Hinman (see [4: V]). Concerning this problem, we shall consider several problems about sections, envelopes and degrees of type 2 objects.

A type 2 object is a function from ω^ω into ω, where ω is the set of non-negative integers. For any tupe 2 object F, $1\text{-sc}(F)$ and $1\text{-env}(F)$ are defined by

$$1\text{-sc}(F) = \{\alpha \in \omega^\omega : \alpha \text{ is recursive in } F\}$$

$$1\text{-env}(F) = \{P \subseteq \omega : P \text{ is semirecursive in } F\}.$$

Let F and G be type 2 objects. $F \leq G$ means that F is recursive in G. As usual, the degrees of type 2 objects are introduced through the equivalence relation $\equiv$ defined by $F \equiv G$ iff $F \leq G$ and $G \leq F$. $F < G$ represents that $F \leq G$ but $F \not\equiv G$. When F and G are incomparable with respect to $\leq$, we write $F|G$.

Let F be a type 2 object. In §1, we shall discuss some problems on degrees of type 2 objects G with the property:

(*) $\quad 1\text{-sc}(F) = 1\text{-sc}(G)$ and $1\text{-env}(F) = 1\text{-env}(G)$.

In particular, we shall show that there exists an uncountable sequence $G_0 < G_1 < \cdots < G_\nu < \cdots$ $(\nu < \aleph_1)$ of type 2 objects such that $F \equiv G_0$ and each G_ν satisfies (*).

In §2, we shall study the relations between 1-sections and 1-envelopes. By using the forcing method developed by Sacks [6], we shall prove that for any normal type 2 object F there exists a type 2 object G such that 1-sc(F) = 1-sc(G) but 1-env(F) = 1-env(G), where F is said to be normal if the Kleene object E defined by

$$E(\alpha) = \begin{cases} 0 & \text{if } \exists n[\alpha(n) = 0], \\ 1 & \text{otherwise} \end{cases}$$

is recursive in F, (in the case of F = E, such a G was given by Fenstad [1: 5.4.25]. But his example is not correct by Lemma 1.1 of this note).

§1. Let F be an arbitrary type 2 object, and consider the type 2 objects G which satisfies the following condition:

(*) 1-sc(F) = 1-sc(G) and 1-env(F) = 1-env(G).

It is an interesting problem to investigate the structure of the degrees of type 2 objects G under the restriction (*). The following lemma is a slight generalization of Tugué [7: Lemma 1] and Hinman [4: Theorem V.4].

<u>LEMMA 1.1</u>. Let F,G be two type 2 objects and $\varphi(\alpha)$ be a partial functional recursive in F such that

$$\alpha \in 1\text{-sc}(F) \rightarrow \varphi(\alpha)\downarrow \quad \text{and} \quad G(\alpha) = \varphi(\alpha)$$

where "$\downarrow$" means "is defined". Then there exists a primitive recursive function $\pi(e)$ such that

(1) $\{\pi(e)\}^F(\langle\vec{x}\rangle) \simeq \{e\}^G(\vec{x})$ for all e and $\vec{x}$.

Thus 1-sc(G) $\subseteq$ 1-sc(F) and 1-env(G) $\subseteq$ 1-env(F).

PROOF. In [4] and [7], (1) was replaced by a weaker condition (1'):

(1') $\{e\}^G(\vec{x})\downarrow \longrightarrow \{\pi(e)\}^F(\langle\vec{x}\rangle)\downarrow$ and $\{\pi(e)\}^F(\langle\vec{x}\rangle) \simeq \{e\}^G(\vec{x})$.

This implies that 1-sc(G) $\subseteq$ 1-sc(F) but does not imply that 1-env(G) $\subseteq$ 1-env(F). In [4] and [7], a primitive recursive function π satisfying (1') was obtained by using the second recursion theorem. The only case we should care about in their proof is when e is an index with $(e)_0 = 8$:

(2) $\{e\}^G(\vec{x}) \simeq G(\lambda t\{(e)_3\}^G(t,\vec{x}))$.

In this case, the value $\pi(e)$ defined in [4] and [7] had the property:

(3) $\lambda t\{\pi((e)_3)\}^F(\langle t,\vec{x}\rangle)$ is total $\longrightarrow \{\pi(e)\}^F(\langle\vec{x}\rangle)$
$\simeq \varphi(\lambda t\{\pi((e)_3)\}^F(\langle t,\vec{x}\rangle)$.

But it could not be proved that

(4) $\{\pi(e)\}^F(\langle\vec{x}\rangle)\downarrow \longrightarrow \lambda t\{\pi((e)_3)\}^F(\langle t,\vec{x}\rangle)$ is total.

Here we use a well-known trick. First replace the equation (2) by

(5) $\{e\}^G(\vec{x}) \simeq G(\lambda t\{(e)_3\}^G(t,\vec{x})) + 0\cdot F(\lambda t\{(e)_3\}^G(t,\vec{x}))$.

Then use the second recursion theorem as in [4] and [7], and we can obtain a primitive recursive function π which satisfies (3) and (4) for any index e with $(e)_0 = 8$.

PROPOSITION 1.2. Let F be a type 2 object. Then there exist $\aleph_1$ type 2 objects $G_\nu(\nu < \aleph_1)$ such that

(i) $1\text{-sc}(G_\nu) = 1\text{-sc}(F)$ and $1\text{-env}(G_\nu) = 1\text{-env}(F)$ for all $\nu < \aleph_1$

(ii) $F \equiv G_0 < G_1 < \cdots < G_\nu < \cdots \ (\nu < \aleph_1)$.

PROOF. We define G_ν by induction on ν.

Stage 0. We simply put

$$G_0(\alpha) = \langle 0, F(\alpha)\rangle$$

for all $\alpha \in \omega^\omega$. Obviously $G_0 \equiv F$.

Stage $\nu + 1$. Suppose that G_ν is already defined. Take $\beta_\nu \in \omega^\omega$ so that neither β_ν nor $\{\beta_\nu\}$ is recursive in G_ν. Define $G_{\nu+1}$ by

$$G_{\nu+1}(\alpha) = \begin{cases} \langle 0, F(\alpha)\rangle & \text{if } \alpha \in 1\text{-sc}(F), \\ \langle 1, G_\nu(\alpha)\rangle & \text{if } \alpha = \beta_\nu. \\ \langle 2, G_\nu(\alpha)\rangle & \text{otherwise.} \end{cases}$$

Stage ρ (limits). Let $\nu_0 < \nu_1 < \cdots < \nu_n < \cdots < \rho$ $(n \in \omega)$ be a sequence of successor ordinals such that $\rho = \sup_{n<\omega} \nu_n$. We put

$$G_\rho(\alpha) = G_{\nu_{\alpha(0)}}(\lambda x \alpha(x+1)).$$

From our definition, it is easy to see that

(a) if ν is a successor ordinal or $\nu = 0$, then $G_\nu(\alpha) = \langle 0, F(\alpha)\rangle$ for all $\alpha \in 1\text{-sc}(F)$, and also that

(b) if ν is a limit ordinal, then $G_\nu(\alpha) = \langle 0, F(\lambda x\alpha(x+1))\rangle$ for all $\alpha \in$ 1-sc(F).

In view of Lemma 1.1 and avove (a), (b), we observe that 1-sc(G_ν) $\subset$ 1-sc(F) and 1-env(G_ν) $\subseteq$ 1-env(F) for all $\nu < \aleph_1$. On the other hand, the values $F(\alpha)$ for $\alpha \in$ 1-sc(F) are obtained from $G_\nu(\alpha)$ by

$$F(\alpha) = \begin{cases} (G_\nu(\alpha))_1 & \text{if } \nu \text{ is a successor ordinal or } \nu = 0, \\ (G_\nu(\langle 0\rangle ^\frown \alpha))_1 & \text{if } \nu \text{ is a limit ordinal.} \end{cases}$$

We use Lemma 1.1 again and see that 1-sc(F) $\subseteq$ 1-sc(G_ν) and 1-env(F) $\subset$ 1-env(G_ν) for $\nu < \aleph_1$. Thus we have (i).

To see that $G_\nu \leq G_{\nu+1}$, it is enough to observe that

$$G_\nu(\alpha) = \begin{cases} (G_{\nu+1}(\alpha))_1 & \text{if } \nu \text{ is a successor ordinal or } \nu = 0, \\ & \text{or } \nu \text{ is limit and } (G_{\nu+1}(\alpha))_0 \neq 0 \\ G_{\nu+1}(\lambda x\alpha(x+1)) & \text{if } \nu \text{ is limit and } (G_{\nu+1}(\alpha))_0 = 0. \end{cases}$$

It holds that $G_\nu \not\geq G_{\nu+1}$ since $\{\beta_\nu\}$ is recursive in $G_{\nu+1}$ but not recursive in G_ν. Thus we have that $G_\nu < G_{\nu+1}$. This completes the proof.

Next we shall discuss about incomparable degrees under the restriction (*). According to Lemma 1.1, for any type 2 object F, there are $2^{2^{\aleph_0}}$ type 2 objects G with the property (*) such that $F \leq G$, and thus there are at least two type 2 objects G_1, G_2 such that $F \leq G_i$, 1-sc(G_i) = 1-sc(F), 1-env(G_i) = 1-env(F) for $i = 1,2$ and that $G_1 | G_2$. In [4], Hinman showed that there exists a set $\mathcal{G} = \{G_i : i \in I\}$ of cardinality $2^{\aleph_0}$ such that $1\text{-sc}(G_i) = \Delta^0_1$ for all $i \in I$ and $G_i | G_j$ if $i \neq j$. Later, Grilliot [3] showed that there exists such a set $\mathcal{G}$ of $2^{2^{\aleph_0}}$.

Let F be an arbitrary type 2 object. In the present situation, it is an interesting problem whether or not there is a set $\mathcal{G}$ of cardinality $2^{2^{\aleph_0}}$ which has the property mentioned above. By a similar proof to that of Grilliot [3], we can show that the answer to this problem is affirmative if F is a continuous object, where we say that F is continuous if F is recursive in some real $F \in \omega^\omega$.

PROPOSITION 1.3. If F is a continuous type 2 object, then there exists a set $\mathcal{G} = \{G_i : i \in I\}$ of type 2 objects such that

(i) the cardinality of $\mathcal{G}$ is $2^{2^{\aleph_0}}$,

(ii) $F \leq G_i$ for all $i \in I$,

(iii) $1\text{-sc}(G_i) = 1\text{-sc}(F)$ and $1\text{-env}(G_i) = 1\text{-env}(F)$ for all $i \in I$,

(iv) $i,j \in I$ and $i \neq j \rightarrow G_i | G_j$.

PROOF. Let $f \in \omega^\omega$ be a real such that F is recursive in f. Take reals $\alpha_\nu \in \omega^\omega (\nu < 2^{\aleph_0})$ so that

$$\mu,\nu < 2^{\aleph_0} \text{ and } \mu \neq \nu \rightarrow \alpha_\mu \text{ is not recursive in } \alpha_\nu \text{ and } f.$$

Then obviously, $\alpha_\nu \notin 1\text{-sc}(F)$ for all $\nu < 2^{\aleph_0}$ since otherwise α_ν would be recursive in f. For any subset A of $\{\alpha_\nu : \nu < 2^{\aleph_0}\}$, we define a type 2 object G_A by

$$G_A(\alpha) = \begin{cases} \langle 0, F(\alpha)\rangle & \text{if } \alpha \in A, \\ \langle 1, F(\alpha)\rangle & \text{if } \alpha \notin A. \end{cases}$$

Clearly, $F \leq G_A$. From Lemma 1.1, we see that $1\text{-sc}(G_A) = 1\text{-sc}(F)$ and $1\text{-env}(G_A) = 1\text{-env}(F)$.

Claim: if $A,B \subseteq \{\alpha_\nu : \nu < 2^{\aleph_0}\}$ and $G_A \leq G_B$, then $A \subset B$.

Let I be an almost disjoint subset of $P(\{\alpha_\nu : \nu < 2^{\aleph_0}\})$, the power set of $\{\alpha_\nu : \nu < 2^{\aleph_0}\}$, of cardinality $2^{2^{\aleph_0}}$, and let $\mathcal{G} = \{G_A : A \in I\}$. Then $\mathcal{G}$ has the desired properties by our claim.

Now we shall prove the claim. Suppose that $G_A \leq G_B$ but not $A \subseteq B$, and choose a real $\gamma \in A - B$. Then $\langle G_A,\gamma\rangle$ is effectively discontinuous at γ in the sense of Grillion [3]. Therefore $\langle G_A,\gamma\rangle$ is normal, and hence $\langle G_B,\gamma\rangle$ is also normal. This means that $\langle G_B,\gamma\rangle$ is effectively discontinuous at some point. However, noticing that $1\text{-sc}(\langle G_B,\gamma\rangle) \subseteq \Delta^0_1(f,\gamma)$ and $B \cap \Delta^0_1(f,\gamma) = \phi$, we see that $\langle G_B,\gamma\rangle$ is not effectively discontinuous at any point. This is a contradiction. Thus we have that $A \subseteq B$. This completes the proof of the proposition.

The Kleene object E is given by

$$E(\alpha) = \begin{cases} 0 & \text{if } \exists x[\alpha(x) = 0], \\ 1 & \text{otherwise.} \end{cases}$$

For any $\beta \in \omega^\omega - \Delta^1_1$, we put

$$G_\beta(\alpha) = \begin{cases} \langle 0,E(\alpha)\rangle & \text{if } \alpha = \beta, \\ \langle 1,E(\alpha)\rangle & \text{otherwise.} \end{cases}$$

Then $1\text{-sc}(G_\beta) = 1\text{-sc}(E) = \Delta^1_1$ and $1\text{-env}(G_\beta) = 1\text{-env}(E) = \Pi^1_1$. Let γ be another real such that $\gamma \notin \Delta^1_1$. And suppose that $G_\beta \leq G_\gamma$. Then $\{\beta\}$ is recursive in G_β since $\alpha \in \{\beta\} \leftrightarrow (G(\alpha))_0 = 0$, and hence $\{\beta\}$ is recursive in E and γ. Hence $\{\beta\} \in \Delta^1_1(\gamma)$ and thus $\beta \in \Delta^1_1(\gamma)$. Thus we have that $G_\beta \leq G_\gamma \rightarrow \beta \in \Delta^1_1($ from which we can get several results about the degrees of type 2 objects G such that $1\text{-sc}(G) = 1\text{-sc}(E)$ and $1\text{-env}(G) = 1\text{-env}(E)$. For example, there is a set $\mathcal{G} = \{G_i : i \in I\}$ of cardinality $2^{\aleph_0}$ which satisfies the conditions (ii)-(iv) in Proposition 1.3 with $F = E$ since as is well-known there are $2^{\aleph_0}$ reals whose hyperdegrees are mutually incomparable. However, the author does not know whether or not there is such a set $\mathcal{G}$ of cardinality $2^{2^{\aleph_0}}$. By a similar proof to that of 1.2, we get a sequence $G_0 < G_1 < \cdots < G_\nu < \cdots < E_1$ $(\nu < \omega_1^{E_1})$ of type 2 objects such that $G_0 \equiv E$ and $1\text{-sc}(G_\nu) = 1\text{-sc}(E)$, $1\text{-env}(G_\nu) = 1\text{-env}(E)$ for all $\nu < \omega_1^{E_1}$, where E_1 is the Tugué object and $\omega_1^{E_1}$ is the first ordinal which is not recursive in E_1.

§2. In this section, we shall consider the following two problems.

(i) Does 1-sc(F) = 1-sc(G) imply 1-env(F) = 1-env(G)?

(ii) Does 1-env(F) = 1-env(G) imply 1-sc(F) = 1-sc(G)?

If F is normal, then the answer to (ii) is evidently affirmative. To discuss the case where F is not normal, we need the following lemma.

LEMMA 2.1. If F is not normal, then there exists a real f recursive in F such that 1-sc(F) $\subseteq \Delta^0_2(f)$ and 1-env(F) = $\Pi^1_1(f)$.

This lemma is well-known. For the proof, see, for example, Wainer [8].

PROPOSITION 2.2. If F is not normal and 1-sc(F) = 1-sc(G), then 1-env(F) = 1-env(G).

PROOF. let f be a real recursive in F as in 2.1. Then f is recursive in G since 1-sc(F) = 1-sc(G). Therefore, $\Pi^1_1(f) \subset$ 1-env(G), and thus we have that 1-env(F) $\subseteq$ 1-env(G). Similarly we can obtain the inverse inclusion.

PROPOSITION 2.3. If F is not normal, then there exists a type 2 object G such that 1-env(F) = 1-env(G) but 1-sc(F) $\neq$ 1-sc(G).

PROOF. Take a real f as in Lemma 2.1. Let g be a real such that $\Pi^1_1(f) = \Pi^1_1(g)$ and $g \notin \Delta^0_2(f)$. We define a type 2 object G by $G(\alpha) = g(\alpha(0))$. Then $g \notin$ 1-sc(F), $g \in$ 1-sc(G) and 1-env(G) = $\Pi^1_1(g)$. Thus we have that 1-sc(F) $\neq$ 1-sc(G) and 1-env(F) = 1-env(G).

When F is normal, the problem (i) will be answered negatively, which is stated as follows.

PROPOSITION 2.4. For any normal type 2 object F, there exists a type 2 object G such that $1\text{-sc}(F) = 1\text{-sc}(G)$ but $1\text{-env}(F) \neq 1\text{-env}(G)$.

We shall give a proof of this proposition in the case of $F = E$ for notational simplicity. So we assume that $F = E$ hereafter.

To obtain a type 2 object G mentioned in the proposition, we need to examine the forcing method used in the proof of the Plus One Theorem given by Sacks [6].

Here we review the forcing method of Sacks in brief. For any partial functional form ω^{ω} into ω, a hierarchy $\{T^p_\sigma\}$ is defined inductively as follows.

Stage 0. $T^p_0 = \{1\}$. 1 is an index for T^p_0. T^p_0 is total.

Stage $\sigma + 1$. If T^p_σ has an index, T^p_σ is total, $2^e \notin T^p_\sigma$, and $\{e\}^{T^p_\sigma}(m)$ is defined for all m, then 2^e is an index for $T^p_{\sigma+1}$. $T^p_{\sigma+1}$ is total if it has an index and $p(\lambda m\{e\}^{T^p_\sigma}(m))$ is defined whenever 2^e is an index for $T^p_{\sigma+1}$.

$$T^p_{\sigma+1} = T^p_\sigma \cup \{2^e : 2^e \text{ is an index for } T^p_{\sigma+1}\} \cup \{<3^e, m, n> : 2^e \text{ is an index for } T^p_{\sigma+1} \text{ and } \{e\}^{T^p_\sigma}(m) = n\} \cup \{<5^e, n> : 2^e \text{ is an index for } T^p_{\sigma+1} \text{ and } p(\lambda m\{e\}^{T^p_\sigma}(m)) = n\}.$$

Stage σ (limits). 7^e is an index for T^p_σ if 2^e is an index for $T^p_{\delta+1}$ for some $\delta < \sigma$ and $\lambda m\{e\}^{T^p_\delta}(m)$ is the representing function of a set R of indices such that

$$\sigma = \sup_{m \in R} |m|$$

where $|m| = \nu$ means that m is an index for T^p_ν

T^p_σ is total if it has an index.

$T^p_\sigma = \bigcup_{\delta<\sigma} T^p_\delta \cup \{7^e : 7^e$ is an index for $T^p_\sigma\}$.

p is said to generate T^p_σ if T^p_σ has an index and is total. The following three facts hold for this hierarchy.

Fact 1. If q is an extension of p and p generates T^p_σ, then q generates T^q_σ and $T^p_\sigma = T^q_\sigma$.

Fact 2. If $\sigma < \tau$ and p generates T^p_σ and T^p_τ, then T^p_σ has lower Turing degree than T^p_σ.

Fact 3. If p is total and $S \subseteq \omega$, then S is recursive in p and E iff S is recursive in some T^p_σ generated by p.

T^p_σ is called the maximum of p if T^p_σ has an index but is not total. We use $I(p)$ to denote the ordinal σ such that T^p_σ is the maximum of p if it exists. It is easy to see that $I(p)$ is a successor ordinal.

Let M be a countable abstract 1-section. See [6] for the definition of abstract 1-section. What we need here in our proof is the fact that L_{ω_1} is an abstract 1-section, where ω_1 is the first non-recursive ordinal and L_{ω_1} is the set of constructible sets constructed at stage $< \omega_1$. p is called a forcing condition if it meets the following requirements.

(1) $p \in M$ and p has the maximum,

(2) $\alpha \in \mathrm{dom}(p)$ iff α is recursive in some T^p_δ with $\delta < \sigma$ where $\sigma + 1 = I(p)$.

We use $\wp$ to denote the set of forcing conditions.

Sacks defined the forcing language $\mathcal{L}(M)$ and the forcing relation "$p \Vdash \psi$". A type 2 object G is said to be generic if for any sentence ψ of $\mathcal{L}(M)$ there exists a forcing condition $p \subseteq G$ such that either $p \Vdash \psi$ or $p \Vdash \neg\psi$ holds. He proved that if G is generic, then $1\text{-sc}(\langle E,G\rangle) = M \cap \omega^{\omega}$.

Now we shall return to the proof of 2.4.

PROOF OF PROPOSITION 2.4. We take $M = L_{\omega_1}$. Let Q be a Π^1_1 set of unique notations for recursive ordinals.
This means that Q is a Π^1_1 subset of ω and there is a recursive linear ordering $R \subseteq \omega \times \omega$ such that Q is the well-founded part of R and the order type of $\langle Q, R \upharpoonright Q \times Q\rangle$ is ω_1 (see Gandy [2]). For each $a \in Q$ we use $\|a\|$ to denote the order type of the set $\{b \in Q : R(b,a)\}$.

Let $A_n (n \in \omega)$ be an enumeration of all Π^1_1 subsets of Q, and let $\psi_n (n \in \omega)$ be an enumeration of all sentences of the forcing language (L_{ω_1}). We shall construct a sequence $p_0, q_0, p_1, q_1, \ldots, p_n, q_n, \ldots$ of forcing conditions. Take a $p_0 \in \wp$ so that either $p_0 \Vdash \psi_0$ or $p_0 \Vdash \neg\psi_0$ holds. Suppose that $p_n \in \wp$ is already defined and $I(p_n) = \sigma + 1$. Then $T_\sigma^{p_n}$ is generated by p_n but $T_\sigma^{p_n} \notin \mathrm{dom}(p_n)$ where we identify a set with its representing function. So we can obtain an extension q_n of p_n such that $I(q_n) = \sigma + 2$ and

$$q_n(T_\sigma^{p_n}) = \begin{cases} 1 & \text{if} \quad a \in A_n \\ 0 & \text{if} \quad a \notin A_n \end{cases}$$

where a is an element of Q such that $\|a\| = \sigma$. And let p_{n+1} be an extension of q_n such that either $p_{n+1} \Vdash \psi_{n+1}$ or $p_{n+1} \Vdash \neg\psi_{n+1}$ holds.

Let G be a type 2 object which extends all p_n's. We want to show that $\langle E,G\rangle$ is a desired one. Evidently, G is generic. Hence we have that

$$1\text{-sc}(\langle E,G\rangle) = L_{\omega_1} \cap \omega^\omega = \Delta^1_1 = 1\text{-sc}(E).$$

We put $A = \{a \in Q : G(T^G_{\|a\|}) = 0\}$. We can easily find a partial function $\pi(a,x)$ recursive in $\langle E.G\rangle$ such that $\lambda x\pi(a,x)$ is the representing function of $T^G_{\|a\|}$ whenever $a \in Q$ and it is not total if $a \notin Q$. Using this function π, we have that $A = \{a \in \omega : G(\lambda x\pi(a,x)) \simeq 0\}$. Therefore, A is semirecursive in $\langle E,G\rangle$, and thus $A \in 1\text{-env}(\langle E,G\rangle)$. Suppose that $A \in 1\text{-env}(E)$. Then A would be a Π^1_1 subset of Q, and hence it would hold that $A = A_n$ for some $n \in \omega$. Let a be an element of Q such that $I(p_n) = \|a\| + 1$. Then we would have that $a \in A$ iff $a \notin A_n$. This is a contradiction. This completes the proof of the proposition for the Kleene object E.

For an arbitrary normal object F, the proof of the proposition will proceed in the same manner by taking the structure $\langle L_{\omega_1^F}[F], \in, F \upharpoonright L_{\omega_1^F}[F]\rangle$ as a ground model of forcing.

REFERENCES

[1] J. E. Fenstad; General Recursion Theory. An Axiomatic Approach, Springer-Verlag (Berlin) 1980.

[2] R. O. Gandy; Proof of Mostowski's conjecture, Bull. Acad. Polon. Sci. 8(1960), 571-575.

[3] T. J. Grilliot; On effectively discontinuous type 2 objects, Jour. Symb. Logic 36(1971), 245-248.

[4] P. G. Hinman; Ad Astra per Aspera: Hierarchy Schemata in Recursive Function Theory, Ph.D. thesis, University of California, Berkeley, 1966.

[5] S. C. Kleene; Recursive functionals and quantifiers of finite types II, Trans. Amer. Math. Soc. 108(1963), 106-142.

[6] G. E. Sacks; The 1-section of a type n object, in Generalized Recursion Theory, North-Holland (Amsterdam) 1974.

[7] T. Tugué; Predicates recursive in a type 2 object and Kleene hierarchies, Comment. Math. Univ. St. Paul 8(1960), 97-117.

[8] S. S. Wainer; The 1-section of a non-normal type 2 object, in J. E. Fenstad, R. O. Gandy and G. E. Sacks (eds.): Generalized Recursion Theory II, North-Holland 1978, 407-417.

HEYTING VALUED UNIVERSES OF INTUITIONISTIC SET THEORY

by

Gaisi Takeuti and Satoko Titani

Introduction

A Complete Heyting algebra (cHa) is a complete lattice satisfying $\wedge, \bigvee$-distributive law:

$$p \wedge \bigvee_{i \in I} q_i = \bigvee_{i \in I} (p \wedge q_i).$$

For a cHa Ω, Ω and ϕ are denoted by 1 and 0, respectively. Operations $\to$ and $-$ on Ω are defined by

$$(a \to b) = \bigvee\{c \in \Omega \mid a \wedge c \leq b\}, \quad (-a) = (a \to 0).$$

The lattice of open subsets of topological space X, denoted by $\mathcal{O}(X)$, is a cHa and many topological properties can be generalized to those of cHa. For the detail on cHa, refer to Fourman and Scott [1].

The purpose of this paper is to extend the Solovay-Tennenbaum's results on iterated Cohen extensions in [5] to cHa-valued universes, which is closely connected with mathematics. The relation of our results to mathematics will be discussed in the sequel to this paper.

R. J. Grayson presented in [2] a system ZF_I of intuitionistic set theory and construction of its cHa-valued model $V^{(\Omega)}$ within that theory. He also presented a system ZF_I' of intuitionistic set theory with the existence predicate E in his thesis [3], which is equivalent to the former system. In this paper ZF_I stands for his ZF_I' with E.

Chapter I starts with description of the system ZF_I and a model $V^{(\Omega)}$ of ZF_I in Grayson [2], [3]. It is known that if X is a topological space and $\Omega = \mathcal{O}(X)$, then the notion of real number in $V^{(\Omega)}$ is identified with the notion of real valued continuous function on X. Then we consider the notion of cHa and cHa-valued universe $V^{(\Omega)}$ in a fixed cHa-valued universe $V^{(H)}$. Some related subjects in this chapter are discussed in Fourman and Scott [1].

In Chapter 2, we extend the above discussion by the method of Chen's permutation model (cf. [4], [5]). That is, for a cHa Ω and a filter Γ of subgroups of automorphism group of Ω, we construct a sub-universe $V^{(\Gamma)}$ of $V^{(\Omega)}$, which is analogous to the subuniverse $V^{(\Gamma)}$ of a Boolean valued universe in [5]. The followings are examples of Γ such that $V^{(\Gamma)}$ is a model of ZF_I.

1) Let X be a locally compact manifold and $\Omega = \mathcal{O}(X)$. Then an automorphism h^* on Ω corresponds to a homeomorphism $h : X \to X$. Let G be the set $\mathrm{Aut}(\Omega)$ of automorphisms of Ω and Γ be the set of sub-groups F of G such that there exists a compact set Y of X with the following property.

$$f \in F \text{ iff } \forall y \in Y(f(y) = y).$$

Then Γ is a filter of subgroups of G and $V^{(\Gamma)}$ is a model of ZF_I. The notion of real number in $V^{(\Gamma)}$ is identified with the notion of real valued continuous function on X which is componentwise constant outside of a compact subset of X.

2) Let X be a topological space, $\Omega = \mathcal{O}(X)$, G a subgroup of $\mathrm{Aut}(\Omega)$ and

$\Gamma = \{H \subseteq G \mid H$ is a subgroup and $[G : H] < \infty\}$.

3) Let X be a topological space, $\Omega = \mathcal{O}(X)$, and G be a locally compact topological group which is a transformation group on X.

Let

$\Gamma = \{H \subseteq G \mid H$ is a subgroup of G and G/H is compact$\}$.

4) A special case of 2), where X is the upper half plane and G is a discrete subgroup of $SL(2,\mathbb{R})$.

These Γ are filter of subgroups of G and $V^{(\Gamma)}$ is a model of ZF_I.

In Chapter 3, we discuss some topological properties of cHa in $V^{(H)}$.

G.Takeuti, S.Titani

Chapter I. The universe $V^{(\Omega)}$

Our work is based on the system of Grayson [3]. Unfortunately the system in his published version [2] is fairly different from his original system in [3], which we would like to use. Therefore we reproduce his system and properties on his system in order to make the paper more or less self-contained. Therefore §§1-9 in this chapter is mostly due to Grayson.

§1. The system ZF_I.

The system ZF_I of intuitionistic set theory is the first order theory with the following nonlogical symbols and axioms.

(1) Nonlogical symbols, $\in$, $=$, E, where E is a predicate symbol with 1 argument place and Ex is interpreted as 'x exists'.

(2) Nonlogical axioms;

Equality : $u = u$,

$u = v \rightarrow v = u$,

$u = v, \varphi(u) \rightarrow \varphi(v)$,

$(Eu \vee Ev \rightarrow u = v) \rightarrow u = v$.

Extension: $\dot\forall z(z \in u \leftrightarrow z \in v) \wedge (Eu \leftrightarrow Ev) \rightarrow u = v$.

Pair: $\dot\exists z \dot\forall x(x \in z \leftrightarrow x = u \vee x = v)$.

Union: $\dot\exists v \dot\forall x(x \in v \leftrightarrow \dot\exists y \in u(x \in y))$

Power: $\dot\exists v \dot\forall x(x \in v \leftrightarrow x \subseteq u)$

Infinity: $\dot\exists v(\dot\exists x \in v \wedge \dot\forall x \in v \dot\exists y \in v(x \in y))$

Separation: $\dot\exists v \dot\forall x(x \in v \leftrightarrow x \in u \wedge \varphi(x))$

Foundation: $\dot\forall x(\dot\forall y \in x\varphi(y) \rightarrow \varphi(x)) \rightarrow \dot\forall x\varphi(x)$

Replacement: $\dot\exists v(\dot\forall x \in u \dot\exists y\varphi(x,y) \rightarrow \dot\forall x \in u \dot\exists y \in v\varphi(x,y))$

In the above axioms $\dot{\forall}x\cdots$ and $\dot{\exists}x\cdots$ are abbreviations of $\forall x(Ex \to\cdots)$ and $\exists x(Ex \wedge\cdots)$. From now on, we write $\forall x$ and $\exists x$ instead of $\dot{\forall}x$ and $\dot{\exists}x$, since $\forall x$ and $\exists x$ always appear in the form $\dot{\forall}x$ and $\dot{\exists}x$.

If a formula φ is provable in ZF_I, then we write $\vdash \varphi$.

§2. Models of ZF_I

A model of a theory consists of a set Ω of truth values, a universe M and a function $[\![\ ]\!]$ which assigns truth value to each sentence over M. $<\Omega, M, [\![\ ]\!]>$ is a model of ZF_I if the operations $\wedge, \vee, \bigwedge, \bigvee, \to$ and $-$, corresponding to the logical operations $\wedge, \vee, \forall, \exists$, $\to$ and $\neg$, are defined on Ω and satisfy the following conditions.

(1) $\{[\![A]\!] \mid A \text{ is a sentence}\} = \Omega.$

For any sentences A,B and formula $A(a)$ with one variable,

(2) $[\![A \wedge B]\!] = [\![A]\!] \wedge [\![B]\!]$,

(3) $[\![A \vee B]\!] = [\![A]\!] \vee [\![B]\!]$,

(4) $[\![\forall x A(x)]\!] = \bigwedge_{x\in M}(Ex \to [\![A(x)]\!])$,

(5) $[\![\exists x A(x)]\!] = \bigvee_{x\in M}(Ex \wedge [\![A(x)]\!])$,

(6) $[\![A \to B]\!] = [\![A]\!] \to [\![B]\!]$,

(7) $[\![\neg A]\!] = -[\![A]\!]$,

(8) if $\vdash A \leftrightarrow B$, then $[\![A]\!] = [\![B]\!]$.

It is easy to see that if $<\Omega, M, [\![\ ]\!]>$ is a model of ZF_I then Ω is a Heyting algebra. Conversely, if Ω is a cHa then we can define a universe M and function $[\![\ ]\!]$ such that $<\Omega,M,[\![\ ]\!]>$ is a model of ZF_I, as follows.

Let V be a standard universe of ZFC, $V_\alpha^{(\Omega)} \subseteq V$ will be defined for all $\alpha \in \mathrm{Ord}$ by transfinite induction. Assume that $V_\beta^{(\Omega)}$ is defined already for $\beta < \alpha$ and each element u of $V_\beta^{(\Omega)}$ is of the form $<\mathcal{D}(u), |u|, Eu>$, where $\mathcal{D}(u) \subseteq V_\gamma^{(\Omega)}$ for some $\gamma < \beta$, $|u|$ is a function of $\mathcal{D}(u)$ into Ω and $Eu \in \Omega$. For convenience we write $u(x)$ instead of $|u|(x)$. Now we define $V_\alpha^{(\Omega)}$ by

$$V_\alpha^{(\Omega)} = \{u = < \mathcal{D}(u), |u|, Eu > | \ \exists\beta < \alpha(\mathcal{D}(u) \subseteq V_\beta^{(\Omega)})$$

$$|u| : \mathcal{D}(u) \to \Omega \wedge Eu \in \Omega \wedge$$

$$\forall x \in \mathcal{D}(u)(u(x) \leq Eu \wedge Ex)\}$$

Let $V^{(\Omega)} = \bigcup_{\alpha\in \mathrm{Ord}} V_\alpha^{(\Omega)}$.

$[\![\varphi]\!]$ is defined by induction on the number of logical symbols in φ.

An atomic sentence over $V^{(\Omega)}$ is of the form $u = v$, $u \in v$ or Eu, where $u, v \in V^{(\Omega)}$. $[\![u = v]\!]$ and $[\![u \in v]\!]$ are defined by transfinite induction on $\max(\mathrm{rank}(u), \mathrm{rank}(v))$ as follows, where $\mathrm{rank}(u)$ is the rank of u in V

$$[\![u = v]\!] = \bigwedge_{x\in \mathcal{D}(u)} (u(x) \to [\![x \in v]\!]) \wedge \bigwedge_{y\in \mathcal{D}(v)} (v(y) \to [\![y \in u]\!]) \wedge (Eu \leftrightarrow Ev),$$

$$[\![u \in v]\!] = \bigvee_{y\in \mathcal{D}(v)} (v(y) \wedge [\![u = y]\!]),$$

$$[\![Eu]\!] = Eu.$$

Note that for $x \in \mathcal{D}(u)$ and $y \in \mathcal{D}(v)$,

$$\max(\mathrm{rank}(x),\mathrm{rank}(y)) < \max(\mathrm{rank}(u),\mathrm{rank}(v)).$$

Hence,

$$[\![x \in v]\!] = \bigvee_{y\in \mathcal{D}(v)} (v(y) \wedge [\![x = y]\!]),$$

$$[\![u = y]\!] = \bigwedge_{x\in \mathcal{D}(u)} (u(x) \rightarrow [\![x \in y]\!]) \wedge \bigwedge_{t\in \mathcal{D}(y)} (y(t) \rightarrow [\![t \in u]\!]) \wedge (Eu \leftrightarrow Ey)$$

are defined in ealier stage.

For a sentence with logical symbols, $[\![\ ]\!]$ is defined as usual. i.e.,

$$[\![A \wedge B]\!] = ([\![A]\!] \wedge [\![B]\!]).$$

$$[\![A \vee B]\!] = ([\![A]\!] \vee [\![B]\!]).$$

$$[\![A \rightarrow B]\!] = ([\![A]\!] \rightarrow [\![B]\!]).$$

$$[\![\neg A]\!] = ([\![A]\!] \rightarrow 0).$$

$$[\![\exists x A(x)]\!] = \bigvee_{x\in V^{(\Omega)}} Ex \wedge [\![A(x)]\!]$$

$$[\![\forall x A(x)]\!] = \bigwedge_{x\in V^{(\Omega)}} (Ex \rightarrow [\![A(x)]\!]).$$

<u>Theorem 1</u>. If Ω is a cHa, then $\langle\Omega, V^{(\Omega)}, [\![\ ]\!]\rangle$, defined above, is a model of ZF_I.

<u>Proof</u>. cf. [2], [3].

If $u, a, b \in V^{(\Omega)}$, then let $\mathcal{P}^{\Omega}(u)$, $\{a,b\}^{\Omega}$, $\bigcup^{\Omega} u$ and $\{x \in u|\varphi(x)\}^{\Omega}$ be elements of $V^{(\Omega)}$ such that

$$\begin{cases} Eu \le [\![\forall x(x \in \mathcal{P}^{\Omega}(u) \leftrightarrow x \subseteq u)]\!], \\ E\,\mathcal{P}^{\Omega}(u) = Eu; \end{cases}$$

$$\begin{cases} Ea \wedge Eb \le [\![\forall x(x \in \{a,b\}^{\Omega} \leftrightarrow x = a \vee x = b)]\!], \\ E\{a,b\}^{\Omega} = Ea \wedge Eb; \end{cases}$$

$$\begin{cases} Eu \le [\![\forall x(x \in \bigcup^{\Omega} u \leftrightarrow \exists y \in u(x \in y)) , \\ E(\bigcup^{\Omega} u) = Eu; \end{cases}$$

$$\begin{cases} Eu \le [\![\forall x(x \in \{x \in u \mid \varphi(x)\}^{\Omega} \leftrightarrow x \in u \wedge \varphi(x))]\!], \\ E\{x \in u \mid \varphi(x)\}^{\Omega} = Eu. \end{cases}$$

We omit Ω from $\mathcal{P}^{\Omega}(u)$, $\{a,b\}^{\Omega}$, $\cup^{\Omega} u$ and $\{x \in u \mid \varphi(x)\}^{\Omega}$, if there is no confusion.

Class in $V^{(\Omega)}$ is defined as follows. X is a <u>class in $V^{(\Omega)}$</u> if $X = \langle \mathcal{D}(X), |X|, EX \rangle$, $\mathcal{D}(X) \subseteq V^{(\Omega)}$, $|X| : \mathcal{D}(X) \to \Omega$, $EX \in \Omega$ and $(\forall x \in \mathcal{D}(X))(|X|(x) \le Ex \wedge EX)$.

If X,Y are classes in $V^{(\Omega)}$ and $x \in V^{(\Omega)}$, then $[\![X = Y]\!]$, $[\![x \in X]\!]$ are defined by:

$$[\![x \in X]\!] = \bigvee_{t \in \mathcal{D}(X)} [\![x = t]\!] \wedge Et,$$

$$[\![X = Y\,]\!] = \bigwedge_{t \in \mathcal{D}(X)} (X(t) \to [\![t \in Y]\!]) \wedge$$

$$\bigwedge_{t \in \mathcal{D}(Y)} (Y(t) \to [\![t \in X]\!]) \wedge (EX \leftrightarrow EY).$$

It is obious that if X is a class in $V^{(\Omega)}$ and a set in V, then $X \in V^{(\Omega)}$.

§3. Properties of $V^{(\Omega)}$.

Lemma 3.1.

If $u \in V^{(\Omega)}$ and $\varphi(a)$ is a formula over $V^{(\Omega)}$ with one free variable, then

(1) $[\![\exists x \in u\varphi(x)]\!] = \bigvee_{x\in \mathcal{D}(u)} (u(x) \wedge [\![\varphi(x)]\!])$.

(2) $[\![\forall x \in u\varphi(x)]\!] = \bigwedge_{x\in \mathcal{D}(u)} (u(x) \to [\![\varphi(x)]\!])$.

Proof.

(1) $$[\![\exists x \in u\varphi(x)]\!] = \bigvee_{x'\in V^{(\Omega)}} Ex' \wedge [\![x' \in u \wedge \varphi(x')]\!]$$

$$= \bigvee_{x'\in V^{(\Omega)}} \bigvee_{x\in \mathcal{D}(u)} [\![x = x']\!] \wedge u(x) \wedge [\![\varphi(x')]\!]$$

$$= \bigvee_{x\in \mathcal{D}(u)} u(x) \wedge [\![\varphi(x)]\!].$$

(2) $$[\![\forall x \in u\varphi(x)]\!] = \bigwedge_{x'\in V^{(\Omega)}} (Ex' \wedge [\![x' \in u]\!] \to [\![\varphi(x')]\!])$$

$$= \bigwedge_{x'\in V^{(\Omega)}} (\bigvee_{x\in \mathcal{D}(u)} [\![x' = x]\!] \wedge u(x) \to [\![\varphi(x')]\!])$$

$$= \bigwedge_{x'\in V^{(\Omega)}} \bigwedge_{x\in \mathcal{D}(u)} ([\![x' = x]\!] \wedge u(x) \to [\![\varphi(x')]\!])$$

$$= \bigwedge_{x\in \mathcal{D}(u)} (u(x) \to [\![\varphi(x)]\!]).$$

Lemma 3.2.

For $u,v \in V^{(\Omega)}$, $[\![u \in v]\!] \leq Eu \wedge Ev$.

Let 2 be the cHa consisting of 0 and 1, which is a complete subalgebra of any cHa. Then V is equivalent to $V^{(2)}$. We define $\vee : V \to V^{(2)} \subseteq V^{(\Omega)}$ by

$$\mathcal{D}(\check{u}) = \{\check{x} \mid x \in u\},$$

$$x \in u \Rightarrow \check{u}(\check{x}) = 1,$$

$$E\check{u} = 1.$$

Then V is embedded in $V^{(\Omega)}$ by $\vee$.

Lemma 3.4.

For $u, v \in V$,

(1) $u \in v$ iff $[\![\check{u} \in \check{v}]\!] = 1$,

(2) $u = v$ iff $[\![\check{u} = \check{v}]\!] = 1$,

(3) $[\![\forall x \in \check{u}\varphi(x)]\!] = \bigwedge_{x \in u}[\![\varphi(\check{x})]\!]$,

(4) $[\![\exists x \in \check{u}\varphi(x)]\!] = \bigvee_{x \in u}[\![\varphi(\check{x})]\!]$.

Proof.

(1) $[\![\check{u} \in \check{v}]\!] = \bigvee_{y \in v}[\![\check{u} = \check{y}]\!] = 1$ iff $u \in v$.

(2) $[\![\check{u} = \check{v}]\!] = \bigwedge_{x \in u}[\![\check{x} \in \check{v}]\!] \wedge \bigwedge_{y \in v}[\![\check{y} \in \check{u}]\!] = 1$ iff $u = v$.

(3) $[\![\forall x \in \check{u}\varphi(x)]\!] = \bigwedge_{x \in u}(\check{u}(\check{x}) \to [\![\varphi(\check{x})]\!])$

$$= \bigwedge_{x \in u}[\![\varphi(\check{x})]\!].$$

(4) $[\![\exists x \in \check{u}\varphi(x)]\!] = \bigvee_{x \in u}(\check{u}(\check{x}) \wedge [\![\varphi(\check{x})]\!])$

$$= \bigvee_{x \in u}[\![\varphi(\check{x})]\!].$$

Operation $\upharpoonright$.

For $v \in V^{(\Omega)}$ and $p \in \Omega$, define $u \upharpoonright p$ by

$$\mathcal{D}(u \upharpoonright p) = \{x \upharpoonright p \mid x \in \mathcal{D}(u)\}$$

$$x \in \mathcal{D}(u) \Rightarrow (u \upharpoonright p)(x \upharpoonright p) = \bigvee \{u(t) \wedge p \mid t \in \mathcal{D}(u), t \upharpoonright p = x \upharpoonright p\}$$

$$E(u \upharpoonright p) = Eu \wedge p.$$

Then $u \upharpoonright p \in V^{(\Omega)}$ and we have

Lemma 3.4.

(1) $u \upharpoonright Eu = u$

(2) $E(u \upharpoonright p) = Eu \wedge p$

(3) $(u \upharpoonright p) \upharpoonright q = u \upharpoonright p \wedge q$

(4) $[\![u = u \upharpoonright p]\!] \geq p$

(5) $[\![u \in v \upharpoonright p]\!] = [\![u \upharpoonright p \in v]\!] = [\![u \in v]\!] \wedge p$

(6) $[\![u \upharpoonright p = v \upharpoonright p]\!] = p \rightarrow [\![u = v]\!]$

(7) $[\![u = v]\!] = \bigvee \{p \mid [\![u \upharpoonright p = v \upharpoonright p]\!] = 1\}$

Proof.

(1), (2), (3) are obvious from the definition.

(4) Assume $[\![x = x \upharpoonright p]\!] \geq p$ for $x \in V^{(\Omega)}$ with rank $x <$ rank u.

$$[\![u = u \upharpoonright p]\!] = \bigwedge_{x \in \mathcal{D}(u)} (u(x) \rightarrow [\![x \in u \upharpoonright p]\!]) \wedge$$

$$_{x \in \mathcal{D}(u)} (u(x) \wedge p \rightarrow [\![x \upharpoonright p \in u]\!]) \wedge$$

$$Eu \leftrightarrow E(u \upharpoonright p), \text{ where}$$

$$[\![x \in u \upharpoonright p]\!] = \bigvee_{y \in \mathcal{D}(u)} [\![x = y \upharpoonright p]\!] \wedge u(y) \wedge p$$

$$\geq \bigvee_{y \in \mathcal{D}(u)} [\![x = y]\!] \wedge [\![y \upharpoonright p = y]\!] \wedge u(y) \wedge p$$

$$\geq p \wedge u(x), \text{ and}$$

$$[\![x \upharpoonright p \in u]\!] \geq [\![x = x \upharpoonright p]\!] \wedge [\![x \in u]\!] \geq p \wedge u(x).$$

Therefore, $[\![u = u \upharpoonright p]\!] \geq p$.

(5) $$[\![u \in v \restriction p]\!] = \bigvee_{y \in \mathcal{D}(v)} [\![u = y \restriction p]\!] \wedge v(y) \wedge p$$

$$= \bigvee_{y \in \mathcal{D}(v)} [\![u = y]\!] \wedge v(y) \wedge p$$

$$= [\![u \in v]\!] \wedge p.$$

$$[\![u \restriction p \in v]\!] = \bigvee_{y \in \mathcal{D}(v)} [\![u \restriction p = y]\!] \wedge v(y)$$

$$= \bigvee_{y \in \mathcal{D}(v)} [\![u \restriction p = y]\!] \wedge v(y) \wedge Ey$$

$$= \bigvee_{y \in \mathcal{D}(v)} [\![u \restriction p = y]\!] \wedge v(y) \wedge p$$

$$= [\![u \in v]\!] \wedge p.$$

(6) $$[\![u \restriction p = v \restriction p]\!] = \bigwedge_{x \in \mathcal{D}(u)} (u(x) \wedge p \to [\![x \restriction p \in v \restriction p]\!]) \wedge$$

$$\bigwedge_{y \in \mathcal{D}(v)} (v(y) \wedge p \to [\![y \restriction p \in u \restriction p]\!] \wedge$$

$$E(u \restriction p) \leftrightarrow E(v \restriction p)$$

$$= \bigwedge_{x \in \mathcal{D}(u)} (u(x) \wedge p \to [\![x \in v]\!] \wedge p) \wedge$$

$$\bigwedge_{y \in \mathcal{D}(v)} (v(y) \wedge p \to [\![y \in u]\!] \wedge p) \wedge$$

$$(Eu \wedge p \leftrightarrow Ev \wedge p)$$

$$= p \to [\![u = v]\!].$$

(7) Obvious from (6).

Operation $\bigvee$.

Let A be a set and $A \subseteq V^{(\Omega)}$. We say A is compatible if

$$a,b \in A \Rightarrow Ea \wedge Eb \leq [\![a = b]\!].$$

Lemma 3.5.

If $A \subseteq V^{(\Omega)}$ is compatible then there exists a unique element b of $V^{(\Omega)}$ such that

(1) $$Eb = \bigvee\{Ea \mid a \in A\}$$

(2) $$a \in A \Rightarrow Ea \leq [\![a = b]\!],$$

where 'unique' means that if $b, b' \in V^{(\Omega)}$ satisfy (1) and (2) then $[\![b = b']\!] = 1$.

Proof.

Define b by

$$\mathcal{D}(b) = \bigcup_{a \in A} \mathcal{D}(a), \quad Eb = \bigvee\{Ea \mid a \in A\} \quad \text{and}$$

$$x \in \mathcal{D}(b) \Rightarrow b(x) = \bigvee_{a \in A} [\![x \in a]\!].$$

Then for $a \in A$

$$[\![a = b]\!] = \bigwedge_{x \in \mathcal{D}(a)} (a(x) \to [\![x \in b]\!]) \wedge \bigwedge_{y \in \mathcal{D}(b)} (b(y) \to [\![y \in a]\!]) \wedge (Ea \leftrightarrow Eb)$$

$$\geq \bigwedge_{y \in \mathcal{D}(b)} (b(y) \to [\![y \in a]\!]) \wedge (Ea \leftrightarrow Eb), \quad \text{where}$$

$$\begin{aligned} Ea \wedge b(y) &= Ea \wedge \bigvee_{a' \in A} [\![y \in a']\!] \\ &= \bigvee_{a' \in A} Ea \wedge [\![y \in a']\!] \wedge Ea' \\ &\leq \bigvee_{a' \in A} [\![y \in a']\!] \wedge [\![a = a']\!] \\ &\leq [\![y \in a]\!] \end{aligned}$$

Therefore, $Ea \leq (b(y) \to [\![y \in a]\!])$

Also we have $Ea \leq (Ea \leftrightarrow Eb)$. Therefore,

$$Ea \leq [\![a = b]\!].$$

For uniqueness, let $b, b' \in V^{(\Omega)}$ satisfy (1) and (2). Then

$$
\begin{aligned}
[\![b = b']\!] &= (Eb \vee Eb' \to [\![b = b']\!]) \\
&= (\bigvee_{a \in A} Ea \to [\![b = b']\!]) \\
&= \bigwedge_{a \in A} (Ea \to [\![b = b']\!]) \\
&= \bigwedge_{a \in A} [\![b \upharpoonright Ea = b' \upharpoonright Ea]\!] = 1.
\end{aligned}
$$

The element b of $V^{(\Omega)}$ is denoted by $\bigvee A$.

The uniqueness principle

The maximum principle, which is valid in Boolean valued universe, is not valid in $V^{(\Omega)}$. Instead we have:

Lemma 3.6. (Uniqueness principle)

Let $\exists ! x A(x)$ be the formula

$$\exists x A(x) \wedge \forall x \forall y (A(x) \wedge A(y) \to x = y).$$

Then there exists $u \in V^{(\Omega)}$ such that

(1) $[\![A(u)]\!] = [\![\exists ! x A(x)]\!] = Eu,$

(2) $v \in V^{(\Omega)}$, $[\![\exists ! x A(x)]\!] \leq [\![A(v)]\!]$ implies
$Eu \wedge Ev \leq [\![u = v]\!],$

(3) $v \in V^{(\Omega)}$, $[\![\exists ! x A(x)]\!] = Ev \leq [\![A(v)]\!]$ implies
$[\![u = v]\!] = 1.$

Proof.

Let $[\![\exists ! x A(x)]\!] = p$. That is,

$$\bigvee_{x \in V^{(\Omega)}} Ex \wedge [\![A(x) \wedge \forall z \forall y (A(z) \wedge A(y) \to z = y)]\!] = p.$$

Since Ω is a set, we can choose $\{u_\xi | \xi < \alpha\}$ such that

$$\bigvee_{\xi<\alpha} Eu_\xi \wedge [\![A(u_\xi) \wedge \forall z \forall y (A(z) \wedge A(y) \to z = y)]\!] = p.$$

Set $B = \{u_\xi \restriction [\![A(u_\xi) \wedge \forall z \forall y (A(z) \wedge A(y) \to z = y)]\!] | \xi < \alpha\}$. Then B is compatible. Hence $u = \bigvee B \in V^{(\Omega)}$, and u satisfy (1), (2), (3).

§4. Ordinals in $V^{(\Omega)}$.

The formula Ord(x), 'x is an ordinal', is defined by

$Ord(x) = Tr(x) \wedge \forall y \in x Tr(y)$, where

$Tr(x) = \forall y \in x (y \subseteq x)$.

Ord, and $\alpha + 1$ are defined by

$Ord = \{\alpha | Ord(\alpha)\}$ and

$\alpha + 1 = \alpha \cup \{\alpha\}$.

Lemma 4.1.

(1) $\vdash \alpha \in Ord \to \alpha + 1 \in Ord$.

(2) $\vdash A \subseteq Ord \to \cup A \in Ord$.

(3) $\vdash \alpha, \beta \in Ord \wedge \alpha \in \beta \to \alpha + 1 \subseteq \beta$.

(4) $\vdash A \subseteq Ord \wedge \beta \in Ord \to (\cup A \subseteq \beta \leftrightarrow \forall \alpha \in A (\alpha \subseteq \beta))$

(5) If α is an ordinal in V, then $\vdash \check{\alpha} \in Ord$.

Proof.

cf. [2], [3].

As Grayson mentioned in [2], definitions by $\in$-recursion on Ord are justified in ZF_I. That is, he showed that if H is defined in ZF_I and $\vdash \forall x \exists ! y (\langle x, y\rangle \in H)$, then there exists F such that

$$\vdash \forall \alpha \in \mathrm{Ord}\, \exists ! u(\langle\alpha,u\rangle \in F \wedge \langle F \restriction \alpha,u\rangle \in H),$$

where $F \restriction \alpha = \{\langle\beta,u\rangle \in F \mid \beta \in \alpha\}$.

§5. Real numbers in $V^{(\Omega)}$.

Let $\mathbb{Q}$ be the set of rational numbers. Then $\check{\mathbb{Q}}$ is the set of rational numbers in $V^{(\Omega)}$. The formula 'u is a real number' is defined as:

$$\exists L \subseteq \check{\mathbb{Q}}\ \exists U \subseteq \check{\mathbb{Q}}[u = \langle L,U\rangle \wedge P_1(L,U) \wedge \cdots \wedge P_5(L,U)],$$

where

$$P_1(L,U) = \exists r, s \in \check{\mathbb{Q}}(r \in L \wedge s \in U).$$
$$P_2(L,U) = \forall r \in \check{\mathbb{Q}}\ \neg(r \in L \wedge r \in U).$$
$$P_3(L,U) = \forall r \in \check{\mathbb{Q}}(r \in L \leftrightarrow \exists s \in \check{\mathbb{Q}}(r < s \wedge s \in L).$$
$$P_4(L,U) = \forall r \in \check{\mathbb{Q}}(r \in U \leftrightarrow \exists s \in \check{\mathbb{Q}}(s < r \wedge s \in U).$$
$$P_5(L,U) = \forall r,s \in \check{\mathbb{Q}}(r < s \to r \in L \vee s \in U)$$

It is known that if X is a topological space and Ω is the set $\mathcal{O}(X)$ of open subsets of X then a real number u in $V^{(\Omega)}$ is considered as a real valued continuous function on Eu, which assigns $\sup\{r \in \mathbb{Q} \mid x \in [\![\check{r} \in L]\!]\}$ to $x \in Eu$.

§6. Sheaf representation.

If two cHas Ω and Ω' are isomorphic, then the mapping $V^{(\Omega)} \to V^{(\Omega')}$ induced by the isomorphism $\Omega \to \Omega'$ is isomorphism and $V^{(\Omega)}$ and $V^{(\Omega')}$ are identified as a model. Hence we may choose Ω such that

$$\forall a,b \in \Omega\ (\mathrm{rank}\ a = \mathrm{rank}\ b)$$

without loss of generality. Then for $u,v \in V^{(\Omega)}$ rank $\mathcal{D}(u) \leq$ rank $\mathcal{D}(v)$ implies rank $u \leq$ rank v.

A class X in $V^{(\Omega)}$ is said to be <u>definite</u> if

$$x \in \mathcal{D}(X) \Rightarrow X(x) = Ex.$$

<u>Lemma 6.1</u>.

If X is a class in $V^{(\Omega)}$, then there is a definite class in $V^{(\Omega)}$ Y, such that $[\![X = Y]\!] = 1$. If $x \in V^{(\Omega)}$ then there is a definite $y \in V^{(\Omega)}$ such that $[\![x = y]\!] = 1$ and rank $y \leq$ rank x.

<u>Proof</u>.

For a class X in $V^{(\Omega)}$, define Y by

$$\mathcal{D}(Y) = \{t \upharpoonright [\![t \in X]\!] \mid t \in \mathcal{D}(X)\},$$

$$t \in \mathcal{D}(X) \Rightarrow Y(t \upharpoonright [\![t \in X]\!]) = [\![t \in X]\!],$$

$$EY = EX.$$

Then Y is definite and

$$[\![X = Y]\!] = \bigwedge_{t \in \mathcal{D}(X)} (X(t) \to [\![t \in Y]\!]) \wedge$$

$$\bigwedge_{t \in \mathcal{D}(X)} (Y(t \upharpoonright [\![t \in X]\!]) \to [\![t \upharpoonright [\![t \in X]\!] \in X]\!]) \wedge$$

$$(EX \leftrightarrow EY).$$

$$= \bigwedge_{t \in \mathcal{D}(X)} ([\![t \in X]\!] \to [\![t \in X]\!]) = 1.$$

For $x \in V^{(\Omega)}$ and $p \in \Omega$, it is proved, by induction on rank x, that rank $x \upharpoonright p \leq$ rank x. Then the last part of the lemma is obvious.

Definition.

Let $u,v \in V^{(\Omega)}$. v is called sheaf representation (SR) of u if

(1) v is definite,

(2) $Ev = 0 \Rightarrow v = \theta$, where $\theta \in V^{(\Omega)}$ is defined by $\mathcal{D}(\theta) = \phi$ and $E\theta = 0$,

(3) $Ev \neq 0$, $x \in V^{(\Omega)}$, $[\![x \in v]\!] = Ex$,

$$\Rightarrow \exists x' \in \mathcal{D}(v)([\![x = x']\!] = 1),$$

(4) $[\![u = v]\!] = 1$.

u is called sheaf representation (SR) if u is a SR of u itself.

Lemma 6.2.

For any $u \in V^{(\Omega)}$ there exists a SR v of u such that

$$x \in \mathcal{D}(v) \Rightarrow \text{rank } x < \text{rank } u.$$

Proof.

Let $u \in V^{(\Omega)}$. By Lemma 6.1, there exists a definite $u_1 \in V^{(\Omega)}$ such that $[\![u = u_1]\!] = 1$ and rank $u_1 \leq$ rank u. Define u_2 by

$$\mathcal{D}(u_2) = \{x \upharpoonright p \mid x \in \mathcal{D}(u_1),\ p \in \Omega,\ Ex \wedge p \neq 0\} \cup \{\theta\}$$

$$t \in \mathcal{D}(u_2) \Rightarrow u_2(t) = Et,$$

$$Eu_2 = Eu.$$

Then

$$[\![u_1 = u_2]\!] = \bigwedge_{x \in \mathcal{D}(u_1)} (u_1(x) \to [\![x \in u_2]\!]) \wedge \bigwedge_{x \in \mathcal{D}(u_2)} (u_2(x) \to [\![x \in u_1]\!])$$

$$= 1,$$

and rank $u_2 \leq$ rank u_1 by the definition. Now we define v. If $Ev = 0$ then let $v = \theta$. Otherwise, let

$$\mathcal{D}(v) = \{\bigvee A \mid A \subseteq \mathcal{D}(u_2),\ A \text{ is compatible}\},$$

$$\bigvee A \in \mathcal{D}(v) \Rightarrow v(\bigvee A) = \bigvee_{a \in A} [\![a \in u_2]\!] = \bigvee_{a \in A} Ea,$$

$$Ev = Eu.$$

Then v satisfies the conditions as follows.

(i) $A \subseteq \mathcal{D}(u_2) \Rightarrow \text{rank} \bigvee A \leq \text{rank}\ \mathcal{D}(u_2)$, since

$$\mathcal{D}(\bigvee A) = \bigcup_{a \in A} \mathcal{D}(a)$$

$$t \in \mathcal{D}(a),\ a \in A \Rightarrow (\bigvee A)(t) = \bigvee_{a \in A} [\![t \in a]\!]$$

$$E(\bigvee A) = \bigvee_{a \in A} Ea.$$

Hence

$$x \in \mathcal{D}(v) \Rightarrow \text{rank}\ x \leq \text{rank}\ \mathcal{D}(u_2) < \text{rank}\ u_2$$

$$\Rightarrow \text{rank}\ x < \text{rank}\ u.$$

(ii) $[\![u_2 = v]\!] = 1$, for

$$x \in \mathcal{D}(u_2) \Rightarrow x = \bigvee\{x\} \in \mathcal{D}(v) \text{ and}$$

$$u_2(x) = Ex = v(x).$$

$$y \in \mathcal{D}(v) \Rightarrow y = \bigvee A \text{ for some } A \subseteq \mathcal{D}(u_2) \text{ and}$$

$$v(y) = \bigvee_{a \in A} Ea \leq \bigvee_{a \in A} [\![a = \bigvee A]\!] \wedge u_2(a)$$

$$\leq [\![y \in u_2]\!].$$

(iii) Assume that $Ev \neq 0$, $x \in V^{(\Omega)}$ and $[\![x \in v]\!] = Ex$. Then

$$\bigvee_{y \in \mathcal{D}(u_1)} [\![x = y]\!] \wedge Ey = Ex, \text{ and}$$

$$A = \{y \restriction [\![x = y]\!] \mid y \in \mathcal{D}(u_1)\} \text{ is compatible.}$$

Hence, $\bigvee A \in \mathcal{D}(v)$ and

$$[\![\bigvee A = x]\!] \leq \bigvee_{y \in \mathcal{D}(u_1)} [\![x = y]\!] \wedge Ey = Ex = E(\bigvee A)$$

$$[\![\bigvee A = x]\!] = 1.$$

Therefore, v is a SR of u.

Lemma 6.3.

Let $u \in V^{(\Omega)}$ be a SR with $Eu \neq 0$. Then for any $v \in V^{(\Omega)}$ there is $v' \in \mathcal{D}(u)$ such that

$$Ev' = [\![v \in u]\!] \leq [\![v = v']\!].$$

Proof.

Since u is SR and

$$[\![v \restriction [\![v \in u]\!] \in u]\!] = E(v \restriction [\![v \in u]\!]),$$

there is $v' \in \mathcal{D}(u)$ such that

$$[\![v' = v \restriction [\![v \in u]\!]]\!] = 1.$$

Hence $Ev' = [\![v \in u]\!] \leq [\![v = v']\!]$.

Lemma 6.4.

Sheaf representation is unique in the sense that: if u, v are SR and $[\![u = v]\!] = 1$, then

$$\forall x \in \mathcal{D}(u)\, \exists y \in \mathcal{D}(v)\, ([\![x = y]\!] = 1) \quad \text{and}$$

$$\forall y \in \mathcal{D}(v)\, \exists x \in \mathcal{D}(u)\, ([\![x = y]\!] = 1).$$

Proof.

Assume u,v are SR and $[\![u = v]\!] = 1$.

$$x \in \mathcal{D}(u) \Rightarrow u(x) = Ex = [\![x \in v]\!]$$
$$\Rightarrow \exists y \in \mathcal{D}(v)([\![x = y]\!] = 1).$$

Similarly,

$$y \in \mathcal{D}(v) \Rightarrow \exists x \in \mathcal{D}(u)([\![y = x]\!] = 1).$$

Definition (by induction on rank).

$u \in V^{(\Omega)}$ is called hereditary sheaf representation (HSR) of $v \in V^{(\Omega)}$ if

1) u is a sheaf representation of v and

2) $x \in \mathcal{D}(u) \Rightarrow x$ is HSR.

Lemma 6.5.

If $u \in V^{(\Omega)}$, then there exists a unique hereditary sheaf representation of u.

Proof.

We prove the lemma by transfinite induction on the rank of u.

Let u_1 be a sheaf representation of u such that

$$x \in \mathcal{D}(u_1) \Rightarrow \text{rank } x < \text{rank } u.$$

Then, by the induction hypothesis,

$$x \in \mathcal{D}(u_1) \Rightarrow \exists! x'(x' \text{ is HSR of } x).$$

Define v by:

If $Eu = 0$ then $v = \theta$. Otherwise,

$$\mathcal{D}(v) = \{x' \mid x \in \mathcal{D}(u_1)(x' \text{ is HSR of } x)\}$$

$$x \in \mathcal{D}(v) \Rightarrow v(x') = [\![x' \in u]\!]$$

$$Eu' = Eu.$$

Then it is obvious that u' is HSR of u. Uniqueness of HSR of u follows from the induction hypothesis and Lemma 6.4.

Let $\equiv$ be the relation defined on $V^{(\Omega)}$ by

$$u \equiv v \quad \text{iff} \quad [\![u = v]\!] = 1.$$

Then the relation $\equiv$ is an equivalence relation and

$$u \equiv v \Rightarrow Eu = Ev,$$

$$u \equiv v,\ p \in \Omega \Rightarrow u \upharpoonright p \equiv v \upharpoonright p,$$

$$u_1 \equiv v_1, \ldots, u_n \equiv v_n \Rightarrow [\![\varphi(u_1, \ldots, u_n)]\!] = [\![\varphi(v_1, \ldots, v_n)]\!].$$

So we identify u and v such that $u \equiv v$, and denote $V^{(\Omega)}/\equiv$ by $V^{(\Omega)}$, if there is no confusion.

Then if $u \in V^{(\Omega)}$, u' is SR of u and u" is HSR of u, we have

$$\{x \in V^{(\Omega)} \mid [\![x \in u]\!] = Ex\} = \mathcal{D}(u') = \mathcal{D}(u'').$$

Strictly speaking,

$$\{x \in V^{(\Omega)} \mid [\![x \in u]\!] = Ex\}/\equiv\ = \{x/\equiv \mid x \in \mathcal{D}(u')\}$$

$$= \{x/\equiv \mid x \in \mathcal{D}(u'')\},$$

where $x/\equiv$ is the equivalence class of x. We sometimes omit $/\equiv$ as above.

§7. $V^{(\Omega)}$ and sheaves.

Definition.

Set A is called presheaf over a cHa Ω if two functions $E_A : A \to \Omega$ and $\rceil : A \times \Omega \to A$ are defined and satisfy

(1) $a \in A \Rightarrow a \rceil E_A a = a$,

(2) $a \in A,\ p \in \Omega \Rightarrow E_A(a \rceil p) = E_A a \wedge p$,

(3) $a \in A,\ p, q \in \Omega \Rightarrow (a \rceil p) \rceil q = a \rceil p \wedge q$.

A subset $B \subseteq A$ is said to be compatible if

$$a, b \in B \Rightarrow a \rceil E_A b = b \rceil E_A a.$$

Set A is called sheaf over Ω if

(1) $\langle A, E_A, \rceil\rangle$ is a presheaf over Ω,

(2) If $B \subseteq A$ is compatible, then there is $b \in A$ such that

$$E_A b = \bigvee_{a \in B} E_A a,$$
$$a \in B \Rightarrow b \rceil E_A a = a.$$

If $u \in V^{(\Omega)}$, then $\tilde{u} = \{x \in V^{(\Omega)} \mid [\![x \in u]\!] = Ex\}$ (or $\mathcal{D}(u')$ or $\mathcal{D}(u'')$, where u' is SR of u and u'' is HSR of u) is a sheaf with respect to $\restriction$, E, which we call sheaf represented by u.

Conversely, for a given sheaf $\langle A, E_A, \rceil\rangle$ over Ω, there is an element u of $V^{(\Omega)}$ such that the sheaf $\langle \tilde{u}, E, \restriction\rangle$ represented by u is isomorphic to $\langle A, E_A, \rceil\rangle$.

Proof.

Let $\langle A, E_A, \rceil\rangle$ be a given sheaf over Ω. for each $a \in A$, set $a^* = \check{a} \restriction E_A a$. Now we define A^* by

$$\mathcal{D}(A^*) = \{a^* | a \in A\}$$

$$a \in A \Rightarrow A^*(a^*) = Ea^* = E_A a$$

$$EA^* = \bigvee_{a \in A} Ea^*.$$

Also we define $\sim$ on A^* by

$$\mathcal{D}(\sim) = \mathcal{D}(A^* \times A^*)$$

$$a,b \in A \Rightarrow \sim\langle a^*,b^*\rangle = \bigvee\{p \in \Omega | a \urcorner p = b \urcorner p\} \wedge Ea^* \wedge Eb^*$$

$$E\sim = 1.$$

Then

$$[\![\sim \text{ is an equivalence relation on } A^*]\!] = 1.$$

We write $a \sim b$ instead of $\langle a,b\rangle \in \sim$. The equivalence class $\bar{a}$ of a^* is defined by

$$\mathcal{D}(\bar{a}) = \{b^* | b \in A\}$$

$$b \in A \Rightarrow \bar{a}(b^*) = [\![a^* \sim b^*]\!]$$

$$E\bar{a} = Ea^*$$

and the quotient $A^*/\sim$ is defined by

$$\mathcal{D}(A^*/\sim) = \{\bar{a} | a \in A\}$$

$$a \in A \Rightarrow (A^*/\sim)(\bar{a}) = E\bar{a}$$

$$E(A^*/\sim) = EA^*.$$

Then we have

(1) $a \in A,\ p \in \Omega \Rightarrow [\![\bar{a} \upharpoonright p = \overline{a \urcorner p}]\!] = 1,$

(2) $a,\ b \in A \Rightarrow ([\![\bar{a} = \bar{b}]\!] = 1$ iff $a = b),$

(3) $a,\ b \in A \Rightarrow ([\![\bar{a} = \bar{b}]\!] \wedge Ea^* \wedge Eb^* = [\![a^* \sim b^*]\!]),$

(4) $a \in A \Rightarrow [\![\bar{a} \in A^*/\sim]\!] = [\![a^* \in A^*]\!],$

(5) $B \subseteq A,\ \{\bar{a} | a \in B\}$ is compatible
$\Rightarrow [\![\bigvee\{\bar{a} | a \in B\} = \overline{\bigvee B}]\!] = 1.$

Proof of (1) ~ (5).

(1) For $a, b \in A$ and $p \in \Omega$.

$$\bar{a}(b^*) \wedge p = [\![a^* \sim b^*]\!] \wedge p$$
$$= \bigvee \{q \leq Ea^* \wedge Eb^* \wedge p \mid a \urcorner q = b \urcorner q\}$$

$$\overline{(a \urcorner p)}(b^*) = [\![(a \urcorner p)^* \sim b^*]\!]$$
$$= \{q \leq Ea^* \wedge Eb^* \wedge p \mid a \urcorner q = b \urcorner q\}$$
$$= \bar{a}(b^*) \wedge p.$$

Therefore, $[\![\bar{a} \upharpoonright p = \overline{a \urcorner p}]\!] = 1$.

(2) Assume $a, b \in A$ and $[\![\bar{a} = \bar{b}]\!] = 1$. Then

$$Ea^* = \bar{a}(a^*) \leq [\![a^* \in \bar{b}]\!] = \bigvee_{c \in A} [\![a^* = c^*]\!] \wedge \bar{b}(c^*)$$
$$\leq \bar{b}(a^*) \leq [\![a^* \sim b^*]\!].$$

Hence $Ea^* \vee Eb^* \leq [\![a^* \sim b^*]\!]$. It follows that

$$a \urcorner Ea^* \vee Eb^* = b \urcorner Ea^* \vee Eb^*$$

Therefore, $a = b$.

(3) For $a, b \in A$, we have

$$a \urcorner [\![a^* \sim b^*]\!] = b \urcorner [\![a^* \sim b^*]\!].$$

By (1), $[\![\bar{a} \upharpoonright [\![a^* \sim b^*]\!] = \bar{b} \upharpoonright [\![a^* \sim b^*]\!]]\!] = 1$.

Hence $[\![a^* \sim b^*]\!] \leq [\![\bar{a} = \bar{b}]\!]$.

Conversely, since we have

$$[\![\bar{a} \upharpoonright [\![\bar{a} = \bar{b}]\!] = b \upharpoonright [\![\bar{a} = \bar{b}]\!]]\!] = 1.$$
$$a \urcorner [\![\bar{a} = \bar{b}]\!] = b \urcorner [\![\bar{a} = \bar{b}]\!].$$

Therefore, $[\![\bar{a} = \bar{b}]\!] \leq [\![a^* \sim b^*]\!]$.

(4) For $a \in A$,

$$[\![a^* \in A^*]\!] = Ea^* = E\bar{a} = [\![\bar{a} \in A^*/\sim]\!].$$

(5) Assume that $B \subseteq A$ and $\{\bar{a} | a \in B\}$ is compatible. Then

$$(\forall a,b \in B)([\![\bar{a} \upharpoonright Eb = \bar{b} \upharpoonright Ea]\!] = 1).$$

Then $(\forall a,b \in B)(a \urcorner Eb = b \urcorner Ea)$, that is, B is compatible. It follows that $\bigvee B \in A$ and for $a \in B$

$$E\bar{a} \le [\![\bigvee\{\bar{c} | c \in B\} = \bar{a} = \overline{\bigvee B \urcorner E_A a} = \overline{\bigvee B}]\!].$$

Therefore, $E\overline{\bigvee B} = \bigvee_{a \in B} E\bar{a} \le [\![\bigvee_{a \in B} \bar{a} = \overline{\bigvee B}]\!]$, and so

$$[\![\bigvee_{a \in B} \bar{a} = \overline{\bigvee B}]\!] = 1.$$

From the definition of $A^*/\sim$ and (1) ~ (5), it follows that $A^*/\sim$ is a sheaf representation whose domain (sheaf represented by $A^*/\sim$) is isomorphic to A.

We call the above $A^*/\sim$ cannonical representation of the sheaf A.

§8. Functions in $V^{(\Omega)}$.

Ordered pair $\langle u,v \rangle^\Omega$ is defined as usual by

$$\langle u,v \rangle = \{\{u\}^\Omega, \{u,v\}^\Omega\}^\Omega.$$

Then

$$E\langle u,v \rangle^\Omega = Eu \wedge Ev$$

$$Eu_1 \wedge Eu_2 \wedge Ev_1 \wedge Ev_2 \le [\![\langle u_1,u_2 \rangle^\Omega = \langle v_1,v_2 \rangle^\Omega \leftrightarrow u_1 = v_1 \wedge U_2 = v_2]\!]$$

Let $F : X \to Y$, (F is a function from X to Y), be the formula $F \subseteq X \times Y \wedge \forall x \in X \exists! y \in Y(\langle x,y \rangle \in F)$.

Theorem 2.

If X, Y and F are classes in $V^{(\Omega)}$ and $[\![F : X \to Y]\!] = p$, then there exists a unique function $f : \tilde{X} \to \tilde{Y}$ such that for $x, y \in \tilde{X}$

(1) $[\![x \in X]\!] \wedge p \leq [\![\langle x, f(x)\rangle \in F]\!]$,

(2) $[\![x = y]\!] \leq [\![f(x) = f(y)]\!]$,

(3) $Ex \wedge p = Ef(x)$,

(4) $[\![x \in X]\!] \wedge p \leq [\![f(x) \in Y]\!]$,

where $\tilde{X} = \{x \in V^{(\Omega)} \mid [\![x \in X]\!] = Ex\}$, $\tilde{Y} = \{y \in V^{(\Omega)} \mid [\![y \in Y]\!] = Ey\}$, and '$f : \tilde{X} \to \tilde{Y}$ is unique' means that if f and f' satisfy the conditions then $[\![f(x) = f'(x)]\!] = 1$ for every $x \in \tilde{X}$.

Proof.

For $x \in \tilde{X}$, $Ex \wedge p = [\![x \in X]\!] \wedge p \leq [\![\exists! y \in Y(\langle x,y\rangle \in F)]\!]$. By the uniqueness principle, there exists $y \in V^{(\Omega)}$ such that

(a) $Ex \wedge p \leq [\![y \in Y \wedge \langle x,y\rangle \in F]\!]$ and

(b) $y' \in V^{(\Omega)}$, $Ex \wedge p \leq [\![y' \in Y \wedge \langle x,y'\rangle \in F]\!]$

$$\Rightarrow Ex \wedge p \leq [\![y = y']\!].$$

Hence there is a unique HSR $y \in V^{(\Omega)}$ which satisfy (a) and $Ey = Ex \wedge p$. Let the unique y be $f(x)$. Then it is obvious that f is a unique function $\tilde{X} \to V^{(\Omega)}$ and satisfies (1) - (4).

The following theorem is the converse of Theorem 2.

Theorem 3.

If X and Y are classes in $V^{(\Omega)}$, X is definite and F is a function satisfying (1) - (4):

(1) $f : \mathcal{D}(X) \to V^{(\Omega)}$,

(2) $x_1, x_2 \in \mathcal{D}(X) \Rightarrow [\![x_1 = x_2]\!] \leq [\![f(x_1) = f(x_2)]\!]$,

(3) $x \in \mathcal{D}(X) \Rightarrow Ef(x) = Ex \wedge p$,

(4) $x \in \mathcal{D}(x) \Rightarrow [\![x \in X]\!] \wedge p \leq [\![f(x) \in Y]\!]$,

then there exists a unique class F of $V^{(\Omega)}$ such that

(a) $[\![F : X \to Y]\!] \geq p = EF$

(b) $x \in \mathcal{D}(X) \Rightarrow [\![<x,f(x)> \in F]\!] = Ex \wedge p.$

Proof.

Define F by

$$\mathcal{D}(F) = \{<x,f(x)> \mid x \in \mathcal{D}(X)\}$$

$$x \in \mathcal{D}(X) \Rightarrow F<x,f(x)> = Ex \wedge p$$

$$EF = p.$$

(a) For $x \in \mathcal{D}(X)$, we have

$$[\![x \in X]\!] \wedge p \leq [\![f(x) \in Y \wedge <x,f(x)> \in F]\!].$$

Hence $[\![x \in X]\!] \wedge p \leq [\![\exists y \in Y(<x,y> \in F)]\!].$ ...(i)

For $x \in \mathcal{D}(X)$ and $y,y' \in \mathcal{D}(Y)$, we have

$$[\![<x,y> \in F]\!] = \bigvee_{x' \in \mathcal{D}(X)} F(<x',f(x')>) \wedge [\![<x',f(x')> = <x,y>]\!]$$

$$\leq \bigvee_{x' \in \mathcal{D}(X)} [\![x = x']\!] \wedge [\![f(x') = y]\!]$$

$$\leq [\![f(x) = y]\!] \quad \text{by (2).}$$

Hence $[\![<x,y> \in F \wedge <x,y'> \in F]\!] \leq [\![f(x) = y \wedge f(x) = y']\!] \leq [\![y = y']\!].$ ...(ii)

From (i)(ii) it follows that

$$[\![x \in X]\!] \wedge p \leq [\![\exists ! y \in Y(<x,y> \in F)]\!].$$

Therefore, $p \leq [\![\forall x \in X \exists ! y \in Y(<x,y> \in F)]\!].$

(b) is obvious from the definition.

Uniqueness. If F and F' satisfy (a), (b), then for $x \in \mathcal{D}(X)$ and $y \in \mathcal{D}(Y)$, we have

$$\begin{aligned} [\![\langle x,y \rangle \in F]\!] &\leq [\![\langle x,y \rangle \in F \wedge x \in X \wedge \langle x,f(x) \rangle \in F]\!] \\ &\leq [\![y = f'(x) \wedge \langle x,f(x) \rangle \in F']\!] \\ &\leq [\![\langle x,y \rangle \in F']\!], \quad \text{and hence} \\ [\![\langle x,y \rangle \in F]\!] &= [\![\langle x,y \rangle \in F']\!]. \end{aligned}$$

Therefore, $[\![F = F']\!] = 1$.

§9. CHa-morphism.

Definition.

Let $\langle \Omega, \wedge, \bigvee, 1 \rangle$ and $\langle \Omega', \wedge', \bigvee', 1' \rangle$ be cHas. A function $q : \Omega \to \Omega'$ is called cHa-morphism if

(1) $q(1) = 1'$.

(2) $p,q \in \Omega \Rightarrow g(p \wedge q) = g(p) \wedge' f(q)$.

(3) $\{p_i\}_i \subseteq \Omega \Rightarrow g(\bigvee_i p_i) = \bigvee'_i g(p_i)$.

Example.

Let X and Y be topological spaces, $\Omega = \mathcal{O}(Y)$ and $\Omega' = \mathcal{O}(X)$, where $\mathcal{O}(Y)$ (or $\mathcal{O}(X)$) stands for the cHa of open sets of Y (or X). Let $f : X \to Y$ be a continuous function, and let $f^* : \Omega \to \Omega'$ be defined by

$$U \in \Omega \Rightarrow f^*(U) = \{x \in X \mid f(x) \in U\}.$$

Then f^* is a cHa-morphism.

Lemma 9.1.

Let $g : \Omega \to \Omega'$ be a cHa-morphism. Then we have

(1) $\{p_i\}_i \subseteq \Omega \Rightarrow g(\bigwedge_i p_i) \leq \bigwedge_i g(p_i)$

(2) $p \in \Omega \Rightarrow g(\neg p) \leq \neg g(p)$.

(3) $p\ q \in \Omega \Rightarrow g(p \to q) \leq (g(p) \to g(q))$.

<u>Proof</u>. Obvious from the fact that g preserves $\leq$.

If $g : \Omega \to \Omega'$ is cHa-morphism, then we extend g to the mapping $\bar{g} : V^{(\Omega)} \to V^{(\Omega')}$ as follows. For $u \in V^{(\Omega)}$

$$\mathcal{D}(\bar{g}u) = \{\bar{g}x \mid x \in \mathcal{D}(u)\}$$

$$x \in \mathcal{D}(u) \Rightarrow (\bar{g}u)(\bar{g}x) = \{g(u(t)) \mid t \in \mathcal{D}(u),\ \bar{g}(t) = \bar{g}(x)\}$$

$$E\bar{g}u = gEu.$$

<u>Lemma 9.2.</u>

If $u,v \in V^{(\Omega)}$ and $p \in \Omega$, then

(1) $g[\![u \in v]\!] \leq [\![\bar{g}u \in \bar{g}v]\!]$

(2) $g[\![u = v]\!] \leq [\![\bar{g}u = \bar{g}v]\!]$

(3) $[\![\bar{g}\{u,v\} = \{\bar{g}u, \bar{g}v\}]\!] = 1$

$[\![\bar{g} <u,v> = <\bar{g}u,\bar{g}v>]\!] = 1$

(4) $[\![\bar{g}(u \restriction p) = \bar{g}u \restriction \bar{g}p]\!] = 1$.

Proof.

(1) $$g[\![u \in v]\!] = g\bigvee_{y\in \mathcal{D}(v)}[\![u = y]\!] \wedge v(y)$$

$$= \bigvee_{y\in \mathcal{D}(v)} g[\![u = y]\!] \wedge g(v(y))$$

$$\leq \bigvee_{y\in \mathcal{D}(g(v))}[\![\overline{g}u = y]\!] \wedge (\overline{g}v)(y)$$

$$= [\![\overline{g}u \in \overline{g}v]\!]$$

(2) $$g[\![u = v]\!] = g(\bigwedge_{x\in \mathcal{D}(u)} (u(x) \to [\![x \in v]\!]) \wedge$$

$$g(\bigwedge_{y\in \mathcal{D}(v)} (v(y) \to [\![y \in u]\!])) \wedge g(Eu \leftrightarrow Ev),$$ where

$$g(\bigwedge_{x\in \mathcal{D}(u)} (u(x) \to [\![x \in v]\!])$$

$$\leq \bigwedge_{x\in \mathcal{D}(u)} (g(u(x)) \to [\![\overline{g}x \in \overline{g}v]\!])$$

$$\leq \bigwedge_{x\in \mathcal{D}(u)} ((\overline{g}u)(\overline{g}x) \to [\![\overline{g}x \in \overline{g}v]\!])$$

$$g(Eu \leftrightarrow Ev) \leq (E\overline{g}u \leftrightarrow E\overline{g}v).$$

Therefore, $g[\![u = v]\!] \leq [\![\overline{g}u = \overline{g}v]\!]$.

(3) Set $u = \{a,b\}^{\Omega}$. That is,

$$\mathcal{D}(u) = \{a,b\}$$

$$u(a) = u(b) = Ea \wedge Eb$$

$$Eu = Ea \wedge Eb$$

Then $\mathcal{D}(\overline{g}u) = \{\overline{g}a, \overline{g}b\}$,

$(\overline{g}u)(\overline{g}a) = (\overline{g}u)(\overline{g}b) = \overline{g}(a \wedge b) = \overline{g}a \wedge' \overline{g}b$ and

$E\overline{g}u = gEu.$

Therefore, $\overline{g}\{a,b\}^{\Omega} = \{\overline{g}a, \overline{g}b\}^{\Omega'}$, and

$\overline{g}\langle a,b\rangle^{\Omega} = \langle \overline{g}a, \overline{g}b\rangle^{\Omega'}$.

(4) $\mathcal{D}(\overline{g}(u \upharpoonright p)) = \{\overline{g}(x \upharpoonright p) \mid x \in \mathcal{D}(u)\}$

$x \in \mathcal{D}(u) \Rightarrow (\overline{g}(u \upharpoonright p))(\overline{g}(x \upharpoonright p))$

$$= \bigvee_{\substack{t \in \mathcal{D}(u) \\ \overline{g}(x \upharpoonright p) = \overline{g}(t \upharpoonright p)}} g(u(t) \wedge p)$$

$$\leq \bigvee_{t \in \mathcal{D}(u)} [\![\overline{g}(x \upharpoonright p) = \overline{g}(t \upharpoonright p)]\!] \wedge g[\![t \in u]\!] \wedge gp$$

$$= \bigvee_{t \in \mathcal{D}(u)} [\![\overline{g}(x \upharpoonright p) = \overline{g}t \upharpoonright gp]\!] \wedge [\![\overline{g}t \in \overline{g}u]\!] \wedge gp.$$

$\leq [\![\overline{g}(x \upharpoonright p) \in \overline{g}u \upharpoonright gp]\!]$, and

$$(\overline{g}u \upharpoonright gp)(\overline{g}x \upharpoonright gp) = \bigvee_{\substack{t \in \mathcal{D}(u) \\ \overline{g}t \upharpoonright gp = \overline{g}x \upharpoonright gp}} \overline{g}u(\overline{g}t) \wedge gp$$

$$\leq \bigvee_{t \in \mathcal{D}(u)} [\![\overline{g}x \upharpoonright gp = \overline{g}(t \upharpoonright p)]\!] \wedge \overline{g}(u \upharpoonright p)(\overline{g}(t \upharpoonright p))$$

$\leq [\![\overline{g}x \upharpoonright gp \in \overline{g}(u \upharpoonright p)]\!]$

Therefore, $[\![\overline{g}u \upharpoonright gp = \overline{g}(u \upharpoonright p)]\!] = 1.$

§10. Ω in $V^{(H)}$.

In this section we fix a cHa H and consider a cHa within $V^{(H)}$.

For $\Omega, \wedge, \bigvee, 1 \in V^{(H)}$, the formula '$<\Omega, \wedge, \bigvee, 1>$ is a cHa' is the conjuction of the following (a) ~ (h).

(a) $(\wedge : \Omega \times \Omega \to \Omega) \wedge (\bigvee : P(\Omega) \to \Omega) \wedge (1 \in \Omega)$.

(b) $a \in \Omega \to a \wedge a = a$.

(c) $a, b \in \Omega \to a \wedge b = b \wedge a$.

(d) $a, b, c \in \Omega \to a \wedge (b \wedge c) = (a \wedge b) \wedge c$.

(e) $A \subseteq \Omega,\ a \in A \to \bigvee A \geq a$.

(f) $A \subseteq \Omega,\ b \in \Omega,\ \forall a \in A(a \leq b) \to \bigvee A \leq b$.

(g) $A \subseteq \Omega,\ b \in \Omega \to (\bigvee A) \wedge b = \bigvee\{a \wedge b | a \in A\}$.

(h) $a \in \Omega \to a \leq 1$.

where $a \leq b$ stands for $a \wedge b = a$.

Let $u \in V^{(H)}$ and $\tilde{u}$ be the sheaf represented by u, that is, $\tilde{u} = \{x \in V^{(H)} | [\![x \in u]\!] = Ex\}$, where two elements x, y of u satisfying $[\![x = y]\!] = 1$ are identified.

Theorem 4.

Let $\Omega, \wedge, \bigvee, 1 \in V^{(\Omega)}$, $\tilde{\Omega} = \{a \in V^{(\Omega)} | [\![a \in \Omega]\!] = Ea\}$ and $[\![<\Omega, \wedge, \bigvee, 1>$ is a cHa$]\!] = 1$.

(1) There is a function $\tilde{\wedge} : \tilde{\Omega} \times \tilde{\Omega} \to \tilde{\Omega}$ such that for any $a, b, a', b' \in \Omega$,

(i) $[\![a \wedge b = a \tilde{\wedge} b]\!] \geq Ea \wedge Eb$,

(ii) $[\![a = a' \wedge b = b']\!] \leq [\![a \tilde{\wedge} b = a' \tilde{\wedge} b']\!]$,

(iii) $E(a \tilde{\wedge} b) = Ea \wedge Eb$.

(2) There is a function $\tilde{\bigvee}: P(\tilde{\Omega}) \to \tilde{\Omega}$ such that for any $A \subseteq \tilde{\Omega}$,

(i) $[\![\bigvee A^* = \tilde{\bigvee} A]\!] = \bigvee_{a \in A} Ea$,

(ii) $E(\tilde{\bigvee} A) = \bigvee_{a \in A} EA$,

where A^* is an element of $V^{(H)}$ defined by

$$\mathcal{D}(A^*) = A,$$

$$a \in A \Rightarrow A^*(a) = Ea,$$

$$EA^* = \bigvee_{a \in A} Ea.$$

(3) For $a,b,c \in \tilde{\Omega}$, $A \subseteq \tilde{\Omega}$,

$$[\![c = a \wedge b]\!] = [\![c = a \tilde{\wedge} b]\!] \wedge Ea \wedge Eb$$

$$[\![c = \bigvee A^*]\!] = [\![c = \tilde{\bigvee} A]\!] \wedge EA^*.$$

(4) $<\tilde{\Omega}, \tilde{\wedge}, \tilde{\bigvee}, 1>$ is a cHa.

(5) For $p \in H$ let $i(p) = 1 \restriction p$. Then i is an one-to-one cHa-morphism. That is,

$$p,q \in H \Rightarrow i(p \wedge q) = i(p) \tilde{\wedge} i(q)$$

$$A \subseteq H \Rightarrow i(\bigvee A) = \tilde{\bigvee} \{i(p) \mid p \in A\}$$

$$i(1) = 1.$$

$$p,q \in H \Rightarrow (p = q \text{ iff } i(p) = i(q))$$

(6) $a \in \tilde{\Omega}$, $p \in H \Rightarrow [\![a \restriction p = a \tilde{\wedge} i(p)]\!] = 1$

(7) If $A \in V^{(H)}$ and $[\![f : A \to \Omega]\!] = 1$, then

$$[\![\bigvee_{a \in A} f(a) = \tilde{\bigvee}_{a \in \tilde{A}} \tilde{f}(a)]\!] \geq \bigvee_{a \in \tilde{A}} Ea.$$

Proof.

(1) is obvious from Theorm 2 and $\tilde{\Omega} \times \tilde{\Omega} \subseteq (\Omega \times \Omega)^{\sim}$.

(2) For $A \subseteq \tilde{\Omega}$, we have

$$[\![A^* \subseteq \Omega]\!] = 1.$$

Hence $[\![A^* \in P(\Omega)]\!] = EA^*$,

$$A^* \in \widetilde{P(\Omega)}, \text{ and}$$

$$\tilde{\bigvee}(A^*) \in \tilde{\Omega}.$$

Now we define $\tilde{\bigvee} A = \tilde{\bigvee}(A^*)$. Then $\tilde{\bigvee}$ satisfies (2)(i) and (ii).

(3) $[\![c = a \wedge b]\!] = [\![<<a,b>,c> \in \Lambda]\!]$

$$= [\![<<a,b>,c> \in \Lambda]\!] \wedge Ea \wedge Eb$$

$$= [\![<<a,b>,c> \in \Lambda]\!] \wedge [\![<<a,b>, a \tilde{\wedge} b> \in \Lambda]\!]$$

$$\leq [\![c = a \tilde{\wedge} b]\!] \wedge Ea \wedge Eb.$$

Similarly

$$[\![c = \bigvee A^*]\!] = [\![c = \tilde{\bigvee} A]\!] \wedge EA^*$$

(4) is obvious from the definition.

(5) $p \in H \Rightarrow i(p) = 1 \restriction p$ and $[\![1 \restriction p \in \Omega]\!] = p = E(1 \restriction p)$

$$\Rightarrow 1 \restriction p \in \tilde{\Omega}.$$

$p,q \in H \Rightarrow i(p \wedge q) = i(p) \wedge i(q)$, for

$$p \wedge q \leq [\![1 \restriction p = 1]\!]\ [\![1 \restriction q = 1]\!]$$

$$\leq [\![1 \restriction p \wedge 1 \restriction q = 1]\!]$$

$$\leq [\![1 \restriction p \tilde{\wedge} 1 \restriction q = 1]\!], \text{ and hence}$$

$$1 \restriction p \tilde{\wedge} 1 \restriction q = 1 \restriction p \wedge q.$$

$$A \subseteq H \Rightarrow i(\bigvee A) = \tilde{\bigvee}\{i(p) \mid p \in A\}, \quad \text{for:}$$

$$p \in A \Rightarrow p \leq [\![i(p) \leq \tilde{\bigvee}\{i(p) \mid p \in A\}]\!]$$

$$\Rightarrow p \leq [\![1 = \tilde{\bigvee}\{i(p) \mid p \in A\}]\!]$$

Hence $\bigvee A \leq [\![1 = \tilde{\bigvee}\{i(p) \mid p \in A\}]\!]$, and so

$$1 \restriction \bigvee A = \tilde{\bigvee}\{i(p) \mid p \in A\}.$$

$i(1) = 1$ is obvious.

$$p,q \in H \wedge [\![i(p) = i(q)]\!] = 1 \Rightarrow (E(i(p)) \leftrightarrow E(i(q))) = 1$$

$$\Rightarrow p = q.$$

(6) $Ea \wedge p \leq [\![a \wedge i(p) = a \tilde{\wedge} i(p)]\!]$

$$p \leq [\![a = a \restriction p]\!] \wedge [\![1 = 1 \restriction p]\!] \leq [\![a \wedge 1 \restriction p = a \restriction p \wedge 1]\!].$$

Therefore, $p \leq [\![a \tilde{\wedge} 1 \restriction p = a \restriction p]\!]$ and so

$$[\![a \tilde{\wedge} i(p) = a \restriction p]\!] = 1$$

(7) If $A \in V^{(H)}$ and $[\![f : A \to \Omega]\!] = 1$, then we prove

$$[\![\bigvee_{a \in A} f(a) = \tilde{\bigvee}_{a \in \tilde{A}} \tilde{f}(a)]\!] \geq \bigvee_{a \in \tilde{A}} Ea.$$

Set $A' = \{\tilde{f}(a) \mid a \in \tilde{A}\}^*$. Then

$$[\![\bigvee A' = \tilde{\bigvee}_{a \in A} \tilde{f}(a)]\!] \geq EA' \quad \text{and}$$

$$EA' = \bigvee_{a \in \tilde{A}} E\tilde{f}(a) = \bigvee_{a \in \tilde{A}} Ea.$$

Therefore, $[\![\bigvee A' = \tilde{\bigvee}_{a \in \tilde{A}} \tilde{f}(a)]\!] \geq \bigvee_{a \in \tilde{A}} Ea.$

Now it suffices to show that

$$[\![\bigvee_{a \in A} f(a) = \bigvee A']\!] = 1,$$

which is equivalent to

$$[\![\forall a \in A\ f(a) \leq \bigvee A']\!] \wedge [\![\forall x \in A' (x \leq \bigvee_{a \in A} f(a))]\!] = 1.$$

This follows from the definition.

Theorem 5.

Let $\Omega \in V^{(H)}$, $\tilde{\Omega} = \{a \in V^{(H)} | [\![a \in \Omega]\!] = Ea\}$, and $<\Omega, \tilde{\wedge}, \tilde{\bigvee}, 1>$ satisfy the following conditions.

(i) $<\tilde{\Omega}, \tilde{\wedge}, \tilde{\bigvee}, 1>$ is a cHa and $E1 = 1$.

(ii) $E : \tilde{\Omega} \to H$ is a cHa-morphism.

(iii) $a, b, a', b' \in \tilde{\Omega}$

$$\Rightarrow [\![a = a']\!] \wedge [\![b = b']\!] \leq [\![a \tilde{\wedge} b = a' \tilde{\wedge} b']\!].$$

(iv) $A, B \subseteq \tilde{\Omega} \Rightarrow [\![A^* = B^*]\!] \leq [\![\tilde{\bigvee} A = \tilde{\bigvee} B]\!]$.

Then exist $\wedge, \bigvee \in V^{(H)}$ such that

(1) $[\![<\Omega, \wedge, \bigvee, 1>$ is a cHa$]\!] = 1$

(2) $a, b \in \tilde{\Omega} \Rightarrow Ea \wedge Eb \leq [\![a \wedge b = a \tilde{\wedge} b]\!]$,

$A \subseteq \tilde{\Omega} \Rightarrow EA^* \leq [\![\bigvee A^* = \tilde{\bigvee} A]\!]$.

Proof.

Define u by

$$\mathcal{D}(u) = \{<a,b> | a, b \in \tilde{\Omega}\}$$

$$<a,b> \in \mathcal{D}(u) \Rightarrow u<a,b> = Ea \wedge Eb$$

$$Eu = 1$$

Then $[\![u \subseteq \Omega \times \Omega]\!] = 1$, u is definite and satisfies the following conditions

$$\tilde{\wedge} : \mathcal{D}(u) \to \tilde{\Omega} \subseteq V^{(H)}$$

$$<a,b> \in \mathcal{D}(u) \Rightarrow E<a,b> \leq E(a \tilde{\wedge} b)$$

$$<a,b>, <a',b'> \in \mathcal{D}(u) \Rightarrow [\![<a,b> = <a',b'>]\!] \leq [\![a \tilde{\wedge} b = a' \tilde{\wedge} b']\!]$$

$$<a,b> \in \mathcal{D}(u) \Rightarrow [\![<a,b> \in u]\!] \leq [\![a \tilde{\wedge} b \in \Omega]\!].$$

Therefore, by the Theorem 3, there is $\wedge \in V^{(\Omega)}$ such that

$$[\![\wedge : u \to \Omega]\!] = [\![\wedge : \Omega \times \Omega \to \Omega]\!] = 1 \quad \text{and}$$

$$\langle a,b\rangle \in \mathcal{D}(u) \Rightarrow Ea \wedge Eb \leq [\![a \wedge b = a \tilde{\wedge} b]\!].$$

Similarly, we can prove that there exist $\bigvee \in V^{(\Omega)}$ such that

$$[\![\bigvee : P(\Omega) \to \Omega]\!] = 1 \quad \text{and}$$

$$A \subseteq \tilde{\Omega} \Rightarrow E(A^*) \leq [\![\bigvee A^* = \tilde{\bigvee} A]\!].$$

For this case, we define u by

$$\mathcal{D}(u) = \{A^* \mid A \subseteq \tilde{\Omega}\},\ Eu = 1 \quad \text{and}$$

$$A \subseteq \tilde{\Omega} \Rightarrow u(A^*) = EA^*.$$

Then $[\![u = P(\Omega)]\!] = 1$.

Now we prove $[\![\langle\Omega, \wedge, \bigvee, 1\rangle \text{ is a cHa}]\!] = 1$.

(b) $[\![\forall x \in \Omega(x \wedge x = x)]\!] = 1$, for

$$a \in \tilde{\Omega} \Rightarrow [\![a \tilde{\wedge} a = a]\!] = 1$$

Therefore, $[\![a \in \Omega]\!] \leq [\![a \wedge a = a \tilde{\wedge} a]\!]$

$$\leq [\![a \wedge a = a]\!].$$

(c) $[\![\forall x,y \in \Omega(x \wedge y = y \wedge x)]\!] = 1$, for

$$a,b \in \tilde{\Omega} \Rightarrow [\![a \tilde{\wedge} b = b \tilde{\wedge} a]\!] = 1,$$

$$\Rightarrow [\![a \in \Omega \wedge b \in \Omega]\!] \leq [\![a \wedge b = b \wedge a]\!].$$

(d) $[\![\forall x,y,z \in \Omega(x \wedge (y \wedge z) = (x \wedge y) \wedge z)]\!] = 1$, for

$$a,b,c \in \tilde{\Omega} \Rightarrow [\![(a \tilde{\wedge} (b \tilde{\wedge} c)) = (a \tilde{\wedge} b) \tilde{\wedge} c]\!] = 1$$

$$\Rightarrow [\![a \in \Omega \wedge b \in \Omega \wedge c \in \Omega]\!]$$

$$\leq [\![a \wedge (b \wedge c) = (a \wedge b) \wedge c]\!].$$

(e) $[\![\forall B \subseteq \Omega \forall a \in B((\bigvee B) \wedge a = a)]\!] = 1$, for

$B \subseteq \Omega,\ a \in B \Rightarrow [\![(\check{\bigvee} B) \wedge a = a]\!] = 1$

$\Rightarrow EB^* \wedge [\![B^* \subseteq \Omega \wedge a \in B^*]\!]$

$\leq [\![(\bigvee B^*) \wedge a = a]\!]$.

(f) $[\![\forall B \subseteq \Omega \forall b \in \Omega(\forall a \in B(a \wedge b = a) \to (\bigvee B) \wedge b = \bigvee B)]\!] = 1$,

(g) $[\![\forall B \subseteq \Omega \forall b \in \Omega((\bigvee B) \wedge b = \bigvee \{a \wedge b | a \in B\})]\!] = 1$

and

(h) $[\![\forall x \in \Omega(x \leq 1)]\!] = 1$ are proved similarly, q.e.d.

Let $\Omega, \wedge, \bigvee, 1, \Omega', \wedge', \bigvee' 1' \in V^{(H)}$ $\tilde{\Omega}, \tilde{\Omega}'$ be as usual, and

$[\![\langle\Omega, \wedge, \bigvee, 1\rangle$ is a cHa$]\!] = [\![\langle\Omega', \wedge', \bigvee', 1'\rangle$ is a cHa$]\!] = 1$.

Let the formula '$h : \Omega \to \Omega'$ is a cHa-morphism' be defined as the conjuction of

$h : \Omega \to \Omega'$,

$h(1) = 1'$,

$\forall a,b \in \Omega(h(a \wedge b) = h(a) \wedge h(b))$, and

$\forall A \subseteq \Omega(h(\bigvee A) = \bigvee\{h(a) | a \in A\})$.

Then we have

<u>Theorem 6</u>.

(1) If $h \in V^{(H)}$ satisfies

$[\![h : \Omega \to \Omega'$ is a cHa-morphism$]\!] = 1$,

then there exists a cHa-morphism $\tilde{h} : \tilde{\Omega} \to \tilde{\Omega}'$ such that

(i) $\tilde{h} \circ i = i'$ on H,

(ii) $E = E\tilde{h}$ on $\tilde{\Omega}$.

Conversely, if $\tilde{h} : \tilde{\Omega} \to \tilde{\Omega}'$ is a cHa-morphism satisfying (i), (ii), then there exists a unique $h \in V^{(H)}$ such that $[\![h : \Omega \to \Omega'$ is a cHa-morphism$]\!] = 1$

(2) For the above correspondence $h \leftrightarrow \tilde{h}$,

$[\![h : \Omega \to \Omega'$ is a cHa-isomorphism$]\!] = 1$

iff $\tilde{h} : \tilde{\Omega} \to \tilde{\Omega}'$ is a cHa-isomorphism.

Proof.

(1) Assume that

$[\![h : \Omega \to \Omega$ is a cHa-morphism$]\!] = 1$.

Then by Theorem 2 there exists $\tilde{h} : \tilde{\Omega} \to \tilde{\Omega}'$ such that for $a,b \in \tilde{\Omega}$,

$$[\![a \in \Omega]\!] \leq [\![h(a) = \tilde{h}(a)]\!],$$

$$[\![a = b]\!] \leq [\![\tilde{h}(a) = \tilde{h}(b)]\!],$$

$$Ea = Eh(a),$$

$$[\![a \in \Omega]\!] = [\![\tilde{h}(a) \in \Omega']\!].$$

Hence

(a) $\tilde{h} : \tilde{\Omega} \to \tilde{\Omega}$

(b) $\tilde{h}(1) = 1'$ since $[\![h(1) = 1']\!] = 1$.

(c) $a,b \in \tilde{\Omega} \Rightarrow \tilde{h}(a \mathbin{\tilde{\wedge}} b) = \tilde{h}(a) \mathbin{\tilde{\wedge}'} \tilde{h}(b)$, for:

$$Ea \wedge Eb \leq [\![h(a) = \tilde{h}(a)]\!] \wedge [\![h(b) = \tilde{h}(b)]\!]$$

$$\leq [\![h(a \mathbin{\tilde{\wedge}} b) = \tilde{h}(a) \mathbin{\tilde{\wedge}'} \tilde{h}(b)]\!]$$

$$\leq [\![\tilde{h}(a \mathbin{\tilde{\wedge}} b) = \tilde{h}(a) \mathbin{\tilde{\wedge}'} \tilde{h}(b)]\!], \text{ and}$$

$$E(h(a \mathbin{\tilde{\wedge}} b)) = Ea \wedge Eb = E(\tilde{h}(a) \mathbin{\tilde{\wedge}} \tilde{h}(b)).$$

Therefore, $[\![\tilde{h}(a \mathbin{\tilde{\wedge}} b) = \tilde{h}(a) \mathbin{\tilde{\wedge}} \tilde{h}(b)]\!] = 1$.

(d) $A \subseteq \tilde{\Omega} \Rightarrow \tilde{h}(\tilde{\bigvee} A) = \tilde{\bigvee}\{\tilde{h}(a) \mid a \in A\}$, for:

Set $B = \{\tilde{h}(a) \mid a \in A\} \subseteq \tilde{\Omega}'$.

Then

$$a \in A \Rightarrow E\tilde{h}(a) = Ea \leq [\![\tilde{h}(a) \in \{h(a) \mid a \in A^*\}]\!]$$

$$a \in A \Rightarrow E\tilde{h}(a) \leq [\![\tilde{h}(a) \in B^*]\!], \text{ and}$$

$$EB^* = \bigvee_{a \in A} E\tilde{h}(a) = E\{h(a) \mid a \in A^*\}.$$

Therefore, $[\![B^* = \{h(a) \mid a \in A^*\}]\!] = 1$, and so

$$[\![\tilde{h}(\tilde{\bigvee} A) = h(\tilde{\bigvee} A) = h(\bigvee A^*) = \bigvee_{a \in A^*} h(a) = \bigvee' B^*$$

$$= \tilde{\bigvee}' B^* = \tilde{\bigvee}' B]\!] = 1.$$

Therefore, h is cHa-morphism and satisfies (i) and (ii).

Conversely, assume a cHa-morphism $\tilde{h} : \tilde{\Omega} \to \tilde{\Omega}$ satisfies (i) and (ii). Then we have

$$a, b \in \tilde{\Omega} \Rightarrow [\![a = b]\!] \leq [\![\tilde{h}(a) = \tilde{h}(b)]\!], \text{ for:}$$

Set $[\![a = b]\!] = p$. Then we have the following results in order.

$$[\![i(p) \tilde{\wedge} a = i(p) \tilde{\wedge} b]\!] = [\![a \upharpoonright p = b \upharpoonright p]\!] = 1$$

$$\tilde{h} \circ i(p) \tilde{\wedge} \tilde{h}(a) = \tilde{h} \circ i(p) \tilde{\wedge} \tilde{h}(b)$$

$$i'(p) \tilde{\wedge} \tilde{h}(a) = i'(p) \tilde{\wedge} \tilde{h}(b)$$

$$[\![\tilde{h}(a) \upharpoonright p = \tilde{h}(b) \upharpoonright p]\!] = 1$$

$$[\![h(a) = h(b)]\!] \geq p.$$

Therefore, by Theorem 3, there exists a unique $h \in V^{(H)}$ such that

$$[\![h : \Omega \to \Omega']\!] = 1 \text{ and}$$

$$a \in \tilde{\Omega} \Rightarrow Ea \leq [\![\langle a, \tilde{h}(a)\rangle \in h]\!] = [\![h(a) = \tilde{h}(a)]\!].$$

Now we show that

$$[\![h : \Omega \to \Omega' \text{ is a cHa-morphism}]\!] = 1.$$

$[\![\forall x_1, x_2 \in \Omega(h(x_1 \wedge x_2) = h(x_1) \wedge' h(x_2))]\!] = 1$, for:

$$a,b \in \tilde{\Omega} \Rightarrow Ea \wedge Eb \le [\![h(a) = \tilde{h}(a) \wedge h(b) = \tilde{h}(b) \wedge$$
$$(\tilde{h}(a \tilde{\wedge} b) = \tilde{h}(a) \tilde{\wedge}' \tilde{h}(b)) \wedge$$
$$(h(a \wedge b) = \tilde{h}(a \tilde{\wedge} b))]\!]$$
$$\Rightarrow Ea \wedge Eb \le [\![h(a) \wedge' h(b) = h(a \wedge b)]\!].$$

$[\![\forall A \in P(\Omega)[h(\bigvee A) = \bigvee_{a \in A} h(a)]]\!] = 1$, for:

by Theorem 4(7),

$$A \subseteq \tilde{\Omega} \Rightarrow [\![h(\bigvee A^*) = \tilde{h}(\tilde{\bigvee}_{a\in A} a) = \tilde{\bigvee}'_{a \in A}\tilde{h}(a) = \bigvee'_{a \in A^*} h(a)]\!] = 1.$$

Let u be

$$\mathcal{D}(u) = \{A^* | A \subseteq \tilde{\Omega}\},$$
$$A \subseteq \tilde{\Omega} \Rightarrow u(A^*) = EA^*,$$
$$Eu = 1.$$

Then $[\![u = P(\Omega)]\!] = 1.$

Therefore, $[\![\forall A \in P(\Omega)(h(\bigvee A) = \bigvee_{a \in A} h(a))]\!] = 1.$

Since we also have $[\![h(1) = h(1) = 1']\!] = 1$, $[\![h : \Omega \to \Omega$ is a cHa-morphism$]\!] = 1$.

To set the uniqueness of $\tilde{h}$ corresponding to h, let $\tilde{h}'$ be the function $\tilde{h} : \tilde{\Omega} \to \tilde{\Omega}'$ such that $(\forall a \in \tilde{\Omega})(Ea \le [\![h(a) = \tilde{h}'(a)]\!])$. Then for $a \in \tilde{\Omega}$

$$Ea \le [\![h(a) = \tilde{h}(a) \wedge h(a) = \tilde{h}'(a)]\!] \le [\![\tilde{h}(a) = \tilde{h}'(a)]\!].$$

Therefore, $\tilde{h}' = \tilde{h}$. It follows that the correspondence $h \leftrightarrow \tilde{h}$ is one-to-one.

(2) : $[\![h : \Omega \to \Omega'$ is a cHa-isomorphism$]\!] = 1$

iff $\tilde{h} : \tilde{\Omega} \to \tilde{\Omega}$; is a cHa-isomorphism.

Assume $[\![h : \Omega \to \Omega'$ is a cHa-isomrophism$]\!] = 1$.

$a,b \in \Omega$, $\tilde{h}(a) = \tilde{h}(b)$

$\Rightarrow Ea \wedge Eb \leq [\![<a,\tilde{h}(a)> \in h \wedge <b,\tilde{h}(a)> \in h]\!]$

$\leq [\![a = b]\!]$

and $Ea = E\tilde{h}(a) = E\tilde{h}(b) = Eb$

$\Rightarrow a = b$.

Let $a' \in \Omega'$. Then

$$Ea' \leq [\![\exists a \in \Omega(h(a) = a')]\!] \leq \bigvee_{a\in\Omega}[\![\tilde{h}(a) = a']\!] \wedge Ea \leq Ea'$$

Set $A = \{a \restriction [\![\tilde{h}(a) = a']\!] \mid a \in \Omega\}$. Then $\bigvee A \in V^{(\Omega)}$ and

$$[\![\tilde{h}(\tilde{\bigvee} A) = a']\!] \geq \bigvee_{a\in\Omega}[\![\tilde{h}(a) = a']\!] \wedge Ea = Ea'$$

Therefore, $\tilde{h}(\tilde{\bigvee} A) = a'$.

Hence $\tilde{h} : \tilde{\Omega} \to \tilde{\Omega}'$ is a cHa-isomorphism.

Conversely, assume that

$\tilde{h} : \tilde{\Omega} \to \tilde{\Omega}'$ is a cHa-isomorphism.

$a,b \in \tilde{\Omega} \Rightarrow Ea \wedge Eb \wedge [\![h(a) = h(b)]\!]$

$\leq [\![h(a) = \tilde{h}(a) \wedge h(b) = \tilde{h}(b) \wedge h(a) = h(b)]\!]$

$\leq [\![\tilde{h}(a) = \tilde{h}(b)]\!]$

$\leq [\![a = b]\!]$.

The last step $[\![\tilde{h}(a) = \tilde{h}(b)]\!] \leq [\![a = b]\!]$ is proved as follows.

Set $[\![\tilde{h}(a) = \tilde{h}(b)]\!] = p$. Then we have the following results in turn.

$\tilde{h}(a) \restriction p = \tilde{h}(b) \restriction p$

$\tilde{h}(a \restriction p) = \tilde{h}(a \wedge i(p)) = \tilde{h}(a) \wedge i'(p) = \tilde{h}(b) \wedge i'(p) = \tilde{h}(b \restriction p)$

$a \restriction p = b \restriction p$

$p \leq [\![a = b]\!]$.

Therefore, $[\![h : \Omega \to \Omega' \text{ one-to-one}]\!] = 1$.

Proof of $[\![h : \Omega \to \Omega' \text{ onto}]\!] = 1$:

$$a' \in \tilde{\Omega}' \Rightarrow \exists a \in \tilde{\Omega} \text{ such that } \tilde{h}(a) = a'$$
$$\Rightarrow \exists a \in \tilde{\Omega} \text{ such that } Ea = Ea' = [\![\langle a,a'\rangle \in h]\!]$$
$$\Rightarrow [\![a' \in \Omega' \to \exists a \in \Omega(\langle a,a'\rangle \in h)]\!] = 1.$$

Theorem 7.

Let H and C be cHas.

If $i : H \to C$ is a one-to-one cHa-morpnism and there is $\tilde{0} \in C$ satisfying

(a) $\forall p,q \in H(\tilde{0} \wedge i(p) \leq i(q) \to p \leq q)$,

(b) $\forall u \in C \exists p \in H((\tilde{0} \to u) = i(p))$,

then there exist $\Omega, \wedge, \vee, 1 \in V^{(H)}$ such that

$$[\![\langle \Omega, \wedge, \vee, 1\rangle \text{ is a cHa}]\!] = 1 \text{ and } C \cong \tilde{\Omega}.$$

Proof.

Define $E_c : C \to H$ and $\restriction : C \times H \to C$ by

$u \in C \Rightarrow E_c(u) = p$ such that $(\tilde{0} \to u) = i(p)$.

$u \in C,\ p \in H \Rightarrow u \restriction p = u \wedge i(p)$

Then $\langle C, E_c, \restriction\rangle$ is shown to be a sheaf over H, as follows.

$$u \in C \Rightarrow u \upharpoonright E_c u = u \wedge iE_c u = u \wedge (\tilde{0} \to u) = u.$$

$$\begin{aligned} u \in C,\ p \in H \Rightarrow iE_c(u \upharpoonright p) &= (\tilde{0} \to u \wedge i(p)) \\ &= (\tilde{0} \to u) \wedge (\tilde{0} \to i(p)) \\ &= iE_c u \wedge i(p) \\ &= i(E_c u \wedge p), \end{aligned}$$

where $(\tilde{0} \to i(p)) = i(p)$ is proved as follows. Since $\tilde{0} \to i(p) \in C$, there is $q \in H$ such that

$$(\tilde{0} \to i(p)) = i(q).$$

$$\tilde{0} \wedge i(p) = \tilde{0} \wedge i(q)$$

By using (a), $p = q$. Therefore,

$$u \in C,\ p \in H \Rightarrow E_c(u \upharpoonright p) = E_c u \wedge p.$$

$$\begin{aligned} u \in C,\ p,\ q \in H \Rightarrow (a \upharpoonright p) \upharpoonright q &= a \wedge i(p) \wedge i(q) \\ &= a \wedge i(p \wedge q) \\ &= a \upharpoonright p \wedge q. \end{aligned}$$

Therefore, $\langle C, E_c, \upharpoonright \rangle$ is a presheaf over H.

Let $A \subseteq C$ and A be compatible. i.e.,

$$a, b \in A \Rightarrow a \upharpoonright E_c b = b \upharpoonright E_c a.$$

Then

$$\begin{aligned} a \in A \Rightarrow (\bigvee A) \upharpoonright E_c a &= (\bigvee A) \wedge iE_c a \\ &= \bigvee \{b \wedge iE_c a \mid b \in A\} \\ &= \bigvee \{a \upharpoonright E_c b \wedge E_c a \mid b \in A\} \\ &= a. \end{aligned}$$

Therefore, $\langle C, E_C, \lceil \rangle$ is a sheaf over H. Let Ω be the cannonical representation $C^*/\sim$ of the sheaf C defined in §7, and define $\tilde{\wedge}, \tilde{\bigvee}$ by

$$a,b \in C \Rightarrow \bar{a} \,\tilde{\wedge}\, \bar{b} = \overline{a \wedge b},$$

$$B \subseteq C \Rightarrow \tilde{\bigvee}\{\bar{b} \mid b \in B\} = \overline{\bigvee B}.$$

Then $\langle \tilde{\Omega}, \tilde{\wedge}, \tilde{\bigvee}, \tilde{1} \rangle$ is a cHa which is isomorphic to C and satisfies

$$E1 = 1$$

$E : \tilde{\Omega} \to H$ is a cHa-morphism

$$a,b,a',b' \in \Omega \Rightarrow [\![a = a' \wedge b = b']\!] \leq [\![a \,\tilde{\wedge}\, b = a' \,\tilde{\wedge}\, b']\!]$$

$$A,B \subseteq \tilde{\Omega} \Rightarrow [\![A^* = B^*]\!] \leq [\![\tilde{\bigvee} A = \tilde{\bigvee} B]\!],$$

where $\mathcal{D}(A^*) = A$

$$a \in A \Rightarrow A^*(a) = Ea$$

$$EA^* = \bigvee_{a \in A} Ea.$$

By using the Theorem 5, there is $\wedge, \bigvee \in V^{(H)}$ such that

$[\![\langle \Omega, \wedge, \bigvee, 1 \rangle$ is a CHa$]\!] = 1$ and

$$a,b \in \tilde{\Omega} \Rightarrow Ea \quad Eb \leq [\![a \wedge b = a \,\tilde{\wedge}\, b]\!]$$

$$A \subseteq \tilde{\Omega} \Rightarrow EA^* \leq [\![\bigvee A^* = \tilde{\bigvee} A]\!]$$

q.e.d.

Grothendieck topology.

G is called Grothendieck topology if the relation $\leq$ on G, the function $\wedge : G \times G \to G$ and the predicate "...is a covering" on $\mathcal{P}(G)$ are defined and satisfy G1 ~ G6.

(G1) $\{U_j\}_{j \in J}$ is a covering of $U \Rightarrow \forall j \in J(U_j \leq U)$

(G2) $U_1, U_2, U_3 \in G,\ U_1 \leq U_2,\ U_2 \leq U_3 \Rightarrow U_1 \leq U_3$.

(G3) If $U_1, U_2 \in G$, then $U_1 \wedge U_2 \leq U_1$ and $U_1 \wedge U_2 \leq U_2$

(G4) $\{U\}$ is a covering of U.

(G5) If $\{U_j\}_{j\in J}$ is a covering of U and for each $j \in J$ $\{U_{jk}\}_{k\in K_j}$ is a covering of U_j, then $\{U_{jk}\}_{k\in K_j, j\in J}$ is a covering of U.

(G6) If $\{U_j\}_{j\in J}$ is a covering of U and $V \leq U$, then $\{U_j \wedge V\}_{j\in J}$ is a covering of V.

When G is a Grothendick topology, $I \subseteq G$ satisfying the condition (I1) and (I2) is called <u>ideal</u>.

(I1) $U_1, U_2 \in G,\ U_1 \leq U_2 \in I \Rightarrow U_1 \in I$

(I2) If $\{U_j\}_{j\in J}$ is a covering of U, then

$$U \notin I \iff \exists j\in J\ (U_j \notin I)$$

<u>Theorem 8.</u>

Let G be a Grothendick topology and let

$$\mathcal{I} = \{I \mid I \text{ is an ideal of } G\}.$$

For $\{I_k\}_{k\in K} \subseteq \mathcal{I}$, $\bigwedge_k I_k$ and $\bigvee_k I_k$ are defined by

$$\bigwedge_k I_k = \bigcap_k I_k$$
$$\bigvee_k I_k = \{U \mid \exists \{U_j\}_{j\in I}(\{U_j\}_{j\in I} \text{ is a covering of } U \text{ and } \forall j \in J\ \exists k \in K(U_j \in I_k))\}.$$

Then $\langle \mathcal{I}, \wedge, \bigvee, G\rangle$ is a cHa

Proof.

(1) $I_1, I_2 \in \mathcal{I} \Rightarrow I_1 \wedge I_2 \in \mathcal{I}$. For,

$I_1, I_2 \in \mathcal{I},\ U_1 \leq U_2 \in I_1 \wedge I_2 \Rightarrow U_1 \in I_1 \wedge I_2$

If $\{U_j\}_{j \in J}$ is a covering of U, then

$$\begin{aligned} U \notin I_1 \wedge I_2 &\Leftrightarrow U \notin I_1 \text{ or } U \notin I_2 \\ &\Leftrightarrow \exists j(U_j \notin I_1) \text{ or } \exists j(U_j \notin I_2) \\ &\Leftrightarrow \exists j(U_j \notin I_1 \wedge I_2) \end{aligned}$$

(2) $\{I_k\}_k \subseteq \mathcal{I} \Rightarrow \bigvee_k I_k \in \mathcal{I}$. For:

(I1) $V \leq U \in \bigvee_k I_k$

$\Rightarrow \exists \{U_j\}_{j \in J}$ covering of U such that

$\forall j \exists k \{U_j \in I_k\}$

and $V \leq U$

$\Rightarrow \{V_j\}_{j \in J}$ covering of V such that

$\forall j \exists k \{U_j \in I_k\}$

where $V_j = U_j \wedge V$.

$\Rightarrow V \in \bigvee_k I_k$

(I2) Let $\{U_j\}_j$ be a covering of U. Then

$\forall j (U_j \in \bigvee_k I_k) \Rightarrow \forall j \exists \{U_{ji}\}_i$ covering of U_j such that

$\forall i \exists k (U_{ji} \in I_k)$.

$\Rightarrow \{U_{ji}\}_{ij}$ is a covering of U and

$\forall j \forall i \exists k (U_{ij} \in I_k)$

$\Rightarrow U \in \bigvee_k I_k$

$$\begin{aligned} U \in \bigvee_k I_k &\Rightarrow \forall_j\ U_j \leq U \in \bigvee_k I_k \\ &\Rightarrow \forall_j\ U_j \in \bigvee_k I_k \end{aligned}$$

$$U \in \bigvee_k I_k \Leftrightarrow \forall_j\ U_j \in \bigvee_k I_k$$

(3) It is obvious that I satisfies the first five conditions of cHa.

(4) $\forall k(I_i \leq I) \Rightarrow \bigvee_k I_k \leq I$. For,

$\forall k(I_k \leq I)$, $U \in \bigvee_k I_k$

$\Rightarrow \forall k(I_k \leq I)$ and $\exists \{U_j\}_j$ covering of U such that $\forall j \exists k(U_j \in I_k)$.

$\Rightarrow \exists \{U_j\}_j$ covering of U such that $\forall j(U_j \in I)$

$\Rightarrow U \in I$.

(5) $I \wedge (\bigvee_k I_k) = \bigvee_k (I \wedge I_k)$. For,

$U \in I \wedge (\bigvee_k I_k) \iff U \in I$ and $\exists \{U_j\}$ covering of U such that $\forall j \exists k(U_j \in I_k)$

$\iff \exists \{U_j\}$ covering of U such that $\forall j \exists k(U_j \in I \wedge I_k)$

$\iff U \in \bigvee_k (I \wedge I_k)$ q.e.d.

§11. $V^{(\Omega)}$ in $V^{(H)}$.

Let H be a cHa, $\langle \Omega, \wedge, \bigvee, 1\rangle \in V^{(\Omega)}$ and $[\![\langle \Omega, \wedge, \bigvee, 1$ is a cHa$]\!] = 1$. Now we construct $V^{(\Omega)}$ in $V^{(H)}$ and show that $V^{(\Omega)}$ in $V^{(\Omega)}$ is identified with $V^{(\tilde{\Omega})}$.

The class Ord of ordinals is defined in $V^{(H)}$ and definitions by $\in$-recursion on Ord within $V^{(H)}$ is justified (§4). Now let

$$h(S) = \{u = \langle \mathcal{D}_\Omega u, |u|, E_\Omega u\rangle \mid \mathcal{D}_\Omega u \subseteq S \wedge |u| : \mathcal{D}_\Omega u \to \Omega \wedge E_\Omega u \in \Omega \wedge \forall x \in \mathcal{D}_\Omega u \forall y \in \Omega(\langle x,y\rangle \in |u| \to y \leq E_\Omega x \wedge E_x u))\}.$$

Then we define $V^{(\Omega)}$ and $V^{(\Omega)}$ by

$$V_\alpha^{(\Omega)} = \bigcup \{h(V_\beta^{(\Omega)}) \mid \beta \in \alpha)\},$$

$$V^{(\Omega)} = \bigcup \{V_\alpha^{(\Omega)} \mid \alpha \in \mathrm{Ord}\}.$$

We denote the class $\{u \in V^{(H)} \mid [\![u \in V^{(\Omega)}]\!] = Eu\}$ by $(V^{(H)})^{(\Omega)}$.

Rank of $u \in V^{(H)}$ in $V^{(H)}$, $\mathrm{rank}_H u$, is defined by

$\mathrm{rank}_H u = \bigcup\{\mathrm{rank}_H y + 1 \mid y \in u\}$.

For $u, v \in (V^{(H)})^{(\Omega)}$, we define $[\![u = v]\!]_\Omega$ and $[\![u \in v]\!]_\Omega$ by induction on $\max(\mathrm{rank}_H u, \mathrm{rank}_H v)$ as follows.

$$[\![u = v]\!]_\Omega = \bigwedge\{u(x) \to [\![x \in v]\!]_\Omega \mid x \in \mathcal{D}_\Omega(u)\} \wedge \bigwedge\{v(y) \to [\![y \in u]\!]_\Omega \mid y \in \mathcal{D}_\Omega(v)\} \wedge (E_\Omega u \leftrightarrow E_\Omega v).$$

$$[\![u \in v]\!]_\Omega = \bigvee\{[\![u = y]\!]_\Omega \wedge v(y) \mid y \in \mathcal{D}_\Omega v\}.$$

For other set theoretical formula φ, $[\![\varphi]\!]_\Omega$ is defined as usual. Then

$$[\![<\Omega, V^{(\Omega)}, [\![\ ]\!]_\Omega> \text{ is a model of } ZF_I]\!] = 1$$

(cf. [2]).

Now we consider the relation between $V^{(\tilde{\Omega})}$ and $(V^{(H)})^{(\Omega)}$, where $\tilde{\Omega}$ is the sheaf represented by Ω. $[\![\varphi]\!]_{\tilde{\Omega}}$ and $[\![\varphi]\!]$ stand for the truth values of φ in $V^{(\tilde{\Omega})}$ and in $V^{(H)}$, respectively.

Then we have the following theorem.

Theorem 10.

There is an isomorphism Φ such that

(1) $\Phi : V^{(\tilde{\Omega})}/[\![=]\!]_{\tilde{\Omega}} = 1 \cong (V^{(H)})^{(\Omega)}/[\![[\![=]\!]_\Omega = 1]\!] = 1$.

(2) If $\varphi(x_1, \ldots, x_n)$ is a set theoretical formula and $u_1, \ldots, u_n \in V^{(\tilde{\Omega})}$, then

$$[\![[\![\varphi(u_1,\dots,u_n)]\!]_{\tilde{\Omega}} = [\![\varphi(\bar{u}_1,\dots,\bar{u}_n)]\!]_{\Omega}]\!] = 1$$

where:

$[\![=]\!]_{\Omega} = 1$, $[\![[\![=]\!]_{\Omega} = 1]\!] = 1$ are equivalence relations $\sim$ defined by

$a \sim b$ iff $[\![a = b]\!]_{\Omega} = 1$

$a \sim b$ iff $[\![[\![a = b]\!]_{\Omega} = 1]\!] = 1$

respectively, and

$\overline{u_i} = \Phi(u_i)$, $i = 1,\dots,n$.

We write $V^{(\tilde{\Omega})}$, $(V^{(H)})^{(\Omega)}$ instead of $V^{(\tilde{\Omega})}/[\![=]\!]_{\Omega} = 1$ and $(V^{(H)})^{(\Omega)}/[\![[\![=]\!]_{\Omega} = 1]\!] = 1'$ if there is no confusion.

Proof.

(1) For $u \in V^{(\tilde{\Omega})}$ which is HSR, we define $\bar{u} = \Phi(u)$, by the transfinite induction on rank u, as follows.

Assume that $\Phi(x) = \bar{x}$ is defined for $x \in V^{(\tilde{\Omega})}$ with rank $< \alpha$ and the following conditions (1)(2) are satisfied.

(a) $x \in V^{(\tilde{\Omega})}$, rank $x < \alpha$

$\Rightarrow \bar{x} \in (V^{(H)})^{(\Omega)}$, $E_{\Omega}\bar{x} = E_{\tilde{\Omega}}x$, $E\bar{x} = EE_{\tilde{\Omega}}x$.

(b) $x, x' \in V^{(\tilde{\Omega})}$, rank $x < \alpha$, rank $x' < \alpha$

$\Rightarrow [\![i[\![\bar{x} = \bar{x}']\!] \leq [\![x = x']\!]_{\tilde{\Omega}} = [\![\bar{x} = \bar{x}']\!]_{\Omega}]\!] = 1$.

Now for $u \in V^{(\tilde{\Omega})}$ with rank α we define $E_{\Omega}\bar{u}$, $\mathcal{D}_{\Omega}\bar{u}$ and $|\bar{u}| : \mathcal{D}_{\Omega}\bar{u} \to \Omega$ as follows.

$$E_{\Omega}\bar{u} = E_{\tilde{\Omega}}u.$$

$$\begin{cases} \mathcal{D}(\mathcal{D}_\Omega \bar{u}) = \{\bar{x} \mid x \in \mathcal{D}(u)\} \\ x \in \mathcal{D}(u) \Rightarrow (\mathcal{D}_\Omega \bar{u})(\bar{x}) = E\bar{x} \\ E(\mathcal{D}_\Omega \bar{u}) = EE_{\tilde{\Omega}}u. \end{cases}$$

Then $\mathcal{D}_\Omega \bar{u} \in V^{(H)}$ is definite. Define $|\bar{u}| : \mathcal{D}(\mathcal{D}_\Omega \bar{u}) \to \tilde{\Omega}$ by $|\bar{u}|(\bar{x}) = u(x)$. $|\bar{u}|$ satisfies the conditions of Theorem 4. That is,

$$x, x' \in \mathcal{D}(u) \Rightarrow i[\![\bar{x} = \bar{x}']\!] \le [\![x = x']\!]_{\tilde{\Omega}} \le (E_{\tilde{\Omega}}x \leftrightarrow E_{\tilde{\Omega}}x')$$

$$\Rightarrow i[\![\bar{x} = \bar{x}']\!] \le (u(x) \leftrightarrow u(x'))$$

$$\Rightarrow [\![\bar{x} = \bar{x}']\!] \le [\![u(x) = u(x')]\!]$$

$$x \in \mathcal{D}(u) \Rightarrow E\bar{x} = EE_{\tilde{\Omega}}x = Eu(x) = E|\bar{u}|(\bar{x}).$$

$$x \in \mathcal{D}(u) \Rightarrow [\![\bar{x} \in \mathcal{D}_\Omega \bar{u}]\!] = E\bar{x} = Eu(x) = [\![|\bar{u}|(\bar{x}) \in \tilde{\Omega}]\!].$$

Therefore, $[\![|\bar{u}| : \mathcal{D}_\Omega \bar{u} \to \Omega]\!] = EE_{\tilde{\Omega}}u$, and

$$x \in \mathcal{D}(u),\ a \in \tilde{\Omega} \Rightarrow E\bar{x} \wedge Ea \wedge [\![\langle x, a\rangle \in |u|]\!]$$

$$\le [\![a = |\bar{u}|(\bar{x}) = u(x) \le E_{\tilde{\Omega}}u \wedge E_{\tilde{\Omega}}x]\!]$$

Therefore, $\bar{u} = \langle \mathcal{D}_\Omega \bar{u}, |\bar{u}|, E_\Omega \bar{u}\rangle \in (V^{(H)})^{(\Omega)}$.

Now we prove the following conditions are satisfied for $u, u' \in V^{(\tilde{\Omega})}$ with rank $\le \alpha$.

(a) $E\bar{u} = EE_{\tilde{\Omega}}u$, $E_\Omega \bar{u} = E_{\tilde{\Omega}}u$.

(b) $i[\![\bar{u} = \bar{u}']\!] \le [\![u = u']\!]_{\tilde{\Omega}} = [\![\bar{u} = \bar{u}']\!]_\Omega$.

Proof.

(a) $E\bar{u} = E\langle \mathcal{D}_\Omega \bar{u}, |\bar{u}|, E_\Omega \bar{u}\rangle = EE_\Omega \bar{u} = EE_{\tilde{\Omega}}u$.

(b) By using equality axiom, for $x \in \mathcal{D}(u) \cup \mathcal{D}(u')$,

$$[\![\bar{u} = \bar{u}']\!] \leq [\![[\![\bar{x} \in \bar{u}]\!]_\Omega = [\![\bar{x} \in \bar{u}']\!]_\Omega]\!]$$

$$\therefore \quad i[\![\bar{u} = \bar{u}']\!] \wedge [\![\bar{x} \in \bar{u}]\!]_\Omega = i[\![\bar{u} = \bar{u}']\!] \wedge [\![\bar{x} \in \bar{u}']\!]_\Omega$$

$$\therefore \quad i[\![\bar{u} = \bar{u}']\!] \leq ([\![\bar{x} \in \bar{u}]\!]_\Omega \leftrightarrow [\![\bar{x} \in \bar{u}']\!]_\Omega).$$

Since $[\![\bar{u} = \bar{u}']\!] \leq [\![E_\Omega \bar{u} = E_\Omega \bar{u}']\!]$, we have

$$i[\![\bar{u} = \bar{u}']\!] \leq (E_\Omega \bar{u} \leftrightarrow E_\Omega \bar{u}').$$

Therefore, $i[\![\bar{u} = \bar{u}']\!] \leq [\![\bar{u} = \bar{u}']\!]_\Omega$.

For $x \in \mathcal{D}(u) \cup \mathcal{D}(u')$,

$$[\![x \in u]\!]_{\tilde{\Omega}} = \bigvee_{y \in \mathcal{D}(u)} [\![x = y]\!]_{\tilde{\Omega}} \wedge u(y)$$

$$= \bigvee_{\bar{y} \in \mathcal{D}_\Omega(\bar{u})} [\![x = y]\!]_\Omega \wedge |\bar{u}|\ \bar{y})$$

$$= [\![\bar{x} \in \bar{u}]\!]_\Omega.$$

Hence,

$$[\![u = u']\!]_{\tilde{\Omega}} = \bigwedge_{x \in \mathcal{D}(u) \cup \mathcal{D}(u')} [\![x \in u \leftrightarrow x \in u']\!]_{\tilde{\Omega}} \wedge (E_{\tilde{\Omega}} u \leftrightarrow E_{\tilde{\Omega}} u')$$

$$= \bigwedge_{\bar{x} \in \mathcal{D}_\Omega(\bar{u}) \cup \mathcal{D}_\Omega(\bar{u}')} [\![\bar{x} \in \bar{u} \leftrightarrow \bar{x} \in \bar{u}']\!]_\Omega \wedge (E_\Omega u \leftrightarrow E_\Omega u')$$

$$= [\![\bar{u} = \bar{u}']\!]_\Omega.$$

Therefore, $i[\![\bar{u} = \bar{u}']\!] \leq [\![u = u']\!]_{\tilde{\Omega}} = [\![\bar{u} = \bar{u}']\!]_\Omega$.

It follows that, for each $u \in V^{(\Omega)}$, $\bar{u} \in (V^{(H)})^{(\Omega)}$ is defined and the following conditions are satisfied.

$$u, u' \in V^{(\tilde{\Omega})} \Rightarrow [\![i[\![\bar{u} = \bar{u}']\!] \leq [\![u = u']\!]_{\tilde{\Omega}} = [\![\bar{u} = \bar{u}']\!]_\Omega]\!] = 1$$

and

$$[\![[\![u \in u']\!]_{\tilde{\Omega}} = [\![\bar{u} \in \bar{u}']\!]_\Omega]\!] = 1.$$

To see that the mapping

$$\Phi : V^{(\tilde{\Omega})} \Big/ _{[\![=]\!]_{\tilde{\Omega}}=1} \longrightarrow (V^{(H)})^{(\Omega)} \Big/ _{[\![[\![=]\!]_{\Omega}=1]\!]=1}$$

defined by $\Phi(u) = \bar{u}$ is isomorphism, it suffices to show that Φ is onto mapping.

Assume $v \in (V^{(H)})^{(\Omega)}$, i.e., $[\![v \in V^{(\Omega)}]\!] = Ev$. Then

$$[\![\exists \mathcal{D}_\Omega(v)\ \exists |v|\ \exists E_\Omega v [v = \langle \mathcal{D}_\Omega(v), |v|, E_\Omega v\rangle \wedge$$
$$|v| : \mathcal{D}_\Omega(v) \to \Omega \wedge E_\Omega v \in \Omega \wedge$$
$$(\forall x \in \mathcal{D}_\Omega(v)(v(x) \leq E_\Omega v \wedge E_\Omega x)]]\!] = Ev.$$

Therefore, $\exists \mathcal{D}_\Omega(v), |v|, E_\Omega v \in V^{(H)}$ such that

$$[\![v = \langle \mathcal{D}_\Omega(v), |v|, E_\Omega v\rangle \wedge |v| : \mathcal{D}_\Omega(v) \to \Omega \wedge E_\Omega v \in \Omega$$
$$(\forall x \in \mathcal{D}_\Omega(v))(v(x) \leq E_\Omega v \wedge E_\Omega x)]\!] = Ev$$

and

$$E\, \mathcal{D}_\Omega(v) = EE_\Omega v = Ev.$$

We may assume that $\mathcal{D}_\Omega(v), |v|, E_\Omega v$ are HSR.

Assume that for $y \in \mathcal{D}(\mathcal{D}_\Omega(v))$ there is a unique HSR $x \in V^{(\Omega)}$ such that $[\![\Phi(x) = \bar{x} = y]\!] = 1$. And define $u \in V^{(\tilde{\Omega})}$ by

$$\mathcal{D}(u) = \{x \mid [\![\bar{x} = y]\!] = 1 \wedge y \in \mathcal{D}\ \mathcal{D}_\Omega(v)\}$$
$$x \in \mathcal{D}(u) \Rightarrow u(x) = |v|(\bar{x})$$
$$E_{\tilde{\Omega}} u = E_\Omega v.$$

Let u' be the HSR of u. Then

$$E_\Omega \overline{u'} = E_{\tilde{\Omega}} u' = E_{\tilde{\Omega}} u = E_{\tilde{\Omega}} v.$$

$$\bar{x} \in \mathcal{D}(\mathcal{D}_{\Omega}\bar{u}') = \{x \mid [\![x \in u]\!] = Ex \text{ and } x \text{ is HSR}\}$$

$$\Rightarrow |\bar{u}'|(\bar{x}) = u'(x) = [\![x \in u]\!]_{\tilde{\Omega}}$$

$$= \tilde{\bigvee}_{\bar{y} \in \mathcal{D}(\mathcal{D}_{\Omega}(v))} [\![x = y]\!]_{\tilde{\Omega}} \wedge |v|(\bar{y})$$

$$= \tilde{\bigvee}_{\bar{y} \in \mathcal{D}(\mathcal{D}_{\Omega}(v))} [\![\bar{x} = \bar{y}]\!]_{\Omega} \wedge |v|(\bar{y})$$

$$\leq [\![\bar{x} \in v]\!]_{\Omega}$$

$$x \in \mathcal{D}(\mathcal{D}_{\Omega}v) \Rightarrow |v|(x) = u(x) \leq [\![x \in u]\!]_{\tilde{\Omega}} = [\![\bar{x} \in \bar{u}']\!]_{\tilde{\Omega}}$$

$$[\![[\![\overline{u'} = v]\!]_{\Omega} = 1]\!] = 1.$$

Therefore Φ is onto mapping and (1) was proved.

(2) In (1) we proved that

$$u_1, u_2 \in V^{(\tilde{\Omega})} \Rightarrow [\![u_1 = u_2]\!]_{\tilde{\Omega}} = [\![\bar{u}_1 = \bar{u}_2]\!]_{\Omega}$$

$$[\![u_1 \in u_2]\!]_{\tilde{\Omega}} = [\![\bar{u}_1 \in \bar{u}_2]\!]_{\Omega}.$$

That is, for atomic formula φ we have

$$[\![[\![\varphi(u_1,\ldots,u_n)]\!]_{\tilde{\Omega}} = [\![\varphi(\bar{u}_1,\ldots,\bar{u}_n)]\!]_{\Omega}]\!] = 1.$$

If $\varphi(x_1,\ldots,x_n) = \exists x \varphi'(x,x_1,\ldots,x_n)$, then

$$[\![\varphi(u_1,\ldots,u_n)]\!]_{\tilde{\Omega}} = \tilde{\bigvee}_{x \in V^{(\tilde{\Omega})}} (E_{\tilde{\Omega}}x \wedge [\![\varphi'(x,u_1,\ldots,u_n]\!]_{\tilde{\Omega}})$$

$$= \bigvee_{\bar{x} \in V^{(\Omega)}} (E_{\Omega}\bar{x} \wedge [\![\varphi'(\bar{x},\bar{u}_1,\ldots,\bar{u}_n)]\!]_{\Omega})$$

$$= [\![(\exists x \in V^{(\Omega)})(\varphi'(\bar{x},\bar{x}_1,\ldots,\bar{x}_n)]\!]_{\Omega}$$

Other cases are proved similarly.

Definition.

For each HSR u in $V^{(H)}$ we define $\check{u} \in (V^{(H)})^{(\Omega)}$ in $V^{(H)}$ by

$$\check{u} = \langle \mathcal{D}_\Omega \check{u}, |\check{u}|, E_\Omega \check{u} \rangle, \quad \text{where}$$

$$\mathcal{D}_\Omega \check{u} = \{\check{x} \mid x \in u\}^H,$$

$$|\check{u}| = \{\langle \check{x}, E_\Omega \check{x} \rangle \mid \check{x} \in \mathcal{D}_\Omega \check{u}\}^H,$$

$$E_\Omega \check{u} = iEu.$$

Lemma 11.1.

(1) $u \in V^{(H)} \Rightarrow \check{u} \in (V^{(H)})^{(\Omega)}$

(2) $u \in V^{(H)} \Rightarrow E\check{u} = iEu$

(3) $u, u' \in V^{(H)} \Rightarrow i[\![\check{u} = \check{u}']\!] \leq i[\![u = u']\!] = [\![\check{u} = \check{u}']\!]_\Omega$

$$i[\![u \in u']\!] = [\![\check{u} \in \check{u}']\!]_\Omega$$

Proof.

(1) Since $[\![x \in u \to \exists \alpha \in \mathrm{ord}(x \in V_\alpha^{(\Omega)})]\!] = 1$, we have

$$Eu \leq [\![\exists \alpha \in \mathrm{Ord}(\mathcal{D}_\Omega u \subseteq V_\alpha^{(\Omega)})]\!].$$

We also have

$$Eu \leq [\![\, |\check{u}| : \mathcal{D}_\Omega \check{u} \to \Omega]\!] \wedge$$

$$[\![(\forall x \in \mathcal{D}_\Omega \check{u})(\forall a \in \Omega)(\langle x, a\rangle \in |\check{u}| \to a \leq E_\Omega \check{x} \wedge E_\Omega \check{u})]\!] \wedge$$

$$[\![E_\Omega \check{u} \in \Omega]\!]$$

Therefore, $Eu = E\check{u} \leq [\![\check{u} \in (V^{(H)})^{(\Omega)}]\!]$, by using (2)

(2) $E\check{u} = EE_\Omega \check{u} = EiEu = Eu$

(3) For HSRs $u, u' \in V^{(H)}$, by using induction on rank u, and rank u', we have

$$[\![\check{u} = \check{u}']\!] = [\![\,|\check{u}| = |\check{u}'| \wedge \mathcal{D}_\Omega \check{u} = \mathcal{D}_\Omega \check{u}' \wedge E_\Omega \check{u} = E_\Omega \check{u}']\!],$$ where

$$[\![\,|\check{u}| = |\check{u}'|\,]\!] = \bigwedge_{x \in \mathcal{D}(u)} (Ex \to [\![\langle \check{x}, iEx\rangle \in |\check{u}'|\,]\!]) \wedge$$

$$\bigwedge_{y \in \mathcal{D}(u')} (Ey \to [\![\langle \check{y}, iE\check{y}\rangle \in |u|\,]\!]) \wedge (E|\check{u}| \leftrightarrow E|\check{u}'|),$$

$$[\![\langle \check{x}, iEx\rangle \in |\check{u}'|\,]\!] = \bigvee_{t \in \mathcal{D}(u')} [\![\langle \check{x}, iEx\rangle = \langle \check{t}, iEt\rangle]\!] \wedge Et$$

$$\leq \bigvee_{t \in \mathcal{D}(u')} [\![x = t]\!] \wedge Et$$

$$= [\![x \in u']\!],$$ and

$$E|\check{u}| = Eu$$

Therefore, $$[\![\,|\check{u}| = |\check{u}'|\,]\!] \leq \bigwedge_{x \in \mathcal{D}(u)} (Ex \to [\![x \in u']\!]) \wedge \bigwedge_{x' \in \mathcal{D}(u')} (Ex' \to [\![x' \in u']\!]) \wedge (Eu \leftrightarrow Eu')$$

$$\leq [\![u = u']\!].$$

$$[\![\,[\![\check{u} = \check{v}]\!]_\Omega = \bigwedge_{x \in \mathcal{D}_\Omega \check{u}} (\check{u}(x) \to [\![x \in \check{v}]\!]_\Omega) \wedge$$

$$\bigwedge_{y \in \mathcal{D}_\Omega \check{v}} (\check{v}(y) \to [\![y \in \check{u}]\!]_\Omega) \wedge (E_\Omega u \leftrightarrow E_\Omega v)$$

$$= \bigwedge_{x \in \mathcal{D}(u)} (iEx \to [\![\check{x} \in \check{v}]\!]_\Omega) \wedge$$

$$\bigwedge_{y \in \mathcal{D}(v)} (iEy \to [\![\check{y} \in \check{u}]\!]_\Omega) \wedge (iEu \leftrightarrow iEv)$$

$$= i[\![u = v]\!]\,]\!] = 1,$$ since

$\bigwedge\{ip \mid p \in A\} = i(\bigwedge A)$ and $(ip \to iq) = i(p \to q)$.

$$[\![\,[\![\check{u} \in \check{v}]\!]_\Omega = \bigvee_{y \in \mathcal{D}_\Omega \check{v}} [\![\check{u} = y]\!]_\Omega \wedge \check{v}(y)$$

$$= \tilde{\bigvee}_{y \in \mathcal{D}(v)} [\![\check{u} = \check{y}]\!]_\Omega \wedge iEy$$

$$= \tilde{\bigvee}_{y \in \mathcal{D}(v)} i[\![u = y]\!] \wedge iEy$$

$$= i[\![u \in v]\!]\,]\!] = 1$$

Now we extend the cHa-morphism $i : H \to \tilde{\Omega}$ to $i_* : V^{(H)} \to V^{(\tilde{\Omega})}$. Then i_* is the mapping which corresponds to the imbedding $\vee : V^{(H)} \to (V^{(H)})^{(\Omega)}$, as shown in the following theorem.

Theorem 11.

(1) $u \in V^{(H)} \Rightarrow [\![[\![\overline{i_* u} = u]\!]_\Omega = 1]\!] = 1$, where $\overline{i_* u} = \Phi(i_* u)$ is defined in Theorem 10.

(2) If $\varphi(x_1,\ldots,x_n)$ is a set theoretical formula and $u_1,\ldots,u_n \in V^{(H)}$, then

$$[\![[\![\varphi(i_* u_1,\ldots,i_* u_n)]\!]_{\tilde{\Omega}} = [\![\varphi(\check{u}_1,\ldots,\check{u}_n)]\!]_\Omega]\!] = 1$$

Proof.

(1) First we prove $i[\![u = v]\!] = [\![i_* u = i_* v]\!]_{\tilde{\Omega}}$ and

$$i[\![u \in v]\!] = [\![i_* u \in i_* v]\!]_{\tilde{\Omega}}.$$

$$i[\![u = v]\!] = i \bigwedge_{x \in \mathcal{D}(u)} (u(x) \to [\![x \in v]\!]) \wedge i \bigwedge_{y \in \mathcal{D}(v)} (v(y) \quad [\![y \in u]\!]) \wedge$$

$$i(Eu \leftrightarrow Ev),$$ where

$$i \bigwedge_{x \in \mathcal{D}(u)} (u(x) \to [\![x \in v]\!]) = \bigwedge_{x \in \mathcal{D}(u)} (iu(x) \to i[\![x \in v]\!])$$

$$= \bigwedge_{x \in \mathcal{D}(u)} ((i_* u)(i_* x) \to [\![i_* x \in i_* v]\!]),$$

$$i(Eu \leftrightarrow Ev) = (iEu \leftrightarrow iEv) = (E_{\tilde{\Omega}} i_* u \leftrightarrow E_{\tilde{\Omega}} i_* v).$$

Therefore, $i[\![u = v]\!] = [\![i_* u = i_* v]\!]$.

$$i[\![u \in v]\!] = i \bigvee_{y \in \mathcal{D}(v)} [\![u = y]\!] \wedge v(y)$$

$$= \bigvee_{y \in \mathcal{D}(v)} i[\![u = y]\!] \wedge iv(y)$$

$$= [\![i_* u \in i_* v]\!].$$

Let $u \in V^{(H)}$ be HSR and prove (1) by induction on the rank of u.

$$E_\Omega \overline{i_* u} = E_{\tilde\Omega} i_* u = iEu = E_\Omega \check{u}.$$

$$\check{x} \in \mathcal{D}(\mathcal{D}_\Omega u) \Rightarrow$$

$$[\![(\check{u}(\check{x}) \to [\![\check{x} \in \overline{i_* u}]\!]_\Omega) = (E_\Omega \check{x} \to [\![\overline{i_* x} \in \overline{i_* u}]\!]_\Omega)$$

$$= (iE\ x \to [\![i_* x \in i_* u]\!]_{\tilde\Omega})$$

$$= i\ Ex \to i[\![x \in u]\!]$$

$$= i(Ex \to [\![x \in u]\!]) \ = 1]\!] = 1$$

Therefore, $[\![\bigwedge_{x \in \mathcal{D}_\Omega u} (\check{u}(x) \to [\![x \in \overline{i_* u}]\!]_\Omega) = 1]\!] = 1$ ... (i)

$$\overline{i_* x} \in \mathcal{D}(\mathcal{D}_\Omega \overline{i_* u}) \Rightarrow$$

$$[\![(\overline{i_* u}(\overline{i_* x}) \to [\![\overline{i_* x} \in \check{u}]\!]_\Omega)$$

$$= (i_* u(i_* x) \to [\![\check{x} \in \check{u}]\!]_\Omega)$$

$$= (iEx \to i[\![x \in u]\!]) = 1]\!] = 1$$

Therefore, $[\![\bigwedge_{x \in \mathcal{D}_\Omega(\overline{i_* u})} (\overline{i_* u}(x) \to [\![x \in \check{u}]\!]_\Omega) = 1]\!] = 1$ ... (ii)

By (i) and (ii), $[\![[\![\overline{i_* u} = \check{u}]\!]_\Omega = 1]\!] = 1$ for HSR $u \in V^{(H)}$.

Let $u \in V^{(H)}$ and u' be the HSR of u. Then

$$[\![i[\![u = u']\!] \leq [\![i_* u = i_* u']\!]_{\tilde\Omega} = [\![\overline{i_* u} = \overline{i_* u'}]\!]_\Omega = 1]\!] = 1$$

Therefore, $[\![[\![\overline{i_* u} = \overline{i_* u'} = \check{u}' = \check{u}]\!]_\Omega = 1]\!] = 1.$

(2) By using Theorem 10 and (1)

$$[\![[\![\varphi(i_* u_1, \ldots, i_* u_n)]\!]_{\tilde\Omega}$$

$$= [\![\varphi(\overline{i_* u_1}, \ldots, \overline{i_* u_n})]\!]_\Omega$$

$$= [\![\varphi(\check{u}_1, \ldots, \check{u}_n)]\!]_\Omega]\!] = 1,$$ q.e.d.

Chapter II. The universe $V^{(\Gamma)}$

§1. $V^{(\Gamma)}$.

Let H be a cHa. We define a H-valued model $V^{(\Gamma)}$ of ZF_I which is a generalization of $V^{(H)}$. The analogous B-valued model $V^{(\Gamma)}$ for a complete Boolean algebra B is well known as a model for independence proof [4], [5].

Let Aut(H) be the set of automorphisms on H. An automorphism $g \in \mathrm{Aut}(H)$ is extended to the mapping $\bar{g} : V^{(H)} \to V^{(H)}$ as mentioned in Chap. §9. For $g \in \mathrm{Aut}(H)$ and $u \in V^{(H)}$, $\bar{g}u$ was defined by

$$\mathcal{D}(\bar{g}u) = \{\bar{g}x \mid x \in \mathcal{D}(u)\},$$

$$x \in \mathcal{D}(u) \Rightarrow (\bar{g}u)(\bar{g}x) = \bigvee_{\substack{x' \in \mathcal{D}(u) \\ \bar{g}x = \bar{g}x'}} g(u(x')) = g(u(x)).$$

$$E(\bar{g}u) = gEu.$$

Lemma 1.1.

(1) If $g,h \in \mathrm{Aut}(H)$, then $\bar{g}\bar{h} = \overline{gh}$, $\bar{g}^{-1} = \overline{g^{-1}}$.

(2) If $g \in \mathrm{Aut}(H)$ and $u,v \in V^{(H)}$, then

$$g[\![u = v]\!] = [\![\bar{g}u = \bar{g}v]\!],$$

$$g[\![u \in v]\!] = [\![\bar{g}u \in \bar{g}v]\!]$$

(3) If $g \in \mathrm{Aut}(H)$, $u \in V^{(H)}$ and $p \in H$, then

$$[\![\bar{g}(u \restriction p) = \bar{g}u \restriction gp]\!] = 1.$$

Proof.

(1) and (2) are obvious.

(3) By using Lemma I.9.2(*) and (2)

(*) Lemma I.9.2 means Lemma 9.2 in Chapter I.

$$g[\![u = v]\!] \leq [\![\overline{g}u = \overline{g}v]\!] \leq g[\![\overline{g^{-1}}\,\overline{g}u = \overline{g^{-1}}\,\overline{g}v]\!]$$
$$\leq g[\![u = v]\!] .$$

Similarly, $g[\![u \in v]\!] = [\![\overline{g}u \in \overline{g}v]\!]$.

(4) By Lemma I.9.2.

Let $G \subseteq \mathrm{Aut}(H)$ be a subgroup of H. A set Γ of subgroups of G is called <u>filter</u> if

(1) $H_1, H_2 \in \Gamma \Rightarrow H_1 \cup H_2 \in \Gamma$

(2) $H_1 \in \Gamma$, H_2 is a subgroup of G, $H_1 \subseteq H_2 \Rightarrow H_2 \in \Gamma$.

A filter Γ is said to be <u>normal</u> if $G_\Gamma \in \Gamma$, where

$$G_\Gamma = \{g \in G \mid \forall H \subseteq G(H \in \Gamma \text{ iff } gHg^{-1} \in \Gamma)\}$$

Let Γ be a normal filter of subgroups of G. Then we define $V^{(\Gamma)}$ by

$$V^{(\Gamma)} = \{u \in V^{(H)} \mid \mathcal{D}(u) \subseteq V^{(\Gamma)} \wedge G_u \in \Gamma\}, \text{ where}$$
$$G_u = \{g \in G \mid [\![\overline{g}u = u]\!] = 1\}.$$

A class in $V^{(H)}$ is said to be <u>a class in $V^{(\Gamma)}$</u> if

$$\mathcal{D}(X) \subseteq V^{(\Gamma)} \text{ and } \{g \in G \mid [\![\overline{g}X = X]\!] = 1\} \in \Gamma,$$

where $\overline{g}X$ is defined by

$$(\overline{g}X) = \{\overline{g}x \mid x \in X\},$$
$$x \in X \Rightarrow (\overline{g}X)(\overline{g}x) = gX(x)$$
$$E\overline{g}X = gEX.$$

It is obvious that if a class X in $V^{(\Gamma)}$ is a set then $X \in V^{(\Gamma)}$.

Truth value of a formula φ in $V^{(\Gamma)}$, denoted by $[\![\varphi]\!]_\Gamma$, is defined by:

$$[\![u = v]\!]_\Gamma = [\![u = v]\!],$$
$$[\![u \in v]\!]_\Gamma = [\![u \in v]\!],$$
$$[\![\forall x\varphi(x)]\!]_\Gamma = \bigwedge_{x \in V^{(\Gamma)}} [\![\varphi(x)]\!]_\Gamma,$$
$$[\![\exists x\varphi(x)]\!]_\Gamma = \bigvee_{x \in V^{(\Gamma)}} [\![\varphi(x)]\!]_\Gamma.$$

For other logical operations, $[\![\ ]\!]_\Gamma$ is defined as usual.

Lemma 1.2.

(1) $g \in G_\Gamma \Rightarrow \bar{g}\, V^{(\Gamma)} = V^{(\Gamma)}$.

(2) If $\varphi(a_1,\ldots,a_n)$ is a formula without constants and $u_1,\ldots,u_n \in V^{(\Gamma)}$, then

$$g \in G_\Gamma \Rightarrow g[\![\varphi(u_1,\ldots,u_n)]\!]_\Gamma = [\![\varphi(\bar{g}u_1,\ldots,\bar{g}u_n)]\!]_\Gamma.$$

(3) If $\varphi(a)$ is a formula and $u \in V^{(\Gamma)}$, then

$$[\![\forall x \in u\varphi(x)]\!]_\Gamma = \bigwedge_{x \in \mathcal{D}(u)} (u(x) \to [\![\varphi(x)]\!]_\Gamma),$$
$$[\![\exists x \in u\varphi(x)]\!]_\Gamma = \bigvee_{x \in \mathcal{D}(u)} (u(x) \wedge [\![\varphi(x)]\!]_\Gamma).$$

Proof.

(1) is proved by induction on the rank of u.

For $u \in V^{(\Gamma)}$,

$$\mathcal{D}(\bar{g}u) = \{\bar{g}x \mid x \in \mathcal{D}(u)\} \subseteq V^{(\Gamma)} \text{ by the induction hypothesis,}$$

and $G_{\bar{g}u} \in \Gamma$, for,

$$g^{-1}G_{\overline{gu}}g = g^{-1}\{h \in G \mid [\![\overline{hgu} = \overline{gu}]\!] = 1\}g$$

$$= g^{-1}\{h \in G \mid [\![\overline{g^{-1}hgu} = u]\!] = 1\}g$$

$$= \{h' \in G \mid [\![h'u = u]\!] = 1\} \in \Gamma.$$

Since $g \in G_\Gamma$, we have $G_{\overline{gu}} \in \Gamma$, and hence $\overline{gu} \in V^{(\Gamma)}$. Therefore, $\bar{g}\, V^{(\Gamma)} = V^{(\Gamma)}$.

(2) is proved by induction on the number of logical symbols in $\varphi(a_1,\ldots,a_n)$. For an atomic formula φ it is proved in Lemma 5.1.

$$g[\![u = v]\!]_\Gamma = [\![\bar{g}u = \bar{g}v]\!]_\Gamma,$$
$$g[\![u \in v]\!]_\Gamma = [\![\bar{g}u \in \bar{g}v]\!]_\Gamma.$$

$$g[\![\forall x\varphi(x,u_1,\ldots,u_n)]\!]_\Gamma = g \bigwedge_{x\in V^{(\Gamma)}} [\![\varphi(x,u_1,\ldots,u_n)]\!]_\Gamma$$

$$= \bigwedge_{x\in V^{(\Gamma)}} [\![\varphi(\bar{g}x,\bar{g}u_1,\ldots,\bar{g}u_n)]\!]_\Gamma,$$

by the induction hypothesis,

$$= \bigwedge_{x\in V^{(\Gamma)}} [\![\varphi(x,\bar{g}u_1,\ldots,\bar{g}u_n)]\!]_\Gamma$$

$$= [\![\forall x\varphi(x,\bar{g}u_1,\ldots,\bar{g}u_n)]\!]_\Gamma.$$

Similarly

$$g[\![\exists x\varphi(x,u_1,\ldots,u_1)]\!]_\Gamma = [\![\exists x\varphi(x,\bar{g}u_1,\ldots,\bar{g}u_n)]\!]_\Gamma.$$

For other cases, it is obvious.

(3) $$[\![(\quad x \in u)(\varphi(x))]\!]_\Gamma = \bigwedge_{x' \in V^{(\Gamma)}} [\![x' \in u \to \varphi(x')]\!]_\Gamma$$

$$= \bigwedge_{x' \in V^{(\Gamma)}} \Big(\bigvee_{x \in \mathcal{D}(u)} [\![x' = x]\!]_\Gamma \wedge u(x) \to \varphi(x') \Big)$$

$$= \bigwedge_{x' \in V^{(\Gamma)}} \bigwedge_{x \in \mathcal{D}(u)} ([\![x' = x]\!]_\Gamma \wedge u(x) \to \varphi(x'))$$

$$= \bigwedge_{x' \in \mathcal{D}(u)} (u(x) \to \varphi(x)).$$

$$[\![(\exists x \in u)\varphi(x)]\!]_\Gamma = \bigvee_{x' \in V^{(\Gamma)}} [\![x' \in u \wedge \varphi(x')]\!]_\Gamma$$

$$= \bigvee_{x' \in V^{(\Gamma)}} \bigvee_{x \in \mathcal{D}(u)} [\![x' = x]\!]_\Gamma \wedge u(x) \wedge [\![\varphi(x')]\!]_\Gamma$$

$$= \bigvee_{x \in \mathcal{D}(u)} u(x) \wedge [\![\varphi(x')]\!]_\Gamma, \qquad \text{q.e.d.}$$

Let $H_\Gamma = \{p \in H \mid \{g \in G \mid gp = p\} \in \Gamma\}$.

Lemma 1.3.

(1) $u \in V^{(\Gamma)}$, $p \in H_\Gamma \Rightarrow u \upharpoonright p \in V^{(\Gamma)}$

(2) If $\varphi(a_1, \ldots, a_n)$ is a formula without constants and $u_1, \ldots, u_n \in V^{(\Gamma)}$, then

$$[\![\varphi(u_1, \ldots, u_n)]\!]_\Gamma \in H_\Gamma.$$

(3) If φ is a sentence on $V^{(\Gamma)}$ in which every quantifier is restricted, then $[\![\varphi]\!]_\Gamma = [\![\varphi]\!]$.

Proof.

(1) For $u \in V^{(\Gamma)}$ and $p \in H_\Gamma$, we have

$$\mathcal{D}(u \upharpoonright p) = \{x \upharpoonright p \mid x \in \mathcal{D}(u)\} \subseteq V^{(\Gamma)}$$

by the induction hypothesis, and

$$G_u \cap \{g \in G | gp = p\} \subseteq G_{u\restriction p}, \quad \text{for}$$

$$g \in G_u, \ gp = p$$

$$\Rightarrow [\![\bar{g}(u \restriction p) = \bar{g}u \restriction gp = u \restriction p]\!]_\Gamma = 1$$

$$\Rightarrow \ g \in G_{u\restriction p}.$$

Therefore, $G_{u\restriction p} \in \Gamma$,

and hence $u \restriction p \in V^{(\Gamma)}$.

(2) If $g \in G_{u_1} \cap \cdots \cap G_{u_n} \cap G_\Gamma$, then

$$g[\![\varphi(u_1,\ldots,u_n)]\!]_\Gamma = [\![\varphi(\bar{g}u_1,\ldots,\bar{g}u_n)]\!]_\Gamma$$
$$= [\![\varphi(u_1,\ldots,u_n)]\!]_\Gamma.$$

It follows that

$$G_{u_1} \cap \cdots \cap G_{u_n} \cap G_\Gamma \subseteq \{g \in G | g[\![\varphi(u_1,\ldots,u_n)]\!]_\Gamma = [\![\varphi(u_1,\ldots,u_n)]\!]_\Gamma\}.$$

Since $G_{u_1} \cap \cdots \cap G_{u_n} \cap G_\Gamma \in \Gamma$, we have

$$\{g \in G | g[\![\varphi(u_1,\ldots,u_n)]\!]_\Gamma = [\![\varphi(u_1,\ldots,u_n)]\!]_\Gamma\} \in \Gamma.$$

Therefore,

$$[\![\varphi(u_1,\ldots,u_n)]\!]_\Gamma \in H_\Gamma.$$

(3) By Lemma 1.2 (3).

§2. Γ-sheaf representation.

Definition.

$v \in V^{(\Gamma)}$ is a Γ-sheaf representation (Γ-SR) of $u \in V^{(\Gamma)}$ if

(1) v is definite

(2) $Eu = 0 \Rightarrow v = \theta$, where $\mathcal{D}(\theta) = \phi$, $E\theta = 0$.

(3) $Eu \neq 0$, $x \in V^{(\Gamma)}$, $[\![x \in u]\!] = Ex$

$\Rightarrow \exists x' \in \mathcal{D}(u)([\![x = x']\!] = 1)$.

Lemma 2.1.

For $u \in V^{(\Gamma)}$ there exists a Γ-SR $v \in V^{(\Gamma)}$ of u such that

$x \in \mathcal{D}(v) \Rightarrow \text{rank } x < \text{rank } u$

Proof.

We prove this lemma in the same way as Lemma I.6.2.

(i) Define u_1 by

$$\mathcal{D}(u_1) = \{x \restriction [\![x \in u]\!] \mid x \in \mathcal{D}(u)\}$$

$$x \in \mathcal{D}(u_1) \Rightarrow u_1(x \restriction [\![x \in u]\!]) = [\![x \in u]\!]$$

$$Eu_1 = E_u.$$

Then $[\![u_1 = u]\!] = 1$, $u_1 \in V^{(\Gamma)}$ and $\text{rank } u_1 \leq \text{rank } u$. $u_1 \in V^{(\Gamma)}$ follows because $[\![x \in u]\!] \in H_\Gamma$ by Lemma 1.3.

(ii) Define u_2 by

$$\mathcal{D}(u_2) = \{x \restriction p \mid x \in \mathcal{D}(u_1),\ p \in H_\Gamma\}$$

$$x \in \mathcal{D}(u_1),\ p \in H_\Gamma \Rightarrow u_2(x \restriction p) = Ex \wedge p$$

$$Eu_2 = Eu.$$

Then $[\![u_2 = u_1]\!] = 1$, $u_2 \in V^{(\Gamma)}$ and rank $u_2 \leq$ rank u_1.

(iii) Define v by

$$\mathcal{D}(v) = \{\bigvee A \mid A \subseteq \mathcal{D}(u_2), \text{ A is compatible, and } \bigvee A \in V^{(\Gamma)}\}$$

$$A \in \mathcal{D}(v) \Rightarrow v(\bigvee A) = E(\bigvee A) = \bigvee_{a\in A} Ea.$$

$$Ev = Eu_2.$$

then $[\![v = u_2]\!] = 1$, for,

$$\bigvee A \in \mathcal{D}(v) \Rightarrow v(\bigvee A) = \bigvee_{a\in A} Ea = \bigvee_{a\in A} [\![a = \bigvee A \wedge a \in u_2]\!]$$
$$\leq [\![\bigvee A \in u_2]\!].$$

$$x \in \mathcal{D}(u_2) \Rightarrow u_2(x) = Ex = E(\bigvee\{x\})$$
$$\leq [\![x \in v]\!].$$

Hence $[\![v = u]\!] = 1$, $v \in V^{(\Gamma)}$ and $x \in \mathcal{D}(v) \Rightarrow$ rank $x <$ rank u. $v \in V^{(\Gamma)}$ follows from the facts: $\mathcal{D}(v) \subseteq V^{(\Gamma)}$, $[\![v = u]\!] = 1$ and $u \in V^{(\Gamma)}$.

Further, v is Γ-SR since

$$[\![x \in v]\!] = Ex,\ x \in V^{(\Gamma)} \Rightarrow \bigvee_{y\in \mathcal{D}(u)} [\![x = y]\!] \wedge u(y) = Ex$$
$$\Rightarrow [\![x = \bigvee_{y\in \mathcal{D}(u)} y \restriction [\![x = y]\!]]\!] = 1,$$
$$\text{and } \bigvee_{y\in \mathcal{D}(u)} y \restriction [\![x = y]\!] \in \mathcal{D}(v).$$

Γ-hereditary sheaf representation (Γ-HSR).

Definition.

$v \in V^{(\Gamma)}$ is Γ-HSR of $u \in V^{(\Gamma)}$ if

(1) v is Γ-SR,

(2) $x \in \mathcal{D}(v) \Rightarrow x$ is Γ-HSR.

Lemma 2.2.

For $u \in V^{(\Gamma)}$ there exists a unique Γ-HSR $v \in V^{(\Gamma)}$ of u.

Proof.

By Lemma 2.1, there exists a Γ-SR of u, say u_1. We define v by induction on the rank of u. Let

$$\mathcal{D}(v) = \{x' \mid x' \text{ is the } \Gamma\text{-HSR of } x \text{ and } x \in \mathcal{D}(u_1)\}$$

$$x' \in \mathcal{D}(v) \Rightarrow v(x') = Ex'$$

$$Ev = Eu.$$

Then obviously $v \in V^{(\Gamma)}$, $[\![u = v]\!] = 1$ and v is Γ-HSR.

Proof of uniqueness:

Let $u, u' \in V^{(\Gamma)}$, Γ-HSR and $[\![u = u']\!] = 1$. If $Eu = 0$, then $Eu' = 0$ and so $u = \theta = v$. If $Eu \neq 0$, then

$$x \in \mathcal{D}(u) \Rightarrow x \in V^{(\Gamma)}, \quad x \text{ is } \Gamma\text{-HSR and } [\![x \in u]\!] = Eu,$$

$$\Rightarrow \exists x' \in \mathcal{D}(u')([\![x = x']\!] = 1),$$

$$\Rightarrow x = x' \in \mathcal{D}(u').$$

Hence $\mathcal{D}(u) = \mathcal{D}(u')$, which results $u = u'$.

Lemma 2.3.

If $u \in V^{(\Gamma)}$ is Γ-HSR and $g \in G_\Gamma$, then $\overline{g}u$ is Γ-HSR.

Proof.

Assume that $u \in V^{(\Gamma)}$ is Γ-HSR and $g \in G_\Gamma$.

(i) $\overline{g}u \in V^{(\Gamma)}$ by Lemma 1.2(1).

(ii) $E(\overline{g}u) = 0 \Rightarrow gEu = 0 \Rightarrow Eu = 0 \Rightarrow u = \theta$

$$\Rightarrow \overline{g}u = \theta.$$

(iii) $\overline{g}u$ is definite, for:

$$x \in \mathcal{D}(u) \Rightarrow (\overline{g}u)(\overline{g}x) = gu(x) = gEx = E\overline{g}x.$$

(iv) $E(\overline{g}u) \neq 0,\ x \in V^{(\Gamma)},\ [\![x \in \overline{g}u]\!] = Ex$

$$\Rightarrow Eu \neq 0,\ \overline{g^{-1}}x \in V^{(\Gamma)}, [\![\overline{g^{-1}}x \in u]\!] = E\overline{g^{-1}}x,$$

$$\Rightarrow \exists x' \in \mathcal{D}(u)([\![\overline{g^{-1}}x = x']\!] = 1),$$

$$\Rightarrow \exists y' \in \mathcal{D}(\overline{g}u)([\![x = y']\!] = 1).$$

(v) $x \in \mathcal{D}(u) \Rightarrow \overline{g}x$ is Γ-HSR by induction hypothesis, q.e.d.

§3. Properties of $V^{(\Gamma)}$.

Theorem 1.

$\langle H, V^{(\Gamma)}, [\![\ ]\!]_\Gamma\rangle$ is a model of ZF_I.

Proof.

Equality.

$$[\![u_1 = v_1 \quad \cdots \quad u_n = v_n \quad \varphi(u_1,\dots,u_n)]\!]_\Gamma \leq [\![\varphi(v_1,\dots,v_n)]\!]_\Gamma$$

is proved by induction on the number of logical symbols in φ, since $[\![u = v]\!]_\Gamma = [\![u = v]\!]$, $[\![u \in v]\!]_\Gamma = [\![u \in v]\!]$ and $V^{(H)}$ is a model of ZF_I. Other axioms of equality are obvious.

Extension.

$$[\![\forall x(x \in u \leftrightarrow x \in v) \wedge (Eu \leftrightarrow Ev)]\!]_\Gamma$$
$$= [\![\forall x(x \in u \leftrightarrow x \in v) \wedge (Eu \leftrightarrow Ev)]\!]$$
$$= [\![u = v]\!] = [\![u = v]\!]_\Gamma.$$

Pair.

$$\mathcal{D}\{u,v\}^H = \{u,v\} \subseteq V^{(\Gamma)}$$
$$g \in G_\Gamma \cap G_u \cap G_v \Rightarrow [\![\bar{g}\{u,v\}^H = \{u,v\}^H]\!] = 1.$$
$$\{u,v\}^H \in V^{(\Gamma)} \quad \text{and}$$
$$[\![\forall x(x \in \{u,v\}^H \leftrightarrow x = u \vee x = v)]\!]_\Gamma$$
$$= [\![\forall x(x \in \{u,v\}^H \leftrightarrow x = u \vee x = v)]\!] = 1.$$

We write $\{u,v\}^H = \{u,v\}^\Gamma$.

Foundation.

Let $[\![\forall x(\forall y \in x\, \varphi(y)) \to \varphi(x))]\!] = p$ and $u \in V^{(\Gamma)}$. By the induction on the rank of $u \in V^{(\Gamma)}$, we have

$$p \leq \bigwedge_{t \in \mathcal{D}(u)} [\![\varphi(t)]\!] = [\![\forall t \in u\, \varphi(t)]\!]$$

Hence $p \leq [\![\varphi(u)]\!]$.

Comprehension.

Let $\varphi(a)$ be a formula on $V^{(\Gamma)}$ with one free variable a, and $u \in V^{(\Gamma)}$. Define v by

$$Ev = Eu$$

$$\mathcal{D}(v) = \mathcal{D}(u)$$

$$x \in \mathcal{D}(v) \Rightarrow v(x) = [\![x \in u \wedge \varphi(x)]\!]_\Gamma .$$

Then

$$v \in V^{(\Gamma)} \text{ and}$$

$$[\![\forall x(x \in v \leftrightarrow x \in u \wedge \varphi(x)]\!]_\Gamma = 1.$$

We write

$$v = \{x \in u | \varphi(x)\}^\Gamma.$$

Replacement.

Let $u \in V^{(\Gamma)}$ and $[\![\forall y \in u \exists x \, \varphi(y,x)]\!]_\Gamma = p$. Then there is an ordinal α such that

$$y \in \mathcal{D}(u) \Rightarrow p \wedge u(y) \leq \bigvee\{[\![\varphi(y,x)]\!]_\Gamma | \text{rank } x < \alpha\}.$$

$$\Rightarrow p \wedge u(y) \leq \bigvee\{[\![\varphi(y,x)]\!]_\Gamma | x \in V^{(\Gamma)} \cap V_\alpha^{(H)}\}$$

Define v by

$$\mathcal{D}(v) = V^{(\Gamma)} \cap V_\alpha^{(H)}$$

$$x \in \mathcal{D}(v) \Rightarrow v(x) = Ex$$

$$Ev = 1.$$

Then we have

$$p \leq [\![\forall y \in u \exists x \in v \, \varphi(y,x)]\!]_\Gamma.$$

Power set.

If $u \in V^{(\Gamma)} \cap V_{\alpha}^{(H)}$, then define v by

$$\mathcal{D}(v) = V^{(\Gamma)} \cap V_{\alpha+1}^{(H)}$$

$$t \in \mathcal{D}(v) \Rightarrow v(t) = [\![t \subseteq u]\!] \wedge Et \wedge Eu$$

$$Ev = Eu.$$

v satisfies

$v \in V^{(\Gamma)}$ and

$$Eu \leq [\![\forall x(x \in v \leftrightarrow x \subseteq u]\!]_{\Gamma}.$$

We denote Γ-HSR of v by $\mathcal{P}^{\Gamma}(u)$.

Union.

For $u \in V^{(\Gamma)}$ define v by

$$\mathcal{D}(v) = \cup_{x \in \mathcal{D}(u)} \mathcal{D}(x)$$

$$t \in \mathcal{D}(v) = v(t) = \bigvee_{x \in \mathcal{D}(u)} [\![t \in x \wedge x \in u]\!]_{\Gamma}$$

$$Ev = Eu.$$

Then $[\![\forall x(x \in v \leftrightarrow \exists y \in u(x \in y))]\!]_{\Gamma} = 1$.

We denote v by $\cup^{\Gamma} u$.

Infinity.

$$[\![\exists x(x \in \check{\omega}) \wedge \forall x \in \omega \exists y \in \omega(x \in y)]\!]_{\Gamma} = 1.$$

Hence $[\![\exists u(\exists x(x \in u) \wedge \forall x \in u \exists y \in u(x \in y)]\!]_{\Gamma} = 1$.

Lemma 3.1. (Uniqueness principle).

If φ (a) is a formula on $V^{(\Gamma)}$, then there exists a $u \in V^{(\Gamma)}$ such that

(1) $[\![\varphi(u)]\!]_\Gamma \geq [\![\exists! x\, \varphi(x)]\!]_\Gamma = Eu$,

(2) $v \in V^{(\Gamma)}$, $[\![\varphi(v)]\!]_\Gamma \wedge Ev \geq [\![\exists! x\, \varphi(x)]\!]_\Gamma$

$$\Rightarrow [\![u = v]\!]_\Gamma \geq Eu, \text{ and}$$

(3) $v \in V^{(\Gamma)}$, $[\![\varphi(v)]\!]_\Gamma \geq [\![\exists! x\, \varphi(x)]\!]_\Gamma = Ev$

$$\Rightarrow [\![u = v]\!]_\Gamma = 1.$$

Proof.

Let

$$A = \{v \in V^{(\Gamma)} \mid v \text{ is a } \Gamma\text{-HSR and } Ev \leq [\![\exists! x\, \varphi(x) \wedge \varphi(v)]\!]_\Gamma\}.$$

Then (i) A is a set, for

$\{Eu \mid u \in A\}$ is a set and

$u, v \in A$, $Eu = Ev$.

$$\Rightarrow Eu = Ev \leq [\![\exists! x\, \varphi(x) \wedge \varphi(u) \wedge \varphi(v)]\!]_\Gamma$$
$$\leq [\![u = v]\!]_\Gamma$$
$$\Rightarrow u = v.$$

(ii) A is compatible, for

$$u, v \in A \Rightarrow Eu \wedge Ev \leq [\![\exists! x\, \varphi(x) \wedge \varphi(u) \wedge \varphi(v)]\!]_\Gamma$$
$$\leq [\![u = v]\!]_\Gamma.$$

Set $u = \bigvee A$, where

$$\mathcal{D}(\bigvee A) = \cup_{v \in A} \mathcal{D}(v)$$
$$x \in \mathcal{D}(\bigvee A) \Rightarrow (\bigvee A)(x) = \bigvee_{v \in A} [\![x \in v]\!]$$
$$E(\bigvee A) = \bigvee_{v \in A} Ev$$

Then we will show that $u \in V^{(\Gamma)}$.

Since we have

$$\mathcal{D}(u) = \cup_{v \in A} \mathcal{D}(v) \subseteq V^{(\Gamma)},$$

it suffices to show that

$$G_u = \{g \in G \mid [\![\bar{g}u = u]\!]_\Gamma = 1\} \in \Gamma.$$

Let $u_1, \ldots, u_n$ be constants in $\varphi(a)$. Then

$$g \in G_\Gamma \cap G_{u_1} \cap \cdots \cap G_{u_n}, \ v \in A$$

$$\Rightarrow gEv \leq g[\![\exists! x\, \varphi(x) \wedge \varphi(v)]\!]_\Gamma$$

$$\Rightarrow E\bar{g}v \leq [\![\exists! x\, \varphi(x) \wedge \varphi(\bar{g}v)]\!]_\Gamma$$

$$\Rightarrow \bar{g}v \in A, \text{ since } \bar{g}v \text{ is } \Gamma\text{-HSR}.$$

It follows that

$$g \in G_\Gamma \cap G_{u_1} \cap \cdots \cap G_{u_n} \Rightarrow \bar{g}A = A$$

$$\Rightarrow [\![\bar{g}(\bigvee A) = \bigvee A]\!]_\Gamma = 1$$

Therefore $G_u \in \Gamma$.

Now we prove that u defined above satisfies (1)(2) and (3).

(1) $$Eu = \bigvee_{v \in A} Ev \leq \bigvee_{v \in A} [\![\exists! x\, \varphi(x) \wedge \varphi(v)]\!]_\Gamma \leq [\![\exists! x\, \varphi(x)]\!]_\Gamma.$$

Since $y \in V^{(\Gamma)}$, $g \in G_\Gamma \cap G_y \cap G_{u_1} \cap \cdots \cap G_{u_n}$

$\Rightarrow g[\![\exists! x\, \varphi(x) \wedge \varphi(y)]\!]_\Gamma = [\![\exists! x\, \varphi(x) \wedge \varphi(y)]\!]_\Gamma$, we have

$y \in V^{(\Gamma)} \Rightarrow [\![\exists! x\, \varphi(x) \wedge \varphi(y)]\!]_\Gamma \in H_\Gamma$.

Therefore, if $y \in V^{(\Gamma)}$ then $y \upharpoonright [\![\exists! x\, \varphi(x) \wedge \varphi(y)]\!] \in V^{(\Gamma)}$ and the Γ-HSR of $y \upharpoonright [\![\exists! x\, \varphi(x) \wedge \varphi(y)]\!]$ is in A. It follows that

$$[\![\exists! x\, \varphi(x)]\!]_\Gamma = [\![\exists! x\, \varphi(x)]\!]_\Gamma \wedge [\![\exists y\, \varphi(y)]\!]_\Gamma$$
$$= \bigvee_{y \in V^{(\Gamma)}} [\![\exists! x\, \varphi(x) \wedge \varphi(y)]\!]_\Gamma \wedge Ey = Eu.$$

(2) and (3) are obvious from the definition of $\exists! x\, \varphi(x)$, q.e.d.

§4. Another definition of $V^{(\Gamma)}$.

Let $V^{(\Gamma)'}$ be the subclass of $V^{(H)}$ defined by

$$V^{(\Gamma)'} = \{u \in V^{(H)} \mid \mathcal{D}(u) \subseteq V^{(\Gamma)'} \wedge u \text{ is definite } \wedge \{g \in G \mid \bar{g}\, \mathcal{D}(u) = \mathcal{D}(u)\} \in \Gamma\}$$

If $u \in V^{(\Gamma)'}$ then $G_u = \{g \in G \mid [\![\bar{g}u = u]\!] = 1\} \in \Gamma$, since $\bar{g}\, \mathcal{D}(u) = \mathcal{D}(u) \Rightarrow [\![\bar{g}u = u]\!] = 1$. Therefore, $u \in V^{(\Gamma)}$. Conversely, if $u \in V^{(\Gamma)}$ and u is Γ-HSR, then $u \in V^{(\Gamma)'}$. Therefore,

$$V^{(\Gamma)'} / [\![=]\!] = 1 = V^{(\Gamma)} / [\![=]\!] = 1.$$

§5. Functions in $V^{(\Gamma)}$.

Let X,Y be classes in $V^{(\Gamma)}$,

$$\tilde{X} = \{x \in V^{(\Gamma)} \mid [\![x \in X]\!] = Ex\},$$
$$\tilde{Y} = \{y \in V^{(\Gamma)} \mid [\![y \in Y]\!] = Ey\}.$$

Theorem 2.

If h is a class in $V^{(\Gamma)}$ and $[\![h : X \to Y]\!]_\Gamma = p$, then there is a function $\tilde{h} : \tilde{X} \to \tilde{Y}$ such that

(1) $x \in \tilde{X} \Rightarrow E\tilde{h}(x) = Ex \wedge p = [\![\langle x, \tilde{h}(x)\rangle \in h]\!]_\Gamma$.

(2) $x, x' \in \tilde{X} \Rightarrow [\![x = x']\!]_\Gamma \leq [\![\tilde{h}(x) = \tilde{h}(x')]\!]_\Gamma$.

(3) $\{g \in G \mid \bar{g}\tilde{h} = \tilde{h}\bar{g}\} \in \Gamma$, where $\bar{g}\tilde{h} = \tilde{h}\bar{g}$ means that $\forall x \in \tilde{X}(\bar{g}\tilde{h}(x) = \tilde{h}\bar{g}(x))$.

Proof.

Let h be a class in $V^{(\Gamma)}$ and

$$[\![h : X \to Y]\!]_\Gamma = [\![\forall x \in X \ \exists! y \in Y(\langle x,y\rangle \in h)]\!]_\Gamma = p.$$

By the uniqueness principle, if $x \in \tilde{X}$ then there exists a unique Γ-HSR $u \in V^{(\Gamma)}$ such that

$$Eu = Ex \wedge p = [\![\exists! y \in Y(\langle x,y\rangle \in h)]\!]_\Gamma = [\![\langle x,u\rangle \in h]\!]_\Gamma.$$

Set $\tilde{h}(x) = u$. Then $\tilde{h}$ satisfies (1) $\sim$ (3), as follows.

(1) is obvious.

(2) $x, x' \in \tilde{X} \Rightarrow$

$$[\![x = x']\!] \wedge (E\tilde{h}(x) \vee E\tilde{h}(x')) = [\![x = x']\!] \wedge p \wedge Ex$$
$$= [\![x = x' \wedge \exists! y \in Y(\langle x,y\rangle \in h) \wedge \langle x,\tilde{h}(x)\rangle \in h \wedge \langle x',\tilde{h}(x')\rangle \in h]\!]$$
$$\leq [\![\tilde{h}(x) = \tilde{h}(x')]\!]$$

$$x, x' \in \tilde{X} \Rightarrow [\![x = x']\!] \leq [\![\tilde{h}(x) = \tilde{h}(x')]\!].$$

(3) Let $g \in G_\Gamma \cap G_X \cap G_Y \cap G_h \cap \{g' \in G | g'p = p\}$. Then

$$x \in \tilde{X} \Rightarrow g[\![x \in X]\!] = [\![\bar{g}x \in X]\!] = E\bar{g}x$$

$$\Rightarrow \bar{g}x \in \tilde{X}$$

$$\Rightarrow [\![<\bar{g}x, \tilde{h}\bar{g}x> \in h]\!] = E\bar{g}x \wedge p \text{, and}$$

$$[\![<\bar{g}x, \overline{g\tilde{h}x}> \in h]\!] = g[\![<x, \tilde{h}(x)> \in h]\!]$$

$$= E\bar{g}x \wedge p$$

$$\Rightarrow \quad E\bar{g}x \wedge p \leq [\![\tilde{h}\bar{g}x = \overline{g\tilde{h}x}]\!]$$

$$\Rightarrow \quad \tilde{h}\bar{g}x = \overline{g\tilde{h}x}.$$

Therefore,

$$G_\Gamma \cap G_X \cap G_Y \cap G_h \subseteq \{g' \in G | g'p = p\} \subseteq \{g \in G | \tilde{h}\bar{g} = \overline{g\tilde{h}}\},$$

and so,

$$\{g \in G | \tilde{h}\bar{g} = \overline{g\tilde{h}}\} \in \Gamma.$$

Theorem 3.

Assume that classes X, Y in $V^{(\Gamma)}$, $p \in \Omega$ and h satisfy the following conditions

(a) X is definite and $G' = \{g \in G | \bar{g}\, \mathcal{D}(X) = \mathcal{D}(X)\} \in \Gamma$,

(b) $\tilde{h} : \mathcal{D}(X) \to \tilde{Y}$, where $\tilde{Y} = \{y \in V^{(\Gamma)} | [\![y \in Y]\!] = Ey\}$

(c) $p \in H_\Gamma$

(d) $x, x' \in \mathcal{D}(X) \Rightarrow [\![x = x']\!] \leq [\![\tilde{h}(x) = \tilde{h}(x')]\!]$.

(e) $x \in \mathcal{D}(X) \Rightarrow E\tilde{h}(x) = Ex \wedge p$.

(f) $\{g \in G_X | \overline{g\tilde{h}} = \tilde{h}\bar{g} \text{ on } \mathcal{D}(X)\} \in \Gamma$.

Then there exists a class h in $V^{(\Gamma)}$ such that

(1) $Eh = EX \wedge EY \wedge p.$

(2) $[\![h : X \to Y]\!]_\Gamma \geq p.$

(3) $x \in \mathcal{D}(X) \Rightarrow [\![<x,\tilde{h}(x)> \in h]\!] = Ex \wedge p.$

(4) If h' is a class in $V^{(\Gamma)}$ satisfying (1) $\sim$ (3), then

$$[\![h = h']\!] = 1.$$

Proof.

Define h by

$$\mathcal{D}(h) = \{<x,\tilde{h}(x)> \mid x \in \mathcal{D}(X)\}.$$

$$x \in \mathcal{D}(X) \Rightarrow h<x,h(x)> = Ex \wedge p.$$

$$Eh = EX \wedge EY \wedge p.$$

Then

$$\mathcal{D}(h) \subseteq V^{(\Gamma)} \text{ by the definition.}$$

Let

$$G_0 = G_\Gamma \cap G' \cap G_X \cap G_Y \cap \{g \in G_X \mid \overline{g}\tilde{h} = \tilde{h}\overline{g} \text{ on } \mathcal{D}(X) \text{ and } gp = p\}.$$

$$\begin{aligned} g \in G_0 \Rightarrow h<x,\tilde{h}(x)> &= Ex \wedge p = g(E\overline{g^{-1}x} \wedge p) \\ &= gh<\overline{g^{-1}}x,\tilde{h}\overline{g^{-1}}x> \\ &\leq g[\![<\overline{g^{-1}}x,\tilde{h}\overline{g^{-1}}x> \in h]\!] \\ &= [\![<x,\tilde{h}(x)> \in \overline{g}h]\!] \end{aligned}$$

$$\begin{aligned} \overline{g}h<\overline{g}x,\overline{g}h(x)> &= gh<x,h(x)> \\ &= E\overline{g}x \wedge p \\ &\leq [\![<\overline{g}x,\tilde{h}\overline{g}x> \in h]\!] \end{aligned}$$

$$E\overline{g}h = gEh = g(EX \wedge EY \wedge p) = Eh$$

Therefore, $g \in G_0$ implies $[\![\bar{g}h = h]\!] = 1$, and so $h \in V^{(\Gamma)}$.

Now we prove that h satisfies (2) $\sim$ (4).

(2) Let $x \in \mathcal{D}(X)$.

$$\begin{aligned} Ex \wedge p &\leq [\![<x,h(x)> \in h]\!] \wedge [\![\tilde{h}(x) \in Y]\!] \\ &\leq [\![\exists y \in Y(<x,y> \in h)]\!]. \end{aligned}$$

$y, y' \in \mathcal{D}(Y) \Rightarrow$

$$\begin{aligned} &Ex \wedge p \wedge [\![<x,y> \in h \wedge <x,y'> \in h]\!] \\ &\leq \bigvee_{t \in \mathcal{D}(X)} [\![<x,y> = <t,\tilde{h}(t)>]\!] \wedge Et \wedge p \wedge \\ &\qquad \bigvee_{t' \in \mathcal{D}(X)} [\![<x,y'> = <t',h(t')>]\!] \wedge Et' \wedge p \\ &\leq [\![y = y']\!] \end{aligned}$$

Hence $\quad Ex \wedge p \leq [\![\exists ! y \in Y \ (<x,y> \in h)]\!]$.

Therefore, we have $p \leq [\![h : X \to Y]\!]_\Gamma$.

(3) is obvious.

(4) Assume that h, h' are Γ-classes satisfying (1) $\sim$ (3). Then

$$\begin{aligned} t \in \mathcal{D}(h) \Rightarrow h(t) &\leq [\![\exists x \in X \exists y \in Y(t = <x,y> \in h)]\!]_\Gamma \wedge p \\ &\leq \bigvee_{\substack{x \in \mathcal{D}(X) \\ y \in \mathcal{D}(Y)}} [\![t = <x,y> \in h]\!]_\Gamma \wedge Ex \wedge p \\ &\leq \bigvee_{\substack{x \in \mathcal{D}(X) \\ y \in \mathcal{D}(Y)}} [\![t = <x,y> \in h \wedge <x,\tilde{h}(x)> \in h \wedge \\ &\qquad\qquad <x,\tilde{h}(x)> \in h']\!] \\ &\leq \bigvee_{x \in \mathcal{D}(X)} [\![t = <x,\tilde{h}(x)> \wedge <x,\tilde{h}(x)> \in h']\!] \\ &\leq [\![t \in h']\!]_\Gamma. \end{aligned}$$

Similarly, $t' \in \mathcal{D}(h') \Rightarrow h'(t') \leq [\![t' \in h]\!]_\Gamma$. Therefore, $[\![h = h']\!]_\Gamma = 1$.

Remark.

Let X be a locally compact manifold and $\Omega = \mathcal{O}(X)$. Then each automorphism h^* on Ω corresponds to a homeomorphism $h : X \to X$. Let G be the set $\mathrm{Aut}(\Omega)$ of automorphisms on Ω and Γ be the set of subgroups F of G such that $F \in \Gamma$ iff

$$\exists Y \subseteq X\ [Y \text{ is compact} \wedge F = \{f \in G \mid \forall y \in Y(f(y) = y)\}]$$

Then obviously Γ is a normal filter of subgroups of G. Hence $V^{(\Gamma)}$ is a model of ZF_I.

Now consider the set $\mathbb{R}^{(\Gamma)}$ of real numbers in $V^{(\Gamma)}$, that is,

$$\mathbb{R}^{(\Gamma)} = \{u \mid u \text{ is a real number}\}^{\Gamma}.$$

For $u \in V^{(\Gamma)}$,

$$[\![u \text{ is a real number}]\!]_{\Gamma} = [\![u \text{ is a real number}]\!].$$

Therefore,

$[\![u = \langle L,U\rangle \in \mathbb{R}^{(\Gamma)}]\!] = Eu$ implies

(1) $\{g^* \in G \mid [\![\overline{g^*} u = u]\!] = 1\} \in \Gamma$ and

(2) f_u is a continuous function on Eu,

where f_u is defined by

$$x \in Eu \Rightarrow f_u(x) = \sup\{r \in \mathbb{Q} \mid x \in [\![r \in L]\!]\}.$$

It follows that for a real number u of $V^{(\Gamma)}$ there exists a compact subset Y of X such that f_u is a continuous function on Eu and constant on every connected subset of $Eu - Y$.

§6. cHa in $V^{(\Gamma)}$.

Theorem 4.

Let $\Omega, \wedge, \bigvee, 1 \in V^{(\Gamma)}$ satisfy

$[\![<\Omega, \wedge, \bigvee, 1> \text{ is a cHa}]\!]_\Gamma = 1.$

Let

$\tilde{\Omega} = \{a \in V^{(\Gamma)} \mid [\![a \in \Omega]\!] = Ea\},$

$\tilde{\wedge} : \widetilde{\Omega \times \Omega} \to \tilde{\Omega}$ and

$\tilde{\bigvee} : \widetilde{P_\Gamma(\Omega)} \to \tilde{\Omega}$

be induced functions from $\wedge$ and $\bigvee$ in Theorem 2. Then

(1)(i) $a,b \in \tilde{\Omega} \Rightarrow [\![a \wedge b = a \tilde{\wedge} b]\!] = Ea \wedge Eb = E(a \tilde{\wedge} b)$, where $a \wedge b = a \tilde{\wedge} b$ is abbreviation of $<<a,b>, a \tilde{\wedge} b> \in \wedge$.

(ii) $a,b,a',b' \in \tilde{\Omega} \Rightarrow [\![<a,b> = <a',b'>]\!] \leq [\![a \tilde{\wedge} b = a' \tilde{\wedge} b']\!]$.

(iii) $a,b \in \tilde{\Omega},\ g \in G_\Gamma \cap G_{<\Omega,\wedge,\bigvee,1>} \Rightarrow \bar{g}(a \wedge b) = \bar{g}a \wedge \bar{g}b$.

(2)(i) $A \subseteq \tilde{\Omega},\ A^* \in V^{(\Gamma)} \Rightarrow [\![<A^*, \tilde{\bigvee} A> \in \bigvee]\!] = EA^* = E(\tilde{\bigvee} A)$, where A^* is defined by $\mathcal{D}(A^*) = A$, $EA^* = \bigvee_{a \in A} Ea$, and $a \in A \Rightarrow A^*(a) = Ea$.

(ii) $A, B \in P(\tilde{\Omega}),\ A^*, B^* \in V^{(\Gamma)} \Rightarrow [\![A^* = B^*]\!] \leq [\![\tilde{\bigvee} A = \tilde{\bigvee} B]\!]$

(iii) $A \subseteq \tilde{\Omega},\ A^* \in V^{(\Gamma)},\ g \in G_\Gamma \cap G_{<\Omega,\wedge,\bigvee,1>}$

$$\Rightarrow \bar{g}(\tilde{\bigvee} A) = \tilde{\bigvee} \bar{g}A.$$

(3) $a,b,c \in \tilde{\Omega} \Rightarrow [\![a \wedge b = c]\!] = [\![a \tilde{\wedge} b = c]\!] \wedge Ea \wedge Eb$.

$A \subseteq \tilde{\Omega},\ A^* \in V^{(\Gamma)} \Rightarrow [\![\bigvee A^* = c]\!] = [\![\tilde{\bigvee} A = c]\!] \wedge E(\tilde{\bigvee} A)$.

(4) If $A \subseteq \tilde{\Omega}$ and $A^* \in V^{(\Gamma)}$, then

$a \in A \Rightarrow a \leq \tilde{\bigvee} A$ and

$b \in \tilde{\Omega}\ \ (\forall a \in A)(a \leq b) \Rightarrow \tilde{\bigvee} A \leq b.$

(5) If $u, f \in V^{(\Gamma)}$, u is definite, and $[\![f : u \to \Omega]\!]_\Gamma = p$, then

$$[\![\bigvee_{x \in u} f(x) = \widetilde{\bigvee}_{x \in \mathcal{D}(u)} \tilde{f}(u)]\!] \geq p.$$

(6) If $A \subseteq \tilde{\Omega}$, $A^* \in V^{(\Gamma)}$ and $b \in \tilde{\Omega}$, then

$$\{a \tilde{\wedge} b \mid a \in A\}^* \in V^{(\Gamma)} \text{ and}$$
$$(\tilde{\bigvee} A) \tilde{\wedge} b = \widetilde{\bigvee} \{a \tilde{\wedge} b \mid a \in A\}.$$

(7) Let $i : H_\Gamma \to \tilde{\Omega}$ be the function defined by $i(p) = 1 \restriction p$. Then i is an one-to-one $\wedge, \bigvee$ -morphism. That is,

$$p, q \in H_\Gamma \Rightarrow i(p \wedge q) = i(p) \tilde{\wedge} i(q).$$
$$A \subseteq H_\Gamma, \ \bigvee A \in H_\Gamma \Rightarrow i(\bigvee A) = \tilde{\bigvee}_{p \in A} i(p).$$

(8) $a \in \Omega$, $p \in H_\Gamma \Rightarrow [\![a \restriction p = a \tilde{\wedge} i(p)]\!] = 1.$

Proof.

(1) If $a, b \in \tilde{\Omega}$, then $[\![\langle a,b \rangle \in \Omega \times \Omega]\!] = E\langle a,b \rangle$, therefore $\langle a,b \rangle \in \widetilde{\Omega \times \Omega}$. Hence, by using Theorem 2, we have (i)(ii)(iii).

(2) If $A \subseteq \tilde{\Omega}$ and $A^* \in V^{(\Gamma)}$ then $A^* \in \widetilde{P_\Gamma(\Omega)}$, for

$$[\![A^* \in P_\Gamma(\Omega)]\!] = [\![A^* \subseteq \Omega]\!] \wedge EA^* = EA^*.$$

Hence, by Theorem 2, we have (i)(ii)(iii).

(3) If $a, b, c \in \tilde{\Omega}$, then

$$\begin{aligned}
[\![a \wedge b = c]\!] &= [\![\langle\langle a,b \rangle, c\rangle \in \wedge]\!] \\
&= [\![\langle\langle a,b \rangle, c\rangle \in \wedge . \wedge . \langle\langle a,b \rangle, a \tilde{\wedge} b\rangle \in \wedge]\!] \\
&= [\![c = a \tilde{\wedge} b . \wedge . \langle\langle a,b \rangle, a \tilde{\wedge} b\rangle \in \wedge]\!] \\
&= [\![c = a \tilde{\wedge} b]\!] \wedge Ea \wedge Eb.
\end{aligned}$$

If $A \subseteq \Omega$ and $A^* \in V^{(\Gamma)}$, then

$$[\![\langle A^*, c\rangle \in \bigvee]\!] = [\![\langle A^*, c\rangle \in \bigvee_{,\wedge,} \langle A^*, \tilde{\bigvee} A\rangle \in \bigvee]\!]$$

$$= [\![c = \tilde{\bigvee} A]\!] \wedge E(\tilde{\bigvee} A).$$

(4) If $A \subseteq \tilde{\Omega}$ and $A^* \in V^{(\Gamma)}$, then

$$[\![A^* \subseteq \Omega]\!] = 1 \quad \text{and}$$

$$[\![A^* \subseteq \Omega]\!] \leq [\![\langle A^*, b\rangle \in \bigvee \leftrightarrow (\forall a \in A^*)(a \leq b) \wedge$$

$$(\forall c \in \Omega)[(\forall a \in A^*)(a \leq c) \to b \leq c]]\!]$$

$$\leq [\![(\forall a \in A^*)(a \leq \tilde{\bigvee} A) \wedge$$

$$(\forall c \in \Omega)[(\forall a \in A^*)(a \leq c) \to \tilde{\bigvee} A \leq c]]\!].$$

(5) Let $A = \{\tilde{f}(x) \mid x \in \mathcal{D}(u)\}$ and let $\mathcal{D}(A^*) = A$, $EA^* = \bigvee_{a\in A} Ea$ and $(a \in A \Rightarrow A^*(a) = Ea)$ as usual. Then we have $A^* \in V^{(\Gamma)}$, for:

$$\mathcal{D}(A^*) = A \subseteq V^{(\Gamma)},$$

$$g \in G_\Gamma \cap G_u \cap G_{\langle \Omega,\wedge,\bigvee,1\rangle} \cap G_f,\ x \in \mathcal{D}(u)$$

$$\Rightarrow (\bar{g}A^*)(\bar{g}\tilde{f}(x)) = gA^*(\tilde{f}(x)) = gEx = [\![\bar{g}x \in u]\!]$$

$$= \bigvee_{t\in \mathcal{D}(u)} [\![\bar{g}x = t]\!] \wedge Et$$

$$\leq \bigvee_{t\in \mathcal{D}(u)} [\![\tilde{f}\bar{g}x = \tilde{f}t]\!] \wedge E\tilde{f}t$$

$$\leq [\![\overline{g}\tilde{f}(x) \in A^*]\!]$$

Similarly, $A^*(\tilde{f}(x)) \leq [\![\tilde{f}(x) \in \bar{g}A^*]\!]$.

Hence $g \in G_\Gamma \cap G_u \cap G_{\langle \Omega,\wedge,\bigvee,1\rangle} \cap G_f \Rightarrow [\![\bar{g}A^* = A^*]\!] = 1$, and so $A^* \in V^{(\Gamma)}$.

Therefore, $[\![\bigvee A^* = \tilde{\bigvee}_{x\in \mathcal{D}(u)} \tilde{f}(x)]\!] \geq \bigvee_{x\in \mathcal{D}(u)} E\tilde{f}(x)$.

Now it suffices to show that

$$[\![\bigvee_{x\in u} f(x) = \bigvee A^*]\!] = 1.$$

That is,

$$[\![(\forall x \in u)(f(x) \leq \bigvee A^*) \wedge (\forall u \in A^*)(u \leq \bigvee_{x\in u} f(x))]\!] = 1.$$

This is an immediate consequence of the definition of f.

<u>Theorem 5.</u>

Let Ω, $1 \in V^{(\Gamma)}$ and $\tilde{\Omega} = \{a \in V^{(\Gamma)} | [\![a \in \Omega]\!] = Ea\}$. Assume that the functions

$$\tilde{\wedge} : \tilde{\Omega} \times \tilde{\Omega} \to \tilde{\Omega} \quad \text{and}$$
$$\tilde{\bigvee} : \{A \subseteq \tilde{\Omega} | A^* \in V^{(\Gamma)}\} \to \tilde{\Omega}$$

Satisfy the conditions:

(1) $<\tilde{\Omega}, \tilde{\wedge}, \tilde{\bigvee}, 1>$ is a $V^{(\Gamma)}$-cHa, i.e, $<\Omega, \tilde{\wedge}, \tilde{\bigvee}, 1>$ is a lattice satisfying

$$A \subseteq \tilde{\Omega},\ A^* \in V^{(\Gamma)} \Rightarrow \tilde{\bigvee} A \in \tilde{\Omega}$$
$$A \subseteq \tilde{\Omega},\ A^* \in V^{(\Gamma)},\ b \in \tilde{\Omega} \Rightarrow (\tilde{\bigvee} A) \wedge b = \tilde{\bigvee}_{a\in A} a \tilde{\wedge} b$$

(2) $a,b,a',b' \in \tilde{\Omega} \Rightarrow [\![<a,b> = <a',b'>]\!] \leq [\![a \tilde{\wedge} b = a' \tilde{\wedge} b']\!]$,

$A,B \subseteq \tilde{\Omega},\ A^*,B^* \in V^{(\Gamma)} \Rightarrow [\![A^* = B^*]\!] \leq [\![\tilde{\bigvee} A = \tilde{\bigvee} B]\!]$.

(3) $E : \tilde{\Omega} \to H$ is cHa-morphism i.e.,

$$E(1) = 1$$
$$a,b \in \tilde{\Omega} \Rightarrow E(a \tilde{\wedge} b) = Ea \wedge Eb$$
$$A \subseteq \tilde{\Omega},\ A^* \in V^{(\Gamma)} \Rightarrow E(\tilde{\bigvee} A) = \bigvee_{a\in A} Ea.$$

(4) $\{g \in G_\Gamma \mid \overline{g} \in \mathrm{Aut}(\tilde{\Omega})\} \in \Gamma$.

Then there exist $\Lambda, \bigvee \in V^{(\Gamma)}$ such that

$$[\![<\Omega,\Lambda,\bigvee,1> \text{ is a cHa}]\!]_\Gamma = 1,$$

$$a,b \in \tilde{\Omega} \Rightarrow [\![a \wedge b = a \tilde{\wedge} b]\!] = Ea \wedge Eb,$$

$$A \subseteq \tilde{\Omega},\ A^* \in V^{(\Gamma)} \Rightarrow [\![\bigvee A^* = \tilde{\bigvee} A]\!] = E(\tilde{\bigvee} A).$$

Proof.

Define X by

$$\mathcal{D}(X) = \{<a,b> \in \tilde{\Omega} \times \tilde{\Omega} \mid a,b \text{ are } \Gamma\text{-HSR}\}$$

$$<a,b> \in \mathcal{D}(X) \Rightarrow X<a,b> = E<a,b>$$

$$EX = 1$$

Then $[\![X = \Omega \times \Omega]\!] = 1$ and X is definite. By Theorem 3, there exists $\Lambda \in V^{(\Gamma)}$ such that

$$[\![\Lambda : \Omega \times \Omega \to \Omega]\!]_\Gamma = 1$$

$$<a,b> \in \mathcal{D}(X) \Rightarrow [\![<<a,b>, a \tilde{\wedge} b> \in h]\!] = E<a,b>.$$

Define Y by

$$\mathcal{D}(Y) = \{A^* \in V^{(\Gamma)} \mid A \subseteq \tilde{\Omega}, \wedge, A^* \text{ is } \Gamma\text{-HSR}\},$$

$$A^* \in \mathcal{D}(Y) \Rightarrow Y(A^*) = EA^*,$$

$$EY = 1.$$

Then $[\![Y = P_\Gamma(\Omega)]\!] = 1$.

By Theorem 3, there exists $\bigvee \in V^{(\Gamma)}$ such that

$$[\![\bigvee : P_\Gamma(\Omega) \to \Omega]\!]_\Gamma = 1$$

$$A^* \in \mathcal{D}(Y) \Rightarrow [\![<A^*, \tilde{\bigvee} A> \in \bigvee]\!] = EA^*.$$

The proof of $[\![<\Omega,\wedge,\bigvee,1> \text{ is a cHa}]\!] = 1$ is similar to Theorem I.5.(*)

Theorem 6.

Let

$<\Omega,\wedge,\bigvee,1>, <\Omega',\wedge',\bigvee',1'> \in V^{(\Gamma)}$ and

$[\![<\Omega,\wedge,\bigvee,1> \text{ is a cHa}]\!]_\Gamma = [\![<\Omega',\wedge',\bigvee',1'> \text{ is a cHa}]\!]_\Gamma = 1.$

(I) If $h \in V^{(\Gamma)}$ and $[\![h : \Omega \to \Omega' \text{ is a cHa-morphism}]\!]_\Gamma = p$, then there is $\tilde{h} : \tilde{\Omega} \to \tilde{\Omega}'$ such that

(1) $a \in \tilde{\Omega} \Rightarrow Ea \wedge p = [\![<a,\tilde{h}(a)> \in h]\!]$,

(2) $a,b \in \tilde{\Omega} \Rightarrow [\![a = b]\!] \leq [\![\tilde{h}(a) = \tilde{h}(b)]\!]$,

(3) $\tilde{h} : \tilde{\Omega} \to \tilde{\Omega}'$ is $\tilde{\wedge},\widetilde{\bigvee}$-morphism,

(4) $g \in H_\Gamma \Rightarrow \tilde{h}i(q) = i'(p \wedge q)$, where $i(q) = 1 \restriction q$ and $i'(q) = 1' \restriction q$,

(5) $a \in \tilde{\Omega} \Rightarrow E\tilde{h}(a) = Ea \wedge p$,

(6) $\{g \in G \mid g\tilde{h} = \widetilde{hg}\} \in \Gamma$.

If $[\![h : \Omega \to \Omega' \text{ is an isomorphism}]\!]_\Gamma = p$, then $\tilde{h} \cap \widetilde{\Omega \restriction p} \times \widetilde{\Omega' \restriction p}$ is an isomorphism.

(II) Conversely, if $p \in H_\Gamma$ and $\tilde{h} : \tilde{\Omega} \to \tilde{\Omega}'$ satisfies (2) $\sim$ (6), then there is a unique $h \in V^{(\Gamma)}$ such that

$[\![h : \Omega \to \Omega' \text{ is a cHa-morphism}]\!]_\Gamma = Eh = p$, and

$a \in \tilde{\Omega} \Rightarrow Ea \wedge p = [\![<a,\tilde{h}(a)> \in h]\!]_\Gamma$.

If $\tilde{h} \cap \widetilde{\Omega \restriction p} \times \widetilde{\Omega' \restriction p}$ is isomorphism, then

$[\![h : \Omega \to \Omega' \text{ is an isomorphism}]\!]_\Gamma = Eh = p.$

(*) Theorem I.5 means Theorem 5 in Chapter I.

Proof.

(I) By using Theorem 2, there exists $\tilde{h} : \tilde{\Omega} \to \tilde{\Omega}'$ satisfying (1)(2)(5)(6).

We prove $\tilde{h}$ satisfies (3) and (4).

(3) $a,b \in \tilde{\Omega} \Rightarrow Ea \wedge Eb \wedge p = [\![<a,\tilde{h}(a)> \in h \wedge <b,\tilde{h}(b)> \in h]\!]$

$$\leq [\![<a \tilde{\wedge} b, \tilde{h}(a) \tilde{\wedge} \tilde{h}(b)> \in h]\!]$$

$$\Rightarrow [\![\tilde{h}(a \tilde{\wedge} b) = \tilde{h}(a) \tilde{\wedge} \tilde{h}(b)]\!] = 1$$

$A \subseteq \tilde{\Omega},\ A^* \in V^{(\Gamma)} \Rightarrow p \wedge EA^* \leq \bigvee_{a \in A^*} h(a) = \widetilde{\bigvee}_{a \in A} \tilde{h}(a)]\!],$

$$\Rightarrow p \wedge EA^* \leq [\![h(\bigvee_{a \in A^*} a) = \widetilde{\bigvee}_{a \in A} \tilde{h}(a)]\!],$$

$$\Rightarrow [\![\tilde{h}(\widetilde{\bigvee} A) = \widetilde{\bigvee}_{a \in A} \tilde{h}(a)]\!] = 1.$$

by using Theorem 4(5).

(4) Since $[\![<1,1'> \in h]\!] \geq p$, we have the following results in turn.

$[\![<1 \restriction q, 1' \restriction p \wedge q> \in h]\!] \geq p \wedge q,$

$[\![\tilde{h}(1 \restriction q) = 1' \restriction p \wedge q]\!] \geq p \wedge q,$

$[\![\tilde{h}(1 \restriction q) = 1' \restriction p \wedge q]\!] = 1.$

If $[\![h : \Omega \to \Omega'$ is an isomorphism$]\!]_{\Gamma} = p$, then

$$a,b \in \widetilde{\Omega \restriction p} \Rightarrow [\![\tilde{h}(a) = \tilde{h}(b)]\!] \wedge p \leq [\![a = b]\!]$$

$$\Rightarrow [\![\tilde{h}(a) = \tilde{h}(b)]\!] \leq [\![a = b]\!].$$

$c \in \widetilde{\Omega' \restriction p} \Rightarrow Ec = Ec \wedge p \leq [\![\exists x \in \Omega(<x,c> \in h)]\!]$ and

$$Ec \wedge p \leq [\![\forall x,y \in \Omega(<x,c> \in h \wedge <y,c> \in h \to x = y)]\!]$$

$$\Rightarrow Ec \leq [\![\exists ! x \in \Omega(<x,c> \in h)]\!]$$

$$\Rightarrow \exists x \in \tilde{\Omega} \text{ such that } [\![\tilde{h}(x) = c]\!] = 1.$$

Therefore, $\tilde{h} \cap \widetilde{\Omega \restriction p} \times \widetilde{\Omega' \restriction p}$ is an isomorphism.

(II) Assume that $p \in H_\Gamma$ and $\tilde{h} : \tilde{\Omega} \to \tilde{\Omega}'$ satisfies (2) ~ (6). Then, by Theorem 3, there exists a unique $h \in V^{(\Gamma)}$ such that

$$[\![h : \Omega \to \Omega']\!]_\Gamma \geq p = Eh, \quad \text{and}$$

$$x \in \mathcal{D}(\Omega) \Rightarrow [\![<x,\tilde{h}(x)> \in h]\!] = Ex \wedge p.$$

$$\tilde{h}(1) = 1' \restriction p \quad \text{and hence} \quad [\![<1,1'> \in h]\!] \geq p,$$

$$a,b \in \tilde{\Omega} \Rightarrow Ea \wedge Eb \wedge p \leq [\![h(a \wedge b) = \tilde{h}(a \wedge b)$$
$$= \tilde{h}(a) \;\tilde{\wedge}'\; \tilde{h}(b)$$
$$= h(a) \wedge' h(b)]\!],$$

$$A \subseteq \tilde{\Omega},\ A^* \in V^{(\Gamma)} \Rightarrow EA^* \wedge p \leq [\![\tilde{h}(\bigvee A^*) = \tilde{h}(\tilde{\bigvee} A) = \tilde{\bigvee}'_{a \in A} h(a)$$
$$= \bigvee_{a \in A^*} h(a)]\!].$$

Therefore, $[\![h : \Omega \to \Omega' \text{ cHa-morphism}]\!] \geq p$.

If $\tilde{h} \cap \widetilde{\Omega \restriction p} \times \widetilde{\Omega' \restriction p}$ is isomorphism, then we have

$$a,b \in \tilde{\Omega} \Rightarrow p \wedge Ea \wedge Eb \wedge [\![h(a) = h(b)]\!] \leq [\![a = b]\!], \quad \text{and}$$

$$c \in \tilde{\Omega}' \Rightarrow \exists a \in \tilde{\Omega} \text{ such that } \tilde{h}(a \restriction p) = c \restriction p$$
$$\Rightarrow [\![\exists x \in \Omega(<x,c> \in h)]\!]_\Gamma \geq Ec \wedge p.$$

Therefore,

$$[\![h : \Omega \to \Omega' \text{ is isomorphism}]\!]_\Gamma \geq p.$$

Remark.

(1) $a,b \in \tilde{\Omega} \Rightarrow (a \to b) = \bigvee\{c \in \tilde{\Omega} \mid a \tilde{\wedge} c \leq b\} \in \tilde{\Omega}$.

(2) $A \subseteq \tilde{\Omega},\ A^* \in V^{(\Gamma)} \Rightarrow \bigwedge A = \bigvee\{c \in \tilde{\Omega} \mid (\forall a \in A)(c \leq a)\} \in \tilde{\Omega}$.

Proof.

(1) Let $X = \{c \in \tilde{\Omega} | a \tilde{\wedge} c \leq b\}$.

$$g \in G_\Gamma \cap G_\Omega \cap G_a \cap G_b \Rightarrow \bar{g}X = X$$
$$\Rightarrow [\![\bar{g}X^* = X^*]\!] = 1.$$

Hence $X^* \in V^{(\Gamma)}$.

Therefore, $a \to b = \bigvee X \in \tilde{\Omega}$.

(2) Let $X = \{c \in \tilde{\Omega} | (\forall a \in A)(c \leq a)\}$. Then

$$X = \{c \in \tilde{\Omega} | Ec \leq [\![\forall a \in A^*(c \leq a)]\!]\}, \text{ and so}$$
$$g \in G_\Gamma \cap G_\Omega \cap G_{A^*} \Rightarrow \bar{g}X = X$$
$$\Rightarrow [\![\bar{g}X^* = X^*]\!] = 1.$$

Hence $X^* \in V^{(\Gamma)}$.

Therefore, $\bigwedge A = \tilde{\bigvee} X \in \tilde{\Omega}$.

§7. $V^{(\Omega)}$ in $V^{(\Gamma)}$.

Let H be a cHa, G be a subgroup of $\mathrm{Aut}(H)$, Γ be a normal filter of subgroups of G, $\langle\Omega,\wedge,\bigvee,1\rangle \in V^{(\Gamma)}$, $[\![\langle\Omega,\wedge,\bigvee,1\rangle$ is a cHa$]\!] = 1$ and $\tilde{\Omega} = \{a \in V^{(\Gamma)} | [\![a \in \Omega]\!] = Ea\}$ as in §6. In §6 we proved that

(a) $\langle\tilde{\Omega},\tilde{\wedge},\tilde{\bigvee},1\rangle$ is a $V^{(\Gamma)}$-cHa, and

(b) if $g \in G_\Gamma \cap G_{\langle\Omega,\wedge,\bigvee,1\rangle}$, then

(i) $\bar{g} \in \mathrm{Aut}(\tilde{\Omega})$,

(ii) $gE = E\bar{g}$ on $\tilde{\Omega}$, and

(iii) $ig = \bar{g}i$ on $H_\Gamma = \{p \in H | \{g \in G | gp = p\} \in \Gamma\}$.

From now on we write G_Ω instead of $G_{\langle\Omega,\wedge,\bigvee,1\rangle}$.

Let

$$F \in \Gamma \Rightarrow \overline{F} = \{\overline{g} \in \mathrm{Aut}(\tilde{\Omega}) \mid g \in F \cap G_\Gamma \cap G_\Omega\},$$

$$\overline{\Gamma} = \{\overline{F} \mid F \in \Gamma\}.$$

Lemma 7.1.

(1) $G_\Gamma \cap G_\Omega \cong \overline{G}$.

(2) $\overline{\Gamma}$ is a normal filter of subgroups of $\overline{G}$.

Proof.

(1) If $g,h \in G_\Gamma \cap G_\Omega$, then $\overline{gh} = \overline{g}\,\overline{h}$, $\overline{g^{-1}} = \overline{g}^{-1}$ and

$$\overline{g} = \overline{h} \Rightarrow \overline{g}\, i\, p = \overline{h}\, i\, p \quad \text{for any } p \in H$$

$$\Rightarrow i\, g\, p = i\, h\, p \quad \text{for any } p \in H$$

$$\Rightarrow g = h.$$

Therefore, $\varphi : G_\Gamma \cap G_\Omega \to \overline{G}$ defined by $\varphi(g) = \overline{g}$ is isomorphism.

(2) If $\overline{F} \in \overline{\Gamma}$ then $\overline{F}$ is a subgroup of $\overline{G}$, for

$$\overline{g}, \overline{h} \in \overline{F} \Rightarrow g,h \in F$$

$$\Rightarrow \overline{g}\,\overline{h} = \overline{gh} \in \overline{F},\ \overline{g}^{-1} = \overline{g^{-1}} \in \overline{F}.$$

$\overline{\Gamma}$ is obviously a filter of subgroups of $\overline{G}$. Hence it suffices to show that $\overline{\Gamma}$ is normal.

Let $g \in \overline{G}$ and $K \in \overline{\Gamma}$. Then there is $h \in G_\Gamma \cap G_\Omega$ and $K' \in \Gamma$ such that $g = \overline{h}$, $K = \overline{K'}$, $K' \in \Gamma$.

$$gKg^{-1} = \overline{h}\,\overline{K'}\,\overline{h}^{-1}$$

$$= \{\overline{h}\,\overline{k}\,\overline{h}^{-1} \mid k \in K'\}$$

$$= \overline{h\,K'\,h^{-1}} \in \overline{\Gamma}$$

Therefore,

$$\overline{G}_{\overline{\Gamma}} = \{\overline{g} \in \overline{G} \mid \forall K \subseteq \overline{G}(K \in \overline{\Gamma} \text{ iff } gKg^{-1} \in \overline{\Gamma})\}$$
$$= \overline{G} \in \overline{\Gamma}.$$

That is, $\overline{\Gamma}$ is normal, q.e.d.

In order to define $V^{(\overline{\Gamma})}$, first we define $V^{(\tilde{\Omega})}$ inductively, as follows.

$$V_{\alpha}^{(\tilde{\Omega})} = \{u = \langle \mathcal{D}(u), |u|, E_{\tilde{\Omega}}u\rangle \mid \mathcal{D}(u) \subseteq V_{\beta}^{(\tilde{\Omega})},\ \beta < \alpha,$$
$$|u| : \mathcal{D}(u) \to \Omega,\ \forall x \in \mathcal{D}(u)(u(x) \leq E_{\tilde{\Omega}}x \,\hat{\wedge}\, E_{\tilde{\Omega}}u)\},$$
$$V^{(\tilde{\Omega})} = \cup_{\alpha \in \mathrm{ord}} V_{\alpha}^{(\tilde{\Omega})}.$$

Remark.

Since $\tilde{\Omega}$ is not complete, $[\![u = v]\!]_{\tilde{\Omega}}$, $[\![u \in v]\!]_{\tilde{\Omega}}$ are not necessarily defined for $u,v \in V^{(\tilde{\Omega})}$.

For $u \in V^{(\tilde{\Omega})}$ and $\overline{g} \in \overline{G}$ we define $\overline{\overline{g}}u$ as usual. i.e.,

$$\begin{cases} \mathcal{D}(\overline{\overline{g}}u) = \{\overline{\overline{g}}x \mid x \in \mathcal{D}(u)\}, \\ x \in \mathcal{D}(\overline{\overline{g}}u) \Rightarrow (\overline{\overline{g}}u)(\overline{\overline{g}}x) = \tilde{\bigvee}\{\overline{g}u(x') \mid x' \in \mathcal{D}(u),\ \overline{\overline{g}}x' = \overline{\overline{g}}x\}, \\ E\overline{\overline{g}}u = \overline{g}Eu. \end{cases}$$

Then $\overline{\overline{g}}u \in V^{(\tilde{\Omega})}$.

Now we define $V^{(\overline{\Gamma})}$.

$$V^{(\overline{\Gamma})} = \{u \in V^{(\tilde{\Omega})} \mid \mathcal{D}(u) \subseteq V^{(\overline{\Gamma})},\ u \text{ is definite},\ \overline{G}(u) \in \overline{\Gamma}\}$$

where

u is definite iff $\forall x \in \mathcal{D}(u)(u(x) = E_{\tilde{\Omega}}x)$, and

$\overline{G}(u) = \{\overline{g} \in \overline{G} \mid \overline{\overline{g}}\,\mathcal{D}(u) = \mathcal{D}(u)\}$.

For $u \in V^{(\tilde{\Omega})}$ and $a \in \tilde{\Omega}$, $u \restriction a$ is defined as usual. i.e.,

$\mathcal{D}(u \restriction a) = \{x \restriction a \mid x \in \mathcal{D}(u)\}$,

$x \in \mathcal{D}(u) \Rightarrow (u \restriction a)(x \restriction a) = \bigvee\{u(x') \wedge a \mid x' \in \mathcal{D}(u),\ x' \restriction a = x \restriction a\}$,

$E(u \restriction a) = Eu \wedge a$.

Then $u \restriction a \in V^{(\tilde{\Omega})}$.

Lemma 7.2.

Let $V_\alpha^{(\overline{\Gamma})} = V^{(\overline{\Gamma})} \cap V_\alpha^{(\tilde{\Omega})}$.

If $u \in V_\alpha^{(\overline{\Gamma})}$, $a \in \tilde{\Omega}$ and $\overline{g} \in \overline{G}$, then

(1) $u \restriction a \in V_\alpha^{(\overline{\Gamma})}$,

(2) $\overline{\overline{g}}u \in V_\alpha^{(\overline{\Gamma})}$,

(3) $\overline{\overline{g}}(u \restriction a) = \overline{\overline{g}}u \restriction \overline{ga}$.

Proof.

Assume the lemma for $x \in V_\beta^{(\overline{\Gamma})}$, $\beta < \alpha$.

(1) (i) $\mathcal{D}(u \restriction a) \subseteq V_\gamma^{(\overline{\Gamma})}$, $\gamma < \alpha$ by the assumption.

(ii) $\overline{G}(u \restriction a) \in \overline{\Gamma}$, for:

$\overline{g} \in \overline{G}(u) \cap \overline{Ga}$, $x \in \mathcal{D}(u)$

$\Rightarrow \overline{\overline{g}}(x \restriction a) = \overline{\overline{g}}x \restriction \overline{ga} = \overline{\overline{g}}x \restriction a \in \mathcal{D}(u \restriction a)$

$\overline{G}(u) \cap \overline{Ga} \subseteq \overline{G}(u \restriction a)$.

$\overline{G}(u \restriction a) \in \overline{\Gamma}$.

(iii) $u \restriction a$ is definite.

By (i)(ii)(iii), $u \restriction a \in V_\alpha^{(\overline{\Gamma})}$.

(2) (i) $\mathcal{D}(\overline{\overline{g}}u) \subseteq V_{\gamma}^{(\overline{\Gamma})}$, $\gamma < \alpha$, by the assumption.

(ii) $\overline{G}(\overline{\overline{g}}u) \in \overline{\Gamma}$, for:

$$K = \{\overline{k} \in \overline{G} \mid \overline{\overline{k}}\ \mathcal{D}(u) = \mathcal{D}(u)\} = \overline{G}(u) \in \overline{\Gamma}, \text{ and}$$

$\overline{\Gamma}$ is normal and $\overline{G} = \overline{G}_{\overline{\Gamma}}$

Hence, $\overline{g}K\overline{g}^{-1} \in \overline{\Gamma}$.

$$\overline{k} \in K,\ x \in \mathcal{D}(u) \Rightarrow \overline{\overline{gkg^{-1}}}\,\overline{\overline{g}}x = \overline{\overline{gk}}x \in \mathcal{D}(\overline{\overline{g}}u).$$

Hence, $\overline{g}K\overline{g}^{-1} \subseteq \overline{G}(\overline{\overline{g}}u)$.

Therefore, $\overline{G}(\overline{\overline{g}}u) \in \overline{\Gamma}$.

(iii) $\overline{\overline{g}}u$ is definite, for

$$x \in \mathcal{D}(u) \Rightarrow (\overline{\overline{g}}u)(\overline{\overline{g}}x) = \overline{g}Ex = E\overline{\overline{g}}x.$$

By (i)(ii)(iii) $\overline{\overline{g}}u \in V^{(\overline{\Gamma})}$.

(3)
$$\begin{aligned}\mathcal{D}(\overline{\overline{g}}(u \restriction a)) &= \{\overline{\overline{g}}(x \restriction a) \mid x \in \mathcal{D}(u)\} \\ &= \{\overline{\overline{g}}x \restriction \overline{g}a \mid x \in \mathcal{D}(u)\}, \text{ by assumption,} \\ &= \mathcal{D}(\overline{\overline{g}}u \restriction \overline{g}a).\end{aligned}$$

For $x \in \mathcal{D}(u)$,

$$\begin{aligned}\overline{\overline{g}}(u \restriction a)(\overline{\overline{g}}(x \restriction a)) &= \overline{g}(E_{\tilde{\Omega}}x \wedge a) \\ &= (\overline{\overline{g}}u \restriction \overline{g}a)(\overline{\overline{g}}x \restriction \overline{g}a). \\ &= (\overline{\overline{g}}u \restriction \overline{g}a)(\overline{\overline{g}}(x \restriction a)).\end{aligned}$$

Therefore, $\overline{\overline{g}}(u \restriction a) = \overline{\overline{g}}u \restriction \overline{g}a$, g.e.d.

Truth value $[\![\varphi]\!]_{\overline{\Gamma}}$ of a formula φ in $V^{(\overline{\Gamma})}$ is defined as follows.

Lemma 7.3.

Let $u,v \in V^{(\overline{\Gamma})}$. Then

(1) There exists $[\![u = v]\!]_{\overline{\Gamma}} = \bigwedge_{x \in \mathcal{D}(u)} (u(x) \to [\![x \in v]\!]_{\overline{\Gamma}})\, \tilde{\wedge}$

$$\bigwedge_{y \in \mathcal{D}(y)} (v(x) \to [\![x \in u]\!]_{\overline{\Gamma}})\, \tilde{\wedge}\, (E_{\tilde{\Omega}} u \leftrightarrow E_{\tilde{\Omega}} v) \in \tilde{\Omega}$$

(2) There exists $[\![u \in v]\!]_{\overline{\Gamma}} = \tilde{\bigvee}_{y \in \mathcal{D}(v)} [\![u = y]\!]\, \tilde{\wedge}\, E_{\tilde{\Omega}} y \in \tilde{\Omega}$.

(3) $\bar{g} \in \bar{G} \Rightarrow \bar{g}[\![u = v]\!]_{\overline{\Gamma}} = [\![\bar{\bar{g}}u = \bar{\bar{g}}v]\!]_{\overline{\Gamma}}$,

$$\bar{g}[\![u \in v]\!]_{\overline{\Gamma}} = [\![\bar{\bar{g}}u \in \bar{\bar{g}}v]\!]_{\overline{\Gamma}}.$$

Proof.

We prove the lemma by induction on max(rank u, rank v).

(i) If rank x < rank u, then there exists $[\![x \in u]\!]_{\overline{\Gamma}} \in \tilde{\Omega}$ and

$$\bar{g} \in \bar{G} \Rightarrow \bar{g}[\![x \in u]\!]_{\overline{\Gamma}} = [\![\bar{\bar{g}}x \in \bar{\bar{g}}u]\!]_{\overline{\Gamma}}.$$

Proof.

Let $A = \{[\![x = t]\!]_{\overline{\Gamma}} \wedge E_{\tilde{\Omega}} t \mid t \in \mathcal{D}(u)\}$. By induction hypothesis,

$$\bar{g} \in \bar{G}(u) \cap \bar{G}(x) \Rightarrow \bar{g}A = \{[\![\bar{\bar{g}}x = \bar{\bar{g}}t]\!]_{\overline{\Gamma}} \wedge E_{\tilde{\Omega}} \bar{\bar{g}}t \mid t \in \mathcal{D}(u)\} = A$$

$$\Rightarrow [\![gA^* = A^*]\!] = 1.$$

Therefore, $A^* \in V^{(\Gamma)}$. It follows that

$$[\![x \in u]\!]_{\overline{\Gamma}} = \tilde{\bigvee}_{t \in \mathcal{D}(u)} [\![x = t]\!]_{\overline{\Gamma}}\, \tilde{\wedge}\, u(t) = \tilde{\bigvee} A \in \tilde{\Omega}.$$

$$\bar{g} \in \bar{G} \Rightarrow \bar{g} \in \mathrm{Aut}(\Omega)$$

$$\Rightarrow g[\![x \in u]\!]_{\overline{\Gamma}} = [\![\bar{\bar{g}}x \in \bar{\bar{g}}u]\!]_{\overline{\Gamma}}.$$

(ii) $[\![u = v]\!] \in \tilde{\Omega}$ and

$$\bar{g} \in \bar{G} \Rightarrow \bar{g}[\![u = v]\!]_{\overline{\Gamma}} = [\![\bar{\bar{g}}u = \bar{\bar{g}}v]\!]_{\overline{\Gamma}}.$$

Proof.

Let $A = \{E_{\tilde{\Omega}}x \to [\![x \in v]\!]_{\overline{\Gamma}} \mid x \in \mathcal{D}(u)\}$.

$$\begin{aligned}\bar{g} \in \bar{G}(u) \cap \bar{G}(v) \Rightarrow \bar{g}A &= \{\bar{g}E_{\tilde{\Omega}}x \to \bar{g}[\![x \in v]\!]_{\overline{\Gamma}} \mid x \in \mathcal{D}(u)\} \\ &= \{E_{\tilde{\Omega}}\bar{g}x \to [\![\bar{g}x \in \bar{g}v]\!]_{\overline{\Gamma}} \mid x \in \mathcal{D}(u)\} \\ &= \{E_{\tilde{\Omega}}x \to [\![x \in v]\!]_{\overline{\Gamma}} \mid x \in \mathcal{D}(u)\} \\ &= A \\ \Rightarrow [\![gA^* = A^*]\!] &= 1.\end{aligned}$$

Hence, $A^* \in V^{(\Gamma)}$.

Therefore, $\bigwedge_{x \in \mathcal{D}(u)} (E_{\tilde{\Omega}}x \to [\![x \in v]\!]_{\overline{\Gamma}}) \in \tilde{\Omega}$.

Similarly, $\bigwedge_{y \in \mathcal{D}(v)} (E_{\tilde{\Omega}}y \to [\![y \in u]\!]_{\overline{\Gamma}}) \in \tilde{\Omega}$ and

$$(E_{\tilde{\Omega}}u \leftrightarrow E_{\tilde{\Omega}}v) \in \tilde{\Omega}.$$

Therefore, $[\![u = v]\!]_{\overline{\Gamma}} \in \tilde{\Omega}$.

$$\begin{aligned}\bar{g} \in \bar{G} &\Rightarrow \bar{g} \in \mathrm{Aut}(\tilde{\Omega}) \\ &\Rightarrow \bar{g}[\![u = v]\!]_{\overline{\Gamma}} = [\![\bar{\bar{g}}u = \bar{\bar{g}}v]\!]_{\overline{\Gamma}}.\end{aligned}$$

(iii) $[\![u \in v]\!]_{\overline{\Gamma}} \in \tilde{\Omega}$ and

$$\bar{g} \in \bar{G} \Rightarrow \bar{g}[\![u \in v]\!]_{\overline{\Gamma}} = [\![\bar{\bar{g}}u = \bar{\bar{g}}v]\!]_{\overline{\Gamma}}.$$

Proof. Similar to (i), (ii).

Lemma 7.4.

If $[\![\varphi(u)]\!]_{\overline{\Gamma}} \in \tilde{\Omega}$ for every $u \in V^{(\overline{\Gamma})}$, then

$$\tilde{\bigvee}_{u \in V^{(\overline{\Gamma})}} [\![\varphi(u)]\!]_{\overline{\Gamma}} \in \tilde{\Omega} \quad \text{and} \quad \bigwedge_{u \in V^{(\overline{\Gamma})}} [\![\varphi(u)]\!]_{\overline{\Gamma}} \in \tilde{\Omega}.$$

Proof.

Let $A = \{[\![\varphi(u)]\!]_{\bar{\Gamma}} | u \in V^{(\bar{\Gamma})}\}$ and let $u_1,\dots,u_n$ and u be all constants in $\varphi(u)$. Then

$$\bar{g} \in \bar{G}(u_1) \cap \cdots \cap \bar{G}(u_n)$$

$$\begin{aligned} \Rightarrow \bar{g}A &= \{\bar{g}[\![\varphi(u)]\!]_{\bar{\Gamma}} | u \in V^{(\bar{\Gamma})}\} \\ &= \{[\![\varphi(\bar{g}u)]\!]_{\bar{\Gamma}} | u \in V^{(\bar{\Gamma})}\} \\ &= \{[\![\varphi(u)]\!]_{\bar{\Gamma}} | u \in V^{(\bar{\Gamma})}\} \\ &= A \end{aligned}$$

$$\Rightarrow [\![\bar{g}A^* = A^*]\!] = 1.$$

Therefore, $A^* \in V^{(\Gamma)}$. It follows that

$$\tilde{\bigvee}_{u \in V^{(\bar{\Gamma})}} [\![\varphi(u)]\!]_{\bar{\Gamma}} \in \tilde{\Omega} \quad \text{and} \quad \bigwedge_{u \in V^{(\bar{\Gamma})}} [\![\varphi(u)]\!]_{\bar{\Gamma}} \in \tilde{\Omega}.$$

By Lemma 7.3, 7.4, we can define $[\![\ ]\!]_{\bar{\Gamma}}$ as usual. i.e.,

$$[\![u = v]\!]_{\bar{\Gamma}} = \bigwedge_{x \in \mathcal{D}(u)} (u(x) \to [\![x \in v]\!]_{\bar{\Gamma}}) \wedge$$

$$\bigwedge_{y \in \mathcal{D}(v)} (v(y) \to [\![y \in u]\!]_{\bar{\Gamma}}) \wedge$$

$$(E_{\tilde{\Omega}}u \leftrightarrow E_{\tilde{\Omega}}v).$$

$$[\![u \in v]\!]_{\bar{\Gamma}} = \tilde{\bigvee}_{y \in \mathcal{D}(v)} [\![u = y]\!]_{\tilde{\Omega}} \wedge v(y).$$

$$[\![\varphi_1 \vee \varphi_2]\!]_{\bar{\Gamma}} = [\![\varphi_1]\!]_{\bar{\Gamma}} \vee [\![\varphi_2]\!]_{\bar{\Gamma}}.$$

$$[\![\varphi_1 \wedge \varphi_2]\!]_{\bar{\Gamma}} = [\![\varphi_1]\!]_{\bar{\Gamma}} \wedge [\![\varphi_2]\!]_{\bar{\Gamma}}.$$

$$[\![\varphi_1 \to \varphi_2]\!]_{\bar{\Gamma}} = [\![\varphi_1]\!]_{\bar{\Gamma}} \to [\![\varphi_2]\!]_{\bar{\Gamma}}.$$

$$[\![\neg\varphi]\!]_{\bar{\Gamma}} = -[\![\varphi]\!]_{\bar{\Gamma}}.$$

$$[\![\forall x\, \varphi(x)]\!]_{\bar{\Gamma}} = \bigwedge_{x \in V^{(\bar{\Gamma})}} [\![\varphi(x)]\!]_{\bar{\Gamma}}.$$

$$[\![\exists x\, \varphi(x)]\!]_{\bar{\Gamma}} = \tilde{\bigvee}_{x \in V^{(\bar{\Gamma})}} [\![\varphi(x)]\!]_{\bar{\Gamma}}.$$

Theorem 7.

There is an isomorphism

$$\Phi : V^{(\overline{\Gamma})} / [\![=]\!]_{\overline{\Gamma}} = 1 \cong (V^{(\Gamma)})^{(\Omega)} / [\![[\![=]\!]_{\Omega} = 1]\!] = 1.$$

Proof.

Let $u = < \mathcal{D}(u), |u|, E_{\tilde{\Omega}}u > \in V_{\alpha}^{(\overline{\Gamma})}$. Assume that for $\beta < \alpha$, $x \in V_{\beta}^{(\overline{\Gamma})}$,

$$\Phi(x) = \overline{x} = < \mathcal{D}_{\Omega}(\overline{x}), |\overline{x}|_{\Omega}, E_{\Omega}\overline{x} > \in (V^{(\Gamma)})^{(\Omega)}$$

is defined and the following (a) (b) (c) are satisfied.

(a) $x \in V_{\beta}^{(\overline{\Gamma})}, \beta < \alpha \Rightarrow E\overline{x} = EE_{\tilde{\Omega}}x, E_{\Omega}\overline{x} = E_{\tilde{\Omega}}x,$

(b) $x, x' \in V_{\beta}^{(\overline{\Gamma})}, \beta < \alpha \Rightarrow i[\![\overline{x} = \overline{x}']\!] \leq [\![x = x']\!]_{\tilde{\Omega}} = [\![\overline{x} = \overline{x}']\!]_{\Omega}$

(c) $x \in V_{\beta}^{(\overline{\Gamma})}, \beta < \alpha, \overline{g} \in \overline{G} \Rightarrow [\![\overline{gx} = \overline{\overline{g}x}]\!] = 1,$

where Ex, $[\![\ ]\!]$ stand for $E_H x$, $[\![\ ]\!]_H$.

Then we define $\overline{u} = <\mathcal{D}_{\Omega}\overline{u}, |\overline{u}|_{\Omega}, E_{\Omega}\overline{u}>$ by

$$(\mathcal{D}\mathcal{D}_{\Omega}\overline{u}) = \{\overline{x} | x \in \mathcal{D}(u)\},$$

$$x \in \mathcal{D}(u) \Rightarrow \mathcal{D}_{\Omega}(\overline{u})(\overline{x}) = E\overline{x},$$

$$E\, \mathcal{D}_{u}(\overline{u}) = EE_{\tilde{\Omega}}u.$$

$$\mathcal{D}(|\overline{u}|) = \{<\overline{x}, E_{\tilde{\Omega}}x> | x \in \mathcal{D}(u)\}$$

$$|\overline{u}|<\overline{x}, E_{\tilde{\Omega}}x> = E\overline{x}$$

$$E|\overline{u}| = EE_{\tilde{\Omega}}u$$

$$E_{\Omega}\overline{u} = E_{\tilde{\Omega}}u.$$

(1) $\overline{u} \in V^{(\Gamma)}$.

Proof.

(i) $\mathcal{D}(\mathcal{D}_\Omega(\bar{u})) = \{\bar{x} | x \in \mathcal{D}(u)\} \subseteq V^{(\Gamma)}$ by the induction hypothesis.

Let $\bar{g} \in \bar{G}(u)$. Then

$$x \in \mathcal{D}(u) \Rightarrow (\bar{g}\,\mathcal{D}_\Omega(\bar{u}))(\overline{gx}) = gE\bar{x} = \overline{Egx} = E\bar{\bar{gx}}$$
$$\leq [\![\bar{\bar{gx}} \in \mathcal{D}_\Omega(\bar{u})]\!]$$
$$= [\![\overline{gx} \in \mathcal{D}_\Omega(\bar{u})]\!]$$
$$(\bar{u})(\bar{x}) = Ex = E\bar{g}\overline{g^{-1}x} = gE\overline{g^{-1}x}$$
$$\leq g[\![\overline{g^{-1}x} \in \mathcal{D}_\Omega(\bar{u})]\!]$$
$$\leq [\![\bar{x} \in \bar{g}\,\mathcal{D}_\Omega(\bar{u})]\!]$$

Therefore, $[\![\bar{g}\,\mathcal{D}_\Omega(\bar{u}) = \mathcal{D}_\Omega(\bar{u})]\!] = 1$. It follows that $\mathcal{D}_\Omega(u) \in V^{(\Gamma)}$.

(ii) $\mathcal{D}(|\bar{u}|) = \{<\bar{x}, E_{\tilde{\Omega}}x> | x \in \mathcal{D}(u)\} \subseteq V^{(\Gamma)}$.

Let $\bar{g} \in \bar{G}(u)$. Then

$$x \in \mathcal{D}(u) \Rightarrow (\bar{g}|\bar{u}|)(\bar{g}<\bar{x}, E_{\tilde{\Omega}}x>) = gE\bar{x} = \overline{Egx}$$
$$= E(\bar{\bar{gx}}) \leq [\![<\bar{\bar{gx}}, E_{\tilde{\Omega}}\bar{\bar{gx}}> \in |\bar{u}|]\!]$$
$$\leq [\![\bar{g}\;<\bar{x}, E_{\tilde{\Omega}}x> \in |\bar{u}|]\!]$$

Therefore, $[\![\bar{g}|\bar{u}| = |\bar{u}|]\!] = 1$. It follows that $|\bar{u}|_\Omega \in V^{(\Gamma)}$.

By (i), (ii) and $E_\Omega\bar{u} = E_{\tilde{\Omega}}u \in V^{(\Gamma)}$, we have $\bar{u} \in V^{(\Gamma)}$.

(2) $\bar{u} \in (V^{(\Gamma)})^{(\Omega)}$.

Proof.

(i) $[\![\forall x \in \mathcal{D}_\Omega(\bar{u})\, \exists\beta \in \mathrm{Ord}(x \in V_\beta^{(\Omega)})]\!] = 1$ by the definition on $V^{(\Omega)}$.

Therefore,

$$[\![\exists\alpha \in \mathrm{Ord}(\mathcal{D}_\Omega(\bar{u}) \subseteq V_\alpha^{(\Omega)})]\!] = 1.$$

(ii) $x \in \mathcal{D}(u) \Rightarrow E\bar{x} = |u|\langle\bar{x}, E_{\tilde{\Omega}}\bar{x}\rangle \wedge [\![E_{\tilde{\Omega}}x \in \Omega]\!]$

$\Rightarrow E\bar{x} \leq [\![\exists y \in \Omega(\langle\bar{x},y\rangle \in |u|)]\!]$

$x \in \mathcal{D}(u),\ y,y' \in \tilde{\Omega}$

$\Rightarrow E\bar{x} \wedge [\![\langle\bar{x},y\rangle \in |u| \wedge \langle\bar{x},y'\rangle \in |u|]\!] \leq [\![y = y']\!]$.

Therefore, $[\![\,|\bar{u}| : \mathcal{D}_{\Omega}(\bar{u}) \to \Omega]\!] = 1$.

(iii) $[\![\,E_{\Omega}\bar{u} \in \Omega]\!] = EE_{\Omega}\bar{u}$.

(iv) $x \in \mathcal{D}(u),\ y \in \tilde{\Omega} \Rightarrow [\![\langle\bar{x},y\rangle \in |\bar{u}|]\!] \leq [\![y = E_{\tilde{\Omega}}\bar{x} \leq E_{\Omega}\bar{u}]\!]$.

By (i) $\sim$ (iv), we have $[\![\bar{u} \in V^{(\Omega)}]\!] = E\bar{u}$, that is, $\bar{u} \in (V^{(\Gamma)})^{(\Omega)}$.

(3) u satisfies (a)(b)(c).

Proof.

(a) $E\bar{u} = EE_{\tilde{\Omega}}u$, $E_{\Omega}\bar{u} = E_{\tilde{\Omega}}u$ by the definition.

(b) For $x \in V_{\beta}^{(\bar{\Gamma})}$, $\beta < \alpha$,

$$[\![[\![x \in u]\!]_{\bar{\Gamma}} = \tilde{\bigvee}_{y \in \mathcal{D}(u)} [\![x = y]\!]_{\bar{\Gamma}} \mathbin{\tilde{\wedge}} E_{\tilde{\Omega}}y$$

$$= \bigvee_{\bar{y} \in \mathcal{D}_{\Omega}(\bar{u})} [\![\bar{x} = \bar{y}]\!]_{\Omega} \wedge E_{\Omega}\bar{y}$$

$$= [\![\bar{x} \in \bar{u}]\!]_{\Omega}]\!] = 1.$$

Let $u,u' \in V_{\alpha}^{(\Gamma)}$.

For $x \in \mathcal{D}(u) \cup \mathcal{D}(u')$, by using the equality axiom, we have

$$[\![\bar{u} = \bar{u}']\!] \leq [\![[\![\bar{x} \in \bar{u}]\!]_{\Omega} = [\![\bar{x} \in \bar{u}']\!]_{\Omega}]\!].$$

It follows that

$$[\![i[\![\bar{u} = \bar{u}']\!] \wedge [\![\bar{x} \in \bar{u}]\!]_{\Omega} = i[\![\bar{u} = \bar{u}']\!] \wedge [\![\bar{x} \in \bar{u}']\!]_{\Omega}]\!] = 1, \text{ and}$$

$$[\![i[\![\bar{u} = \bar{u}']\!] \leq ([\![\bar{x} \in \bar{u}]\!]_{\Omega} \leftrightarrow [\![\bar{x} \in \bar{u}']\!]_{\Omega}) = ([\![x \in u]\!]_{\bar{\Gamma}} \leftrightarrow [\![x \in u]\!]_{\bar{\Gamma}})]\!] = 1.$$

Since $[\![\bar{u} = \bar{u}']\!] \leq [\![E_{\Omega}\bar{u} = E_{\Omega}\bar{u}']\!]$, we also have

$$i[\![\bar{u} = \bar{u}']\!] \leq (E_{\tilde{\Omega}}u \leftrightarrow E_{\tilde{\Omega}}u') = (E_{\Omega}\bar{u} \leftrightarrow E_{\Omega}\bar{u}').$$

Therefore, $i[\![\bar{u} = \bar{u}']\!] \leq [\![u = u']\!]_{\bar{\Gamma}} = [\![\bar{u} = \bar{u}']\!]_{\Omega}$. Let $\bar{g} \in \bar{G}$.

(i) $\mathcal{D}(\overline{\mathcal{D}_{\Omega}\bar{\bar{g}}u}) = \{\overline{\bar{\bar{g}}x} \mid x \in \mathcal{D}(u)\}$,

$\mathcal{D}(\bar{g}\ \mathcal{D}_{\Omega}u) = \{\overline{gx} \mid x \in \mathcal{D}(u)\}$, and

$(\overline{\mathcal{D}_{\Omega}\bar{\bar{g}}u})(\overline{\bar{\bar{g}}x}) = E\overline{\bar{\bar{g}}x} = E\overline{gx} = (\bar{g}\ \mathcal{D}_{\Omega}u)(\overline{gx})$.

Since $[\![\overline{\bar{\bar{g}}x} = \overline{gx}]\!] = 1$, we have

$$[\![\mathcal{D}_{\Omega}(\overline{\bar{\bar{g}}u}) = \bar{g}\ \mathcal{D}_{\Omega}(\bar{u})]\!] = 1.$$

(ii) $\mathcal{D}(\overline{|\bar{\bar{g}}u|}) = \{\overline{<\bar{\bar{g}}x, E_{\tilde{\Omega}}\bar{\bar{g}}x>} \mid x \in \mathcal{D}(u)\}$

$\mathcal{D}(\bar{g}|\bar{u}|) = \{\overline{<\overline{gx}, \bar{g}E_{\tilde{\Omega}}x>} \mid x \in \mathcal{D}(u)\}$

$$x \in \mathcal{D}(x) \Rightarrow \overline{|\bar{\bar{g}}u|}\,\overline{<\bar{\bar{g}}x, E_{\Omega}\bar{\bar{g}}x>} = E(\overline{\bar{\bar{g}}x}) = E(\overline{gx})$$

$$= \bar{g}|\bar{u}|\,<\overline{gx}, \bar{g}E_{\tilde{\Omega}}x>.$$

Therefore, $[\![\overline{|\bar{\bar{g}}u|} = \bar{g}|\bar{u}|]\!] = 1$

(iii) $E_{\Omega}\overline{\bar{\bar{g}}u} = E_{\tilde{\Omega}}\bar{\bar{g}}u = \bar{g}E_{\Omega}u = \bar{g}E_{\Omega}\bar{u}$.

By (i)(ii)(iii)

$$[\![\overline{\bar{\bar{g}}u} = <\mathcal{D}_{\Omega}(\overline{\bar{\bar{g}}u}), \overline{\bar{\bar{g}}u}, E_{\Omega}\overline{\bar{\bar{g}}u}>$$

$$= <\bar{g}\ \mathcal{D}_{\Omega}U, \bar{g}|\bar{u}|, \bar{g}E_{\Omega}\bar{u}> = \overline{gu}]\!] = 1.$$

Now the function

$$\Phi : V^{(\bar{\Gamma})} \longrightarrow (V^{(\Gamma)})^{(\Omega)}$$

was defined. Φ is isomorphism if

(1) $[\![u = v]\!]_{\overline{\Gamma}} = [\![\bar{u} = \bar{v}]\!]_{\Omega}$ for $u,v \in V^{(\overline{\Gamma})}$.

(2) $[\![u \in v]\!]_{\overline{\Gamma}} = [\![\bar{u} \in \bar{v}]\!]_{\Omega}$ for $u,v \in V^{(\overline{\Gamma})}$.

(3) $\Phi : V^{(\overline{\Gamma})} \to (V^{(\Gamma)})^{(\Omega)} / [\![[\![=]\!]_{\Omega} = 1]\!] = 1$ is onto.

(1) was proved already.

(2) $$[\![[\![u \in v]\!]_{\overline{\Gamma}} = \tilde{\bigvee}_{y \in \mathcal{D}(v)} [\![u = y]\!]_{\overline{\Gamma}} \tilde{\wedge} v(y)$$

$$= \bigvee_{\bar{y} \in \mathcal{D}_{\Omega}(\bar{v})} [\![\bar{u} = \bar{y}]\!]_{\Omega} \wedge E_{\Omega}\bar{y}$$

$$= [\![\bar{u} \in \bar{v}]\!]_{\Omega}]\!] = 1.$$

(3) Let $v \in (V^{(\Gamma)})^{(\Omega)}$.

$$Ev = [\![v \in V^{(\Omega)}]\!]$$

$$\leq [\![\exists\, \mathcal{D}_{\Omega}(v), |v|_{\Omega}, E_{\Omega}v (v = \langle \mathcal{D}_{\Omega}(v), |v|_{\Omega}, E_{\Omega}v\rangle \wedge$$
$$\mathcal{D}_{\Omega}(v) \subseteq V^{(\Omega)} \wedge |v|_{\Omega} : \mathcal{D}_{\Omega}(v) \to \Omega \wedge E_{\Omega}v \wedge$$
$$\forall x \in \mathcal{D}_{\Omega}(v)\, \forall y \in \Omega (\langle x,y\rangle \in |v|_{\Omega} \to y \leq E_{\Omega}x \wedge E_{\Omega}v))$$

Hence, by the uniqueness principle, there exist $\mathcal{D}_{\Omega}(v)$, $|v|_{\Omega}$, $E_{\Omega}v$ such that

$$Ev = E\,\mathcal{D}_{\Omega}(v) \leq [\![\mathcal{D}_{\Omega}(v) \subseteq V^{(\Omega)}]\!]$$

$$Ev = E|v|_{\Omega} \leq [\![\,|v|_{\Omega} : \mathcal{D}_{\Omega}(v) \to \Omega]\!]$$

$$Ev = EE_{\Omega}v \leq [\![E_{\Omega}v \in \Omega]\!]$$

Let $\mathcal{D}_{\Omega}(v)$, $|v|_{\Omega}$, $E_{\Omega}v$ be all Γ-HSR.

$$\mathcal{D}(\mathcal{D}_{\Omega}(v)) \subseteq (V^{(\Gamma)})^{(\Omega)}$$

$$|v|_{\Omega} : \mathcal{D}(\mathcal{D}_{\Omega}(v)) \to \tilde{\Omega}$$

$$E_{\Omega}v \in \tilde{\Omega}$$

Assume that for each $x \in \mathcal{D}(\mathcal{D}_\Omega(v))$ there is $x^* \in V^{(\overline{\Gamma})}$ such that

$$[\![\Phi(x^*) = \overline{x^*} = x]\!] = 1.$$

Define v^* by

$$\mathcal{D}(v^*) = \{\overline{\overline{g}}x^* \upharpoonright [\![\overline{g}x \in v]\!]_\Omega \mid x \in \mathcal{D}\mathcal{D}_\Omega v,\ g \in \overline{G} \cap \overline{G}_v\}$$

$$v^*(\overline{\overline{g}}x^*) = [\![gx \in v]\!]$$

$$E\, v^* = E\, v.$$

Then we show that

(i) $v^* \in V^{(\overline{\Gamma})}$ and

(ii) $[\![\, [\![\overline{v^*} = v]\!]_\Omega = 1]\!] = 1.$

(i) v^* is definite and $(g \in \overline{G} \cap \overline{G}_v \Rightarrow \overline{\overline{g}}\, \mathcal{D}(v^*) = \mathcal{D}(v^*))$ by the definition. Hence $v^* \in V^{(\overline{\Gamma})}$.

(ii) If $x \in \mathcal{D}\mathcal{D}_\Omega(v)$, $g \in \overline{G} \cap \overline{G}_v$, then

$$\overline{v^*(\overline{\overline{g}}x^* \upharpoonright [\![gx \in v]\!]_\Omega)} = E_{\tilde{\Omega}}(\overline{\overline{g}}x^* \upharpoonright [\![gx \in v]\!]\ ) \leq [\![\overline{g}x \in v]\!]_\Omega$$

$$[\![\overline{g}x \in v]\!] \leq [\![\overline{\overline{g}}x^* = gx^* \upharpoonright [\![gx \in v]\!]]\!]_{\overline{\Gamma}}$$

$$= [\![\overline{\overline{\overline{g}}x^*} = \overline{g}x = \overline{\overline{\overline{g}}x^* \upharpoonright [\![\overline{g}x \in v]\!]_\Omega}]\!]_\Omega.$$

Hence $[\![(\forall y \in \mathcal{D}_\Omega v^*)(\overline{v^*}(y) \to [\![y \in v]\!]_\Omega)]\!] = 1.$

We also have

$$[\![(\forall y \in \mathcal{D}_\Omega v)(v(y) \to [\![y \in v^*]\!]_\Omega)]\!] = 1.$$

Therefore, $[\![\, [\![\overline{v^*} = v]\!] = 1]\!] = 1,$ q.e.d.

Remark.

If $u,v \in (V^{(\Gamma)})^{(\Omega)}$ and $\bar{g} \in \bar{G}$, then

$$[\![\bar{g}[\![u \in v]\!]_{\Omega} = [\![\bar{g}u \in \bar{g}v]\!]_{\Omega}]\!] = 1,$$

$$[\![\bar{g}[\![u = v]\!]_{\Omega} = [\![\bar{g}u = \bar{g}v]\!]_{\Omega}]\!] = 1.$$

Proof.

$$x \in \mathcal{D}\mathcal{D}_{\Omega}v \Rightarrow Ex \wedge Ev \leq [\![<x,v(x)> \in |v|]\!]$$

$$E\bar{g}x \wedge E\bar{g}v \leq [\![<\bar{g}x,\bar{g}v(x)> \in |\bar{g}v|]\!]$$

$$E\bar{g}x \wedge E\bar{g}v \leq [\![\bar{g}v(x) = (\bar{g}v)(\bar{g}x)]\!].$$

Therefore, $\bar{g}[\![u \in v]\!]_{\Omega} = \bar{g}\bigvee_{y \in \mathcal{D}_{\Omega}v}[\![u = y]\!]_{\Omega} \wedge v(y)$

$$= \bar{g}\tilde{\bigvee}_{y \in \mathcal{D}\mathcal{D}_{\Omega}v}[\![u = y]\!]_{\Omega} \wedge v(y)$$

$$= \tilde{\bigvee}_{y \in \mathcal{D}\mathcal{D}_{\Omega}v}[\![\bar{g}u = \bar{g}y]\!]_{\Omega} \wedge (\bar{g}v)(\bar{g}x)$$

$$= [\![\bar{g}u \in \bar{g}v]\!]_{\Omega}$$

Similarly, $[\![g[\![u = v]\!]_{\Omega} = [\![\bar{g}u = \bar{g}v]\!]_{\Omega}]\!] = 1.$

§8. $V^{(\Gamma')}$ in $V^{(\Gamma)}$.

We use the same H, $V^{(H)}$, $V^{(\Gamma)}$, $(V^{(\Gamma)})^{(\Omega)}$, $V^{(\bar{\Gamma})}$ as those in §6.

Let $G', \Gamma' \in V^{(\Gamma)}$ satisfy

$[\![G'$ is a subgroup of $\mathrm{Aut}(\Omega)]\!]_{\Gamma} = 1.$

$[\![\Gamma'$ is a normal filter of subgroups of $G']\!]_{\Gamma} = 1.$

If $h \in V^{(\Gamma)}$ and $[\![h \in \mathrm{Aut}(\Omega)]\!]_{\Gamma} = Eh$, then, by Theorem 6, there is $\tilde{h} : \tilde{\Omega} \to \tilde{\Omega}$ satisfying

(1) $\tilde{h} \cap (\widetilde{\Omega \upharpoonright Eh} \times \widetilde{\Omega \upharpoonright Eh}) \in Aut(\widetilde{\Omega \upharpoonright Eh})$.

(2) $a,b \in \tilde{\Omega} \Rightarrow [\![a = b]\!] \leq [\![\tilde{h}a = \tilde{h}b]\!]$.

(3) $\tilde{h}i = i$ on $\{g \in H_\Gamma | g \leq Eh\}$.

(4) $E = E\tilde{h}$ on $\widetilde{\Omega \upharpoonright Eh}$.

(5) $\bar{g} \in \bar{G}_h \Rightarrow \bar{g}\tilde{h} = \tilde{h}\bar{g}$ on $\tilde{\Omega}$.

Set

$$\tilde{G}' = \{h | h \in V^{(\Gamma)} \wedge [\![h \in G']\!]_\Gamma = Eh\}$$

Lemma 8.1.

If $\tilde{h}, \tilde{k} \in \tilde{G}'$, then $\widetilde{hk} = \tilde{h}\tilde{k}$ and $\widetilde{h^{-1}} = \tilde{h}^{-1}$.

Proof.

If $\tilde{h}, \tilde{k} \in G'$, then

$$a \in \tilde{\Omega} \Rightarrow [\![\langle a, \tilde{k}(a) \rangle \in k]\!] = Ea \wedge Ek,$$

$$[\![\langle \tilde{k}(a), \tilde{h}\tilde{k}(a) \rangle \in h]\!] = Ea \wedge Ek \wedge Eh.$$

$$\Rightarrow [\![\langle a, \tilde{h}\tilde{k}(a) \rangle \in hk]\!] = Ea \wedge Ek \wedge Eh$$

$$= Ea \wedge E(hk).$$

$$[\![\widetilde{hk}(a) = \tilde{h}\tilde{k}(a)]\!] = 1.$$

Therefore, $\widetilde{hk} = \tilde{h}\tilde{k}$. Similarly, $\widetilde{h^{-1}} = \tilde{h}^{-1}$.

For $\tilde{h} \in G'$ and $u \in V^{(\bar{\Gamma})}$ we define $\bar{\tilde{h}}u$ by

$$\mathcal{D}(\bar{\tilde{h}}u) = \{\bar{\tilde{h}}x | x \in \mathcal{D}(u)\},$$

$$(\bar{\tilde{h}}u)(\bar{\tilde{h}}x) = \tilde{h}u(x) = \tilde{h}E_{\tilde{\Omega}}x,$$

$$E_{\tilde{\Omega}}\bar{\tilde{h}}u = \tilde{h}E_{\tilde{\Omega}}u.$$

Lemma 8.2.

(1) $\tilde{h} \in \tilde{G}' \Rightarrow \bar{\tilde{h}} : V^{(\bar{\Gamma})} \to V^{(\bar{\Gamma})}$.

(2) $\tilde{h} \in \tilde{G}',\ u \in V^{(\bar{\Gamma})} \Rightarrow E_{\tilde{\Omega}}\bar{\tilde{h}}u \le iEh$.

(3) $\tilde{h} \in \tilde{G}',\ u,v \in V^{(\bar{\Gamma})}$

$$\Rightarrow \tilde{h}[\![u = v]\!]_{\bar{\Gamma}} = [\![\bar{\tilde{h}}u = \bar{\tilde{h}}v]\!]_{\bar{\Gamma}} \wedge iEh,$$

$$\tilde{h}[\![u \in v]\!]_{\bar{\Gamma}} = [\![\bar{\tilde{h}}u \in \bar{\tilde{h}}v]\!]_{\bar{\Gamma}} \wedge iEh.$$

(4) $\tilde{h},\tilde{k} \in \tilde{G}',\ u \in V^{(\bar{\Gamma})} \Rightarrow i[\![h = k]\!] \le [\![\bar{\tilde{h}}u = \bar{\tilde{k}}u]\!]_{\bar{\Gamma}}$

Proof.

(1) Let $\tilde{h} \in \tilde{G}'$ and $u \in V^{(\bar{\Gamma})}$.

$$\bar{g} \in \bar{G}(u) \cap \bar{G}_h,\ x \in \mathcal{D}(u) \Rightarrow \bar{\bar{g}}\bar{\tilde{h}}x = \bar{\tilde{h}}\bar{\bar{g}}x \in \mathcal{D}(\bar{\tilde{h}}x)$$

Therefore, $\bar{G}(\bar{\tilde{h}}u) = \{\bar{g} \in \bar{G} \mid \bar{\bar{g}}\ \mathcal{D}(\bar{\tilde{h}}u) = \mathcal{D}(\bar{\tilde{h}}u)\} \in \bar{\Gamma}$, that is, $\bar{\tilde{h}}u \in V^{(\bar{\Gamma})}$.

(2) $\tilde{h} \in \tilde{G}',\ u \in V^{(\bar{\Gamma})} \Rightarrow E_{\tilde{\Omega}}\bar{\tilde{h}}u = \tilde{h}E_{\tilde{\Omega}}u \le iEh$

(3) Let $\tilde{h} \in \tilde{G}',\ u,v \in V^{(\bar{\Gamma})}$.

Since $\tilde{h}[\![u = v]\!]_{\bar{\Gamma}} = \tilde{h}(\bigwedge_{x \in \mathcal{D}(u)} (E_{\tilde{\Omega}}x \to [\![x \in v]\!]_{\bar{\Gamma}})\ \tilde{\wedge}$

$$\tilde{h}(\bigwedge_{y \in \mathcal{D}(v)} (E_{\tilde{\Omega}}y \to [\![y \in u]\!]_{\bar{\Gamma}})\ \tilde{\wedge}$$

$$\tilde{h}(E_{\tilde{\Omega}}u \leftrightarrow E_{\tilde{\Omega}}v),$$

$$x \in \mathcal{D}(u) \Rightarrow \tilde{h}[\![u = v]\!]_{\bar{\Gamma}}\ \tilde{\wedge}\ E_{\tilde{\Omega}}\bar{\tilde{h}}x \le [\![\bar{\tilde{h}}x \in \bar{\tilde{h}}v]\!]_{\bar{\Gamma}}.$$

Hence $\tilde{h}[\![u = v]\!]_{\bar{\Gamma}} \le \bigwedge_{x \in \mathcal{D}(u)} (E_{\tilde{\Omega}}\bar{\tilde{h}}x \to [\![\bar{\tilde{h}}x \in \bar{\tilde{h}}v]\!]_{\bar{\Gamma}})$

Similarly,

$$\tilde{h}[\![u = v]\!]_{\bar{\Gamma}} \le \bigwedge_{y \in \mathcal{D}(v)} (E_{\tilde{\Omega}}\bar{\tilde{h}}y \to [\![\bar{\tilde{h}}y \in \bar{\tilde{h}}u]\!]_{\bar{\Gamma}})$$

Also $\tilde{h}[\![u = v]\!]_{\bar{\Gamma}} \le (E_{\tilde{\Omega}}hu \leftrightarrow E_{\tilde{\Omega}}hv)$.

Therefore, $\tilde{h}[\![u = v]\!]_{\overline{\Gamma}} \leq [\![\bar{\tilde{h}}u = \bar{\tilde{h}}v]\!]_{\overline{\Gamma}} \,\tilde{\wedge}\, iEh$, and

$$[\![\bar{\tilde{h}}u = \bar{\tilde{h}}v]\!]_{\overline{\Gamma}} \,\tilde{\wedge}\, iEh = \tilde{h}\widetilde{h^{-1}}[\![\bar{\tilde{h}}u = \bar{\tilde{h}}v]\!]_{\overline{\Gamma}}$$
$$\leq \tilde{h}[\![u = v]\!]_{\overline{\Gamma}}.$$

That is, $\tilde{h}[\![u = v]\!]_{\overline{\Gamma}} = [\![\bar{\tilde{h}}u = \bar{\tilde{h}}v]\!]_{\overline{\Gamma}}$.

It follows that

$$\tilde{h}[\![u \in v]\!]_{\overline{\Gamma}} = \tilde{h}\ \widetilde{\bigvee}_{y\in \mathcal{D}(v)} [\![u = y]\!]_{\overline{\Gamma}} \,\tilde{\wedge}\, E_{\tilde{\Omega}}y$$
$$= \widetilde{\bigvee}_{y\in \mathcal{D}(v)} [\![\bar{\tilde{h}}u = \bar{\tilde{h}}y]\!]_{\overline{\Gamma}} \,\tilde{\wedge}\, E_{\tilde{\Omega}}\bar{\tilde{h}}y$$
$$= [\![\bar{\tilde{h}}u \in \bar{\tilde{h}}v]\!]_{\overline{\Gamma}}.$$

(4) Let $\tilde{h},\tilde{k} \in \tilde{G}'$ and $u \in V^{(\overline{\Gamma})}$. If $a \in \tilde{\Omega}$ then we have

$$i[\![h = k]\!] \leq (\tilde{h}a \leftrightarrow \tilde{h}a), \quad \text{for}$$

$$Ea \wedge [\![h = k]\!] = [\![h = k]\!] \wedge [\![\langle a,\tilde{h}a\rangle \in h]\!] \wedge [\![\langle a,\tilde{k}(a)\rangle \in k]\!]$$
$$\leq [\![\tilde{h}a = \tilde{k}a]\!].$$

Hence $\tilde{h}a \,\tilde{\wedge}\, i[\![h = k]\!] = \tilde{k}a \,\tilde{\wedge}\, i[\![h = k]\!]$, and we have

$$i[\![h = k]\!] \leq (\tilde{h}a \leftrightarrow ka).$$

Now assume $i[\![h = k]\!] \leq [\![\bar{\tilde{h}}x = \bar{\tilde{k}}x]\!]_{\overline{\Gamma}}$ for $\forall x \in \mathcal{D}(u)$. Then

$$i[\![h = k]\!] \,\tilde{\wedge}(\bar{\tilde{h}}u(\bar{\tilde{h}}x) = i[\![h = k]\!] \,\tilde{\wedge}\, \tilde{h}E_{\tilde{\Omega}}x$$
$$= i[\![h = k]\!] \,\tilde{\wedge}\, \tilde{k}E_{\tilde{\Omega}}x \,\tilde{\wedge}\, [\![\bar{\tilde{h}}x = \bar{\tilde{k}}x]\!]_{\overline{\Gamma}}$$
$$\leq [\![\bar{\tilde{h}}x \in \bar{\tilde{k}}u]\!]_{\overline{\Gamma}} \quad \text{for } \forall x \in \mathcal{D}(u).$$

Similarly,

$$i[\![h = k]\!] \wedge (\bar{\tilde{k}}u)(\bar{\tilde{k}}x) \leq [\![\bar{\tilde{k}}x \in \bar{\tilde{h}}u]\!]_{\overline{\Gamma}} \quad \text{for } \forall x \in \mathcal{D}(u).$$

Therefore,

$$i[\![h = k]\!] \leq [\![\bar{\tilde{h}}u = \bar{\tilde{k}}u]\!]_{\overline{\Gamma}}.$$

q.e.d.

Theorem 8.

For each $F \in V^{(\Gamma)}$ such that $[\![F \in \Gamma']\!] = EF$, let

$$\tilde{F} = \{\tilde{h} \mid h \in V^{(\Gamma)} \wedge [\![h \in F]\!]_{\Gamma} = Eh\},$$

and define $\tilde{\Gamma}'$, $V^{(\tilde{\Gamma}')}$ by

$$\tilde{\Gamma}' = \{\tilde{F} \mid F \in V^{(\Gamma)} \wedge [\![F \in \Gamma']\!] = EF\},$$
$$V^{(\tilde{\Gamma}')} = \{u \in V^{(\overline{\Gamma})} \mid \mathcal{D}(u) \subseteq V^{(\Gamma')} \wedge \tilde{G}'(u) \in \tilde{\Gamma}'\},$$

where $\tilde{G}'(u) = \{\tilde{h} \in \tilde{G}' \mid [\![\tilde{h}u = u]\!]_{\overline{\Gamma}} \geq iEh\}$.

On the other hand, $(V^{(\Gamma)})^{(\Gamma')}$ is defined by

$$(V^{(\Gamma)})^{(\Gamma')} = \{v \in V^{(\Gamma)} \mid [\![v \in V^{(\Gamma')}]\!]_{\Gamma} = Ev\}.$$

Then the isomorphism

$$\Phi : V^{(\overline{\Gamma})} \longrightarrow (V^{(\Gamma)})^{(\Omega)}$$

defined in §7 induce the isomorphism

$$\Phi : V^{(\tilde{\Gamma}')} \longrightarrow (V^{(\Gamma)})^{(\Gamma')}.$$

Proof.

Assume that $u \in V^{(\tilde{\Gamma}')}$ and

$$x \in \mathcal{D}(u) \Rightarrow \Phi(x) = \overline{x} \in (V^{(\Gamma)})^{(\Gamma')}.$$

We claim that if $t \in V^{(\overline{\Gamma})}$, $h \in V^{(\Gamma)}$ and $[\![h \in G']\!] = Eh$, then $[\![[\![\overline{ht} = ht]\!]_{\Omega} = 1]\!] \geq Eh$, where $\mathcal{D}_{\Omega}(\overline{ht}) = \{\overline{hz} \mid \overline{z} \in \mathcal{D}_{\Omega}(t)\}^{\Gamma}$, $(\overline{ht})(\overline{hz}) = \tilde{h}E_{\Omega}\overline{z}$, and $E_{\Omega}\overline{ht} = \tilde{h}E_{\Omega}\overline{t}$.

Then

$$\tilde{h} \in \tilde{G}'(u) \iff [\![h \in G']\!] = Eh \wedge [\![\overline{\tilde{h}u} = u]\!]_{\overline{\Gamma}} \geq iEh,$$

$$\iff [\![h \in G']\!] = Eh \wedge [\![\tilde{h}u = u]\!]_{\Omega} \geq iEh,$$

$$\iff [\![h \in G']\!] = Eh \wedge [\![[\![hu = u]\!]_{\Omega} = 1]\!] \geq Eh,$$

$$\iff [\![h \in G'_{\overline{u}}]\!] = Eh,$$

$$\iff \tilde{h} \in \tilde{G}'_{\overline{u}} .$$

It follows that

$$\tilde{G}'(u) \in \Gamma' \iff [\![G'_{\overline{u}} \in \Gamma']\!] = EG'_{u} .$$

$$u \in V^{(\tilde{\Gamma}')} \iff \overline{u} \in (V^{(\Gamma)})^{(\Gamma')} .$$

<u>Proof of the claim:</u>

Assume that $(\forall z \in \mathcal{D}(t))([\![[\![\overline{hz} = \overline{\tilde{\overline{h}}z}]\!]_{\Omega} = 1]\!] = 1)$.

Since $\mathcal{D}\mathcal{D}_{\Omega}(\overline{ht}) = \{\overline{hz} \mid z \in \mathcal{D}(t)\}$, and

$\mathcal{D}\mathcal{D}_{\Omega}(\overline{\tilde{\overline{h}}t}) = \{\overline{\tilde{\overline{h}}z} \mid z \in \mathcal{D}(t)\}$

we have $[\![\mathcal{D}_{\Omega}(\overline{ht}) = \mathcal{D}_{\Omega}(\overline{\tilde{\overline{h}}t})]\!] = 1.$

$(\overline{ht})(\overline{hz}) = \tilde{h}E_{\Omega}\overline{z},$

$(\overline{\tilde{\overline{h}}t})(\overline{\tilde{\overline{h}}z}) = E_{\Omega}\tilde{\overline{h}}z = \underline{\tilde{h}E_{\Omega}}\overline{z},$ and

$E_{\Omega}(\overline{ht}) = \tilde{h}E_{\Omega}\overline{t} = E_{\Omega}\tilde{\overline{h}}t$

Therefore, $[\![\overline{ht} = \overline{\tilde{\overline{h}}t}]\!] = 1.$

Chapter III. Topological-properties of H, Ω and $\tilde{\Omega}$

Definitions.

Let $\langle\Omega,\wedge,\bigvee,1\rangle$ be a cHa.

(1) Ω is compact iff

$$(\forall A \subseteq \Omega)(\bigvee A = 1 \rightarrow \exists a_1,\ldots,a_n \in A(a_1 \vee \cdots \vee a_n = 1)),$$

where $\exists a_1,\ldots,a_n \in A(a_1 \vee \cdots \vee a_n = 1)$ is abbreviation of
$\exists n \in \omega \exists f \subseteq n \times A(f : n \rightarrow A \wedge \bigvee_{i \in n} f(i) = 1)$.

(2) Ω is locally compact iff

$$\bigvee \{p \in \Omega \mid \exists q \in \Omega[p \wedge q = 0 \wedge \forall A \subseteq \Omega(\bigvee A \vee q = 1 \rightarrow (\exists a_1,\ldots,a_n \in A)(a_1 \vee \cdots \vee a_n \vee q = 1))]\} = 1.$$

(3) Ω is connected iff

$$\forall p,q \in \Omega[(p \vee q = 1) \wedge (p \wedge q = 0) \rightarrow (p = 0) \vee (q = 0)].$$

(4) Ω is regular iff

$$\forall p \in \Omega(p = \bigvee \{q \in \Omega \mid \exists r \in \Omega((p \wedge r = 1) \quad \wedge \quad (q \wedge r = 0))\}).$$

(5) Ω is normal iff

$$\forall p,q \in \Omega(p \vee q = 1 \rightarrow \exists r,s \in \Omega[(r \wedge s = 0) \wedge (p \vee r = 1) \wedge (q \vee s = 1)]).$$

Let H be a cHa, $\langle\Omega,\wedge,\bigvee,1\rangle \in V^{(H)}$ and

$[\![\langle\Omega,\wedge,\bigvee,1\rangle$ is a cHa$]\!] = 1$. Then
$\tilde{\Omega} = \{a \in V^{(H)} \; [\![a \in \Omega]\!] = Ea\}$ is a cHa as shown in Chapter I.

In this chapter we discuss about the topological properties of H, Ω and $\tilde{\Omega}$.

Theorem 1.

(1) If H is compact and $[\![\Omega$ is compact$]\!] = 1$, then $\tilde{\Omega}$ is compact.

(2) If $\tilde{\Omega}$ is compact, then H is compact.

(3) If H is regular and $\tilde{\Omega}$ is compact, then $[\![\Omega$ is compact$]\!] = 1$.

Proof.

(1) Assume that H is compact, $[\![\Omega$ is compact$]\!] = 1$ and $A \subseteq \tilde{\Omega}$, $\bigvee A = 1$. Then for A^* defined by

$$\mathcal{D}(A^*) = A, \quad EA^* = \bigvee_{a\in A} Ea,$$
$$a \in A \Rightarrow A^*(a) = Ea,$$

we have

$$[\![A^* \subseteq \Omega \wedge \bigvee A^* = 1]\!] = 1.$$

Hence

$$[\![\exists n \in \check{\omega}\, \exists f : n \to A^*(\bigvee_{i\in n} f(i) = 1)]\!] = 1, \text{ and so}$$

$$\bigvee_{n\in\omega} \bigvee_{f\in V^{(H)}} [\![f : n \to A^* \wedge \bigvee_{i\in n} f(i) = 1]\!] = 1$$

Since H is compact, there exist finite sets,

$\{n_1,\dots,n_k\} \subset \omega$,

$\{s_1,\dots,s_k\} \subset \omega$ and

$\{f_{ij} \mid 1 \le i \le k,\ 1 \le j \le s_k\}$ such that

$$\bigvee_{i,j} [\![f_{ij} : n_i \to A^* \wedge \bigvee_{t\in n_i} f_{ij}(t) = 1]\!] = 1.$$

By Theorem I.2, there exists $\widetilde{f_{ij}} : n_i \to \widetilde{A^*}$ such that

$$t \in n_i \Rightarrow [\![\langle \check{t}, \widetilde{f_{ij}}(\check{t})\rangle \in f_{ij}]\!] = Ef_{ij} = E\widetilde{f_{ij}}(\check{t}),$$

and

$$[\![\tilde{\bigvee}_{t\in n_i} \tilde{f}_{ij}(\check{t}) = 1]\!] \geq Ef_{ij},$$

where

$$\tilde{A}^* = \{a \in \tilde{\Omega} | [\![a \in A^*]\!] = Ea\}.$$

It follows that

$$[\![\tilde{\bigvee}_{i,j} \tilde{\bigvee}_{t\in n_i} \tilde{f}_{ij}(\check{t}) = 1]\!] = 1.$$

That is, there exists a finite subset $\{a_1,\ldots,a_n\}$ of $\tilde{A}^*$ such that

$$\tilde{\bigvee}\{a_1,\ldots,a_n\} = 1.$$

For each t with $1 \leq t \leq n$,

$$[\![a_t \in A^*]\!] = \bigvee_{c\in A} [\![a_t = c]\!] \wedge Ec = Ea_t.$$

By the compactness of H, there is $\{c_1,\ldots,c_s\} \subseteq A$ such that

$$([\![a_t = c_1]\!] \wedge Ec_1) \vee \cdots \vee ([\![a_t = c_s]\!] \wedge Ec_s) = Ea_t,$$

Since $[\![a_t = c_r]\!] \wedge Ec_r \leq [\![a_t \leq \tilde{\bigvee}\{c_1,\ldots,c_s\}]\!]$ for $1 \leq r \leq s$, we have $Ea_t \leq [\![a_t \leq \tilde{\bigvee}\{c_1,\ldots,c_s\}]\!]$.

Therefore, $a_t \leq \tilde{\bigvee}\{c_1,\ldots,c_s\}$ for $\{c_1,\ldots,c_s\} \subseteq A$.

Hence there exists a finite subset B of A such that

$$\tilde{\bigvee} B \geq \tilde{\bigvee}\{a_1,\ldots,a_n\} = 1.$$

Therefore, $\tilde{\Omega}$ is compact.

(2) Assume that $\tilde{\Omega}$ is compact, and $G \subseteq H \wedge \bigvee G = 1$. Then we have

$$\tilde{\bigvee}_{g\in G} 1 \upharpoonright g = 1.$$

By the compactness of $\tilde{\Omega}$, there exists a finite subset $\{g_1,\ldots,g_n\}$ of G such that

$$\tilde{\bigvee}\{1 \upharpoonright g_1,\ldots,1 \upharpoonright g_n\} = 1,$$

and hence, $\bigvee\{g_1,\ldots,g_n\} = 1$.

Therefore, H is compact.

(3) Assume that $\tilde{\Omega}$ is compact and H is regular. Let $[\![A \subseteq \Omega \wedge \bigvee A = 1]\!] = p$

Since H is regular,

$$p = \bigvee\{q \in H \mid \exists r \in H(p \vee r = 1 \text{ and } q \wedge r = 0)\}.$$

Let $q,r \in H$, $p \vee r = 1$ and $q \wedge r = 0$. Then

since $[\![\bigvee A = \tilde{\bigvee}\tilde{A} = 1]\!] = p$, where $\tilde{A} = \{a \in V^{(H)} \mid [\![a \in A]\!] = Ea\}$,

we have $(\tilde{\bigvee}\tilde{A}) \upharpoonright p = 1 \upharpoonright p$.

Therefore, $(\tilde{\bigvee}\tilde{A}) \tilde{\vee}\, 1 \upharpoonright r \geq (\tilde{\bigvee}\tilde{A}) \upharpoonright p \,\tilde{\vee}\, 1 \upharpoonright r = 1 \upharpoonright p \,\tilde{\vee}\, 1 \upharpoonright r = 1.$

By the assumption, there exists a finite subset $\{a_1,\ldots,a_n\}$ of $\tilde{A}$ such that

$$\tilde{\bigvee}\{a_1,\ldots,a_n, 1 \upharpoonright r\} = 1.$$

Hence $\tilde{\bigvee}\{a_1,\ldots,a_n\} \geq 1 \upharpoonright q$, and so

$[\![a_1 \vee \cdots \vee a_n = 1]\!] \geq q$

Therefore, $[\![\exists a_1,\ldots,a_n \in A(a_1 \vee \cdots \vee a_n = 1)]\!] \geq q$

It follows that

$[\![\exists a_1,\ldots,a_n \in A(a_1\vee\cdots\vee a_n = 1)]\!] \geq p.$

Since $[\![\check{\phi} \subseteq \Omega]\!] = 1$ and $[\![\check{\phi} \subseteq \Omega]\!] \leq [\![\exists ! u \in \Omega(\bigvee\check{\phi} = u)]\!]$, there exists $u \in \tilde{\Omega}$ such that $[\![\bigvee\check{\phi} = u]\!] = 1$. We denote the u by $\tilde{0}$. Then we have

$$[\![\forall a \in \Omega(\tilde{0} \leq a)]\!] = 1 \text{ and } E\tilde{0} = 1.$$

We say $\tilde{\Omega}$ is $\underline{\tilde{0}\text{-locally compact}}$ if

$$\bigvee\{p\in\tilde{\Omega} \mid \exists q\in\tilde{\Omega}[p \tilde{\wedge} q \leq \tilde{0} \wedge \forall A\subseteq\tilde{\Omega}(\tilde{\bigvee} A \tilde{\vee} q = 1 \rightarrow (\exists a_1,\ldots,a_n \in A)(\tilde{\bigvee}\{a_1,\ldots,a_n\}\tilde{\vee} q = 1))]\} = 1.$$

Theorem 2.

(1) If H is regular and locally compact and $[\![\Omega$ is locally compact$]\!] = 1$, then $\tilde{\Omega}$ is $\tilde{0}$-locally compact.

(2) If H is regular and $\tilde{\Omega}$ is $\tilde{0}$-locally compact, then $[\![\Omega$ is locally compact$]\!] = 1$.

Proof.

1) Assume that H is regular and locally compact and $[\![\Omega$ is locally compact$]\!] = 1$. Then

$$\tilde{\bigvee}\{p\in\tilde{\Omega} \mid [\![\exists q\in\Omega[p\wedge q = \tilde{0} \wedge \forall A\subseteq\Omega(\bigvee A \vee q = 1 \rightarrow \exists N\in\check{\omega}\,\exists f(f:N\rightarrow A \wedge \bigvee_{n\in N} f(n)\vee q = 1))]]\!] \geq Ep\} = 1.$$

Let M be the set of $\langle p,q\rangle \in \tilde{\Omega}\times\tilde{\Omega}$ such that

$$[\![p\wedge q = \tilde{0} \wedge \forall A\subseteq \Omega(\bigvee A \vee q = 1 \rightarrow \exists N\in\check{\omega}\,\exists f(f:N\rightarrow A \wedge \bigvee_{n\in N} f(n) \vee q = 1))]\!] \geq Ep \wedge Eq \neq 0.$$

Then $\tilde{\bigvee}_{\langle p,q\rangle\in M} p \restriction Eq = 1.$

Since H is locally compact,

$$\bigvee\{a \in H \mid \exists b \in H(a \wedge b = 0 \wedge \forall K \subseteq H(\bigvee K \vee b = 1 \rightarrow \exists k_1,\ldots,k_s \in K(k_1 \vee \cdots \vee k_s \vee b = 1)))\} = 1.$$

Let N be the set of $\langle a,b\rangle \in H \times H$ such that

$$a \wedge b = 0 \wedge \forall K \subseteq H(\bigvee K \vee b = 1 \rightarrow \exists k_1,\ldots,k_s \in K(k_1 \vee \cdots \vee k_s \vee b = 1)).$$

Then $\bigvee_{\langle a,b\rangle \in N} a = 1$.

By the regularity of H, for $\langle p,q\rangle \in M$, $\langle a,b\rangle \in N$,

$$Ep \wedge Eq \wedge a = \bigvee\{x \in H \mid \exists y \in H((Ep \wedge Eq \wedge a) \vee y = 1 \text{ and } x \wedge y = 0)\}.$$

Let $L(p,q,a,b)$ be the set of $\langle x,y\rangle \in H \times H$ such that $(Ep \wedge Eq \wedge a) \vee y = 1$ and $x \wedge y = 0$. Then for $\langle x,y\rangle \in L(p,q,a,b)$

(i) $x \leq Ep \wedge Eq \wedge a$ and $b \leq y$

(ii) $(p \restriction x) \wedge (q \vee 1 \restriction y) \leq \tilde{0}$

(iii) $\tilde{\bigvee}_{\langle p,q\rangle \in M, \langle a,b\rangle \in N, \langle x,y\rangle \in L(p,q,a,b)} p \restriction x = 1$.

Proof.

(i) $(Ep \wedge Eq \wedge a) \wedge x = ((Ep \wedge Eq \wedge a) \vee y) \wedge x = x$.

$$b \wedge y = b \wedge ((Ep \wedge Eq \wedge a) \vee y) = b.$$

(ii) $[\![p \wedge q = \tilde{0}]\!] \geq Ep \wedge Eq$ implies $p \restriction x \wedge q \leq \tilde{0}$, and $x \wedge y = 0$ implies $(p \restriction x) \wedge (1 \restriction y) \leq \tilde{0}$. Therefore, $(p \restriction x) \wedge (q \vee 1 \restriction y) \leq \tilde{0}$.

(iii) $\bigvee_{\langle p,q\rangle \in M, \langle a,b\rangle \in N, \langle x,y\rangle \in L(p,q,a,b)} p \restriction x$

$$= \bigvee_{\langle p,q\rangle \in M, \langle a,b\rangle \in N} p \restriction Eq \wedge a$$

$$= \bigvee_{\langle p,q\rangle \in M} p \restriction Eq = 1.$$

Let $\langle p,q\rangle \in M$, $\langle a,b\rangle \in N$ and $\langle x,y\rangle \in L(p,q,a,b)$ and assume $A \subseteq \Omega$ and $\tilde{\bigvee} A \,\tilde{\vee}\, (q \,\tilde{\vee}\, 1 \upharpoonright y) = 1$. Then we have the following results in turn.

$$[\![A^* \subseteq \Omega \wedge (\bigvee A^* \vee (q \vee 1 \upharpoonright y) = 1)]\!] = 1.$$

$$[\![A^* \subseteq \Omega \wedge [\bigvee (A \cup \{1 \upharpoonright y\})^* \vee q = 1]]\!] = 1.$$

$$[\![\exists N \in \check{\omega}\, \exists f (f : N \to A^* \wedge (\bigvee_{n \in N} f(n) \vee 1 \upharpoonright y \vee q = 1))]\!] \geq Ep \wedge Eq.$$

$$\bigvee_{N \in \omega} \bigvee_f [\![f : N \to A^* \wedge (\bigvee_{n \in \check{N}} f(n) \vee q = 1)]\!] \geq Ep \wedge Eq.$$

$$\bigvee_{N \in \omega} \bigvee_f [\![f : N \to A^* \wedge (\bigvee_{n \in \check{N}} f(n) \vee q = 1)]\!] \vee y \vee b = 1.$$

By the same argument as Theorem 1 (1), there is a finite set $\{a_1, \ldots, a_n\}$ of A such that

$$[\![a_1 \vee \cdots \vee a_n \vee q = 1]\!] \vee y = 1.$$

Therefore, $a_1 \tilde{\vee} \cdots \tilde{\vee} a_n \tilde{\vee} q \tilde{\vee} 1 \upharpoonright y = 1$. Hence, by using (ii), (iii) $\tilde{\Omega}$ is $\tilde{0}$-locally compact.

(2) Assume that H is regular and $\tilde{\Omega}$ is $\tilde{0}$-locally compact. Let

$$M = \{\langle p,q\rangle \in \tilde{\Omega} \times \tilde{\Omega} \mid p \wedge q \leq \tilde{0} \wedge \forall A \subseteq \tilde{\Omega} (\tilde{\bigvee} A \vee q = 1 \to \exists \{a_1, \ldots, a_n\} \subseteq A (a_1 \vee \cdots \vee a_n \vee q = 1)\}.$$

We have $\bigvee_{\langle p,q\rangle \in M} p = 1$.

For $\langle p,q\rangle \in M$ and $A \subseteq \tilde{\Omega}$,

$$[\![\tilde{\bigvee} A \vee q = 1]\!] \wedge Ep = \bigvee \{s \in H \mid \exists r ((h \vee r = 1) \wedge (s \wedge r = 0))\},$$

by regularity of H. Let

$$N(p,q,A) = \{\langle s,r\rangle \in H \times H) ([\![\hat{\bigvee} A \vee q = 1]\!] \wedge Ep) \vee r = 1 \text{ and } (s \wedge r = 0)\}.$$

Then

(i) $\bigvee\{s \in H \mid \langle s,r\rangle \in N(p,q,A)\} = [\![\tilde{\bigvee} A \vee q = 1]\!] \wedge Ep.$

(ii) $\langle s,r\rangle \in N(p,q,A) \Rightarrow \tilde{\bigvee} A \tilde{\vee} q \tilde{\vee} 1 \upharpoonright r = 1.$

Proof.

(i) is obvious from regularity of H.

(ii) Let $[\![\tilde{\bigvee} A \vee q = 1]\!] = c$. Then $\tilde{\bigvee} A \tilde{\vee} q \geq 1 \upharpoonright c$.

Therefore,

$$\tilde{\bigvee} A \tilde{\vee} q \tilde{\vee} 1 \upharpoonright r \geq 1 \upharpoonright c \tilde{\vee} 1 \upharpoonright r = 1.$$

Since $\langle p,q\rangle \in M$, there is a finite set $\{a_1,\dots,a_n\} \subseteq A$ such that $a_1 \tilde{\vee} \cdots \tilde{\vee} a_n \tilde{\vee} 1 \upharpoonright r \tilde{\vee} q = 1$, and $a_1 \tilde{\vee} \cdots \tilde{\vee} a_n \tilde{\vee} q \tilde{\vee} 1 \upharpoonright r = 1$ implies $a_1 \tilde{\vee} \cdots \tilde{\vee} a_n \hat{\vee} q \geq 1 \upharpoonright s$. Therefore, for $\langle s,r\rangle \in N(p,q,A)$,

$$[\![\exists a_1,\dots,a_n \subseteq A(\bigvee\{a_1,\dots,a_n\} \vee q = 1)]\!] \geq s.$$

It follows that

$$[\![\hat{\bigvee} A \vee q = 1]\!] \wedge Ep \leq [\![\exists a_1,\dots,a_n \subseteq A(\bigvee\{a_1,\dots,a_n\} \vee q = 1)]\!].$$

Hence $\quad Ep \leq [\![\exists q \in \Omega(\forall A \subseteq \Omega(\bigvee A \vee q = 1 \rightarrow \exists\{a_1,\dots,a_n\}(\bigvee\{a_1,\dots,a_n\} \vee q = 1$

Therefore,

$$[\![\bigvee\{p \in \Omega \mid \exists q \in \Omega[\forall A \subseteq \Omega(\bigvee A \vee q = 1 \rightarrow \exists\{a_1,\dots,a_n\}(\bigvee\{a_1,\dots,a_n\} \vee q = 1))]\}$$
$$\geq \tilde{\bigvee}\{p \in \tilde{\Omega} \mid \langle p,q\rangle \in M\} = 1]\!] = 1.$$

Theorem 3.

(1) If H is connected and $[\![\Omega$ is connected$]\!] = 1$, then $\tilde{\Omega}$ is $\tilde{0}$-connected,

(2) If $\tilde{\Omega}$ is $\tilde{0}$-connected, then H is connected

(3) If for any $h \in H$ $\widetilde{\Omega \upharpoonright h}$ is $\tilde{0}$-connected, then $[\![\Omega$ is connected$]\!] = 1$,

where $\tilde{\Omega}$ is $\tilde{0}$-connected iff

$$p,q \in \tilde{\Omega},\ p \vee q = 1,\ p \wedge q \leq 0 \Rightarrow p \leq \tilde{0} \text{ or } q \leq \tilde{0}.$$

roof.

(1) Assume that H is connected, $\llbracket \Omega$ is connected $= 1\rrbracket$ and $p,q \in \tilde{\Omega},\ p \vee q = 1,\ p \wedge q \leq \tilde{0}$. Set $p' = p \vee \tilde{0}$ and $q' = q \vee \tilde{0}$.

hen

$$\llbracket (p' \vee q' = 1) \wedge (p' \wedge q' = 0) \rrbracket = 1.$$

ince $\llbracket \Omega$ is connected$\rrbracket = 1$,

$$\llbracket (p' = 0) \wedge (q' = 0) \rrbracket = 1.$$

n the other hand,

$$\llbracket (p' = 0) \wedge (q' = 0) \rrbracket \leq \llbracket p' \vee q' = \tilde{0} = 1 \rrbracket = 0.$$

ince H is connected,

$$\llbracket p' = 0 \rrbracket = 0 \text{ or } \llbracket q' = 0 \rrbracket = 0.$$

f $\llbracket p' = \tilde{0} \rrbracket = 0$, then $\llbracket q' = \tilde{0} \rrbracket = 1$, and hence $q \leq \tilde{0}$.

f $\llbracket q' = \tilde{0} \rrbracket = 0$, then $\llbracket p' = \tilde{0} \rrbracket = 1$, and hence $p \leq \tilde{0}$.

2) Assume that $\tilde{\Omega}$ is $\tilde{0}$-connected and

$$h_1, h_2 \in H,\ h_1 \vee h_2 = 1,\ h_1 \wedge h_2 = 0.$$

hen

$$1 \upharpoonright h_1 \vee 1 \upharpoonright h_2 = 1 \text{ and } 1 \upharpoonright h_1 \wedge 1 \upharpoonright h_2 \leq \tilde{0}.$$

y the assumption, $1 \upharpoonright h_1 \leq \tilde{0}$ or $1 \upharpoonright h_2 \leq \tilde{0}$. If $1 \upharpoonright h \leq \tilde{0}$, then . $\leq \llbracket 1 \leq \tilde{0} \rrbracket = 0$. Therefore, $h_1 = 0$ or $h_2 = 0$.

(3) Assume that $(\forall h \in H)(\widetilde{\Omega \upharpoonright h}$ is connected), $p,q \in \tilde{\Omega}$ and $[\![(p \vee q = 1) \wedge (p \wedge q = \tilde{0})]\!] = h$. Then

$$p \upharpoonright h \,\tilde{\vee}\, q \upharpoonright h = 1 \upharpoonright h \quad \text{and} \quad p \upharpoonright h \,\tilde{\wedge}\, q \upharpoonright h \leq \tilde{0}.$$

By the assumption, $p \upharpoonright h \leq \tilde{0}$ or $q \upharpoonright h \leq \tilde{0}$. Therefore, $[\![p = 0]\!] \geq h$ or $[\![q = 0]\!] \geq h$. It follows that,

$$[\![(p \vee q = 1) \wedge (p \wedge q = \tilde{0}) \rightarrow (p = 0) \vee (q = 0)]\!] = 1.$$

REFERENCES

[1] M. P. Fourman and D. S. Scott: Sheaves and Logic. Applications of Sheaves ed. by Fourman, Mulvey and Scott, pp 302-401. Lecture Notes in Mathematics 753, Springer, 1979.

[2] R. J. Grayson: Heyting valued models for intuitionistic set theory. Ibid. pp 402-414.

[3] R. J. Grayson: A sheaf approach to models of set theory. M. Sc. Thesis, Oxford 1975.

[4] D. S. Scott: Lectures on Boolean-valued models for set theory. Lecture notes of the U. C. L. A. Summer Institute on Set Theory, 1967.

[5] R. M. Solovay and S. Tennenbaum: Iterated Cohen extensions and Souslin's problem. Annals of Math., Vol. 94, pp 201-245, 1971.

INDEPENDENCE OF A PROBLEM IN ELEMENTARY ANALYSIS FROM SET THEORY*)

Tosiyuki Tugué and Hisao Nomoto

Introduction

In a correspondence to one of the authors, G. Takeuti has proposed the following problem.

Let A *be a subset of all real numbers* $\mathbb{R}$. *Consider the statement:*

(P) *For any sequence* $\{a_k\}$ *of real numbers, if* $\lim_{k\to\infty} e^{2\pi i a_k t} = 1$ *for every* $t \in A$, *then* $\{a_k\}$ *converges to* 0.

When $A = \mathbb{R}$, *clearly* (P) *holds. What then are the conditions on* A *that the statement* (P) *be true?*

Given A, let us consider the question whether (P) holds or not. We begin with some examples. If A is the set $\mathbb{Q}$ of all rational numbers, then (P) is trivially false when we consider the sequence $\{n!\}$. In general, we have the same answer as $\mathbb{Q}$ for any countable set A, that is, there exists an increasing sequence $\{n_k\}$ of natural numbers such that $\lim_{k\to\infty} e^{2\pi i n_k t} = 1$ for each t of A. Concerning the uncountable set, in addition to $\mathbb{R}$ or intervals, we have examples of nowhere dense null sets for which the answer is Yes. Cantor's set C is such an instance. This fact follows from the following observations:

*) This research was partially supported by Grant-in-Aid for Co-operative Research Proj. No.434007 and Scientific Research Proj. No.546004, Ministry of Education, Japan.

(i) Given a sequence $\{a_k\}$, the Borel set

$$G = \{t \in \mathbb{R} \mid \lim_{k\to\infty} e^{2\pi i a_k t} = 1\}$$

forms a subgroup of the additive group $\mathbb{R}$.

(ii) Cantor set C generates the full group $\mathbb{R}$.

Contrary to these examples, even if A is uncountable, it happens that the answer is No. A simple example is the set of all Liouville's transcendental numbers of the type $\sum_{k=1}^{\infty} \frac{\tau_k}{10^{k!}}$ ($\tau_k = 0, 1, \cdots, 9$), whose cardinality is $2^{\aleph_0}$ [Proposition 1].

Observing these examples, we became interested in obtaining any method to decide whether the statement (P) be true or false for a set A. The result is that there exist sets for which the statement (P) is independent of the ZFC set theory.

In §1, as preliminaries, introducing a class of sets determined by a simple way, we give a criterion of the statement (P) on this class. In §2, we deal with the problem from the viewpoint of cardinality and show that the answer is No for any set A with cardinality $\kappa < 2^{\aleph_0}$, under the assumption of Martin's axiom (Martin and Solovay [8]) and $\aleph_1 < 2^{\aleph_0}$. In §3, we will be concerned with measure and category of sets and reduce the problem to discussions on the class of null and meager sets. §4 is devoted to discussions of independence of the statement (P) from ZFC, especially, on the class of sets of Borel's strong measure 0.

§1. Preliminaries

Let $\mathcal{C}$ be the collection of pairs $\langle A, \{a_k\}\rangle$, A a subset of $\mathbb{R}$ and $\{a_k\}$ a sequence in $\mathbb{R}$, which satisfies the following condition:

(C) $$\lim_{k\to\infty} e^{2\pi i a_k t} = 1 \quad \text{for every} \quad t \in A.$$

Let $p \geqq 2$ be an integer, $\nu = \{n_k\}$ be an increasing sequence in $\mathbb{N}$ and set

$$A_\nu = \{t \mid t = \sum_{k=1}^{\infty} \frac{\tau_k}{p^{n_k}}, \quad \tau_k = 0,1,\cdots,p-1\}.$$

Then we have the following criterion of the statement (P) of A_ν.

<u>PROPOSITION</u> 1. <u>In order that the statement</u> (P) <u>hold for</u> A_ν, <u>it is necessary and sufficient that</u> $\Delta\nu = \{n_{k+1} - n_k\}_{k\in\mathbb{N}}$ <u>be bounded</u>. <u>In particular, if</u> $\Delta\nu$ <u>is unbounded and if</u> $\lim(n_{k_j+1} - n_{k_j}) = \infty$, <u>then</u> $< A_\nu, \{p^{n_{k_j}}\} > \in \mathfrak{C}$.

<u>Proof</u>. For any $t \in A_\nu$, we have

$$\varepsilon_k = \text{the decimal part of } p^{n_k} t = p^{n_k} t - [p^{n_k} t] = p^{n_k} \sum_{j=1}^{\infty} \frac{\tau_{k+j}}{p^{n_{k+j}}}$$

$$\leqq \sum_{j=1}^{\infty} \frac{p-1}{p^{n_{k+j} - n_k}} \leqq \sum_{j=1}^{\infty} \frac{p-1}{p^{n_{k+1} - n_k + j - 1}} = \frac{1}{p^{n_{k+1} - n_k - 1}}.$$

Hence, if $n_{k_j+1} - n_{k_j} \longrightarrow \infty$, it follows $\varepsilon_{k_j} \longrightarrow 0$ and $e^{2\pi i p^{n_{k_j}} t} = e^{2\pi i \varepsilon_{k_j}} \longrightarrow 1$. This shows $<A_\nu, \{p^{n_{k_j}}\}> \in \mathfrak{C}$, hence (P) is false for A_ν.

Let $\Delta\nu$ be bounded and assume $\Delta\nu$ be a repetition of finitely many positive integers $u_1, \cdots, u_r$. Each term n_k of ν is of the form

$$n_k = m + \lambda_{k1} u_1 + \cdots + \lambda_{kr} u_r, \quad k = 1,2,\cdots,$$

where $m = n_1, \lambda_{k1}, \cdots, \lambda_{kr} \in \omega$. Let $\nu_j = \{m + n_{jk}\}$ $(j=1,2,\cdots,r)$ be the subsequence of ν formed by terms of increasing parts of the function $k \mapsto \lambda_{kj}$. For example if $r=2$ and

$$\nu = \{m, m+u_1, \cdots, m+m_1u_1, m+m_1u_1+u_2, \cdots, m+m_1u_1+\ell_1u_2, m+(m_1+1)u_1+\ell_1u_2, \cdots\},$$

then we mean by ν_1 and ν_2 the subsequences

$$\nu_1 = \{m, m+u_1, \cdots, m+m_1u_1, m+(m_1+1)u_1+\ell_1u_2, m+(m_1+2)u_1+\ell_1u_2, \cdots\}$$

and

$$\nu_2 = \{m+m_1u_1+u_2, \cdots, m+m_1u_1+\ell_1u_2, m+(m_1+m_2)u_1+(\ell_1+1)u_2, \cdots\}.$$

Note that ν_j is a finite or infinite sequence according as u_j appears in $\Delta\nu$ for finitely or infinitely many times.

Now, one can show that

$$\{n_{jk}-\ell \mid 0 \leqq \ell \leqq u_j-1,\ 1 \leqq j \leqq r,\ 1 \leqq k\} = \mathbb{N}.$$

Therefore for any $t = \sum_{i=1}^{\infty} \frac{\tau_i}{p^i} \in [0, 1]$, we have

$$t = \sum_{j=1}^{r} \sum_{\ell=0}^{u_j-1} \sum_{k \geqq 1} \frac{\tau_{jk\ell}}{p^{n_{jk}-\ell}}, \text{ where } \tau_{jk\ell} = \tau_i \text{ provided } n_{jk}-\ell = i,$$

so that

$$p^m t = \sum_{j=1}^{r} \sum_{\ell=0}^{u_j-1} p^{m+\ell} t_{j\ell}, \text{ where } t_{j\ell} = \sum_{k \geqq 1} \frac{\tau_{jk\ell}}{p^{m+n_{jk}}} \in A_{\nu_j}.$$

Let G_ν be the additive subgroup of $\mathbb{R}$ generated by A_ν. Then, by the above expression of $p^m t$ together with $A_{\nu_1} \cup \cdots \cup A_{\nu_r} \subseteq A_\nu$, we have $[0, 1] \subseteq G_\nu$, which shows $G_\nu = \mathbb{R}$. Therefore, $\langle A_\nu, \{a_k\}\rangle \in \mathcal{C}$ implies $\langle \mathbb{R}, \{a_k\}\rangle \in \mathcal{C}$, hence it follows that $a_k \longrightarrow 0$.

□

§2. Cardinality of set A

By Proposition 1 we can realize that there are many sets A with cardinal $2^{\aleph_0}$ for which (P) does not hold, as well as the set of Liouville's transcendental numbers of the type mentioned in Introduction. Here, in view of cardinality of A let us consider the case where $|A| < 2^{\aleph_0}$. In this section, we shall use an increasing sequence $\{n_k\}_{k\in\omega}$ of natural numbers as an increasing function $f : \omega \longrightarrow \omega$; $k \longmapsto f_k$.

Given a subset A of $\mathbb{R}$ with cardinality $|A| = \kappa < 2^{\aleph_0}$, let $A = \{t_\alpha \mid \alpha < \kappa\}$.

As a natural extension of the method for the countable case, by transfinite induction we can define a set $\{f^\alpha \mid \alpha < \omega_1\}$ of functions $f^\alpha : \omega \longrightarrow \omega$ such that the following two conditions are satisfied:

(a) for any $\beta < \alpha < \omega_1$, $\{f_k^\alpha\}$ is a subsequence of $\{f_k^\beta\}$ except for its finite terms,

(b) for any $\beta \leq \alpha < \omega_1$ and $t_\beta \in A$, $f_k^\alpha t_\beta - [f_k^\alpha t_\beta]$ converges to a limit s_β which is independent of α $(> \beta)$.

To show this, first we choose a subsequence $\{n_k t_0 - [n_k t_0]\}_{k\in\omega}$ of $\{nt_0 - [nt_0]\}_{n\in\omega}$ such that $n_k t_0 - [n_k t_0] \longrightarrow s_0$, and put $f^0 = \{n_k\}_{k\in\omega}$. As the induction step, for any $\alpha < \omega_1$, suppose that $\{f^\beta \mid \beta < \alpha\}$ be chosen so as to satisfy the conditions (a) and (b). We consider two cases according as α is a successor or limit ordinal.

Case 1. α is a successor ordinal. Let $\alpha = \beta + 1$. The sequence $\{f_n^\beta \cdot t_\alpha\}_{n\in\omega}$ contains a subsequence $\{f_{n_k}^\beta \cdot t_\alpha\}$ such that their decimal parts converge to a limit s_α. Taking such one, define f^α by $f^\alpha(k) = f^\beta(n_k)$. Then f^α satisfies (a) and (b), since f^α is a subsequence of f^β.

Case 2. α is a limit ordinal. We take any increasing sequence $\{\beta_k\}_{k\in\omega}$ of ordinals such that $\alpha = \lim_{k\to\omega} \beta_k$. By the assumption, f^{β_k} is a subsequence of f^{β_i} $(i < k)$ except for its finite terms, say $\{f_0^{\beta_k}, \cdots, f_{n_i}^{\beta_k}\}$, and

$$f_n^{\beta_k} \cdot t_\gamma - [f_n^{\beta_k} \cdot t_\gamma] \longrightarrow s_\gamma, \quad \text{for every } \gamma \leqq \beta_k .$$

Now, define a function $\tilde{f}^\alpha : \omega \longrightarrow \omega$ by

$$\tilde{f}^\alpha(k) = f^{\beta_k}(m_k),$$

where $m_k = \max\{k, n_0, \cdots, n_{k-1}\}+1$.

Let β be any ordinal $< \alpha$. Then, for some k we have $\beta < \beta_k$. By induction hypothesis, f^{β_k} is a subsequence of f^β except for its finite terms. Hence, $\tilde{f}^\alpha$ is a subsequence of f^β except for its finite terms, for $\{\tilde{f}_n^\alpha\}_{n\geqq k}$ is clearly a subsequence of $\{f_n^{\beta_k}\}_{n\in\omega}$ for every k. Again by induction hypothesis $f_n^\beta \cdot t_\beta - [f_n^\beta \cdot t_\beta] \longrightarrow s_\beta$, it follows that $\tilde{f}_n^\alpha \cdot t_\beta - [\tilde{f}_n^\alpha \cdot t_\beta] \longrightarrow s_\beta$.

Finally, considering the sequence $\{\tilde{f}_n^\alpha \cdot t_\alpha - [\tilde{f}_n^\alpha \cdot t_\alpha]\}$ and taking a convergent subsequence of it, say

$$\tilde{f}_{n_k}^\alpha \cdot t_\alpha - [\tilde{f}_{n_k}^\alpha \cdot t_\alpha] \longrightarrow s_\alpha,$$

set $f^\alpha(k) = \tilde{f}^\alpha(n_k)$. f^α clearly satisfies (a) and (b).

Here, we assume Martin's axiom and $\aleph_1 < 2^{\aleph_0}$. Let κ be any cardinal such that $\kappa < 2^{\aleph_0}$.

LEMMA 1 (MA + ¬CH). *For any subset* $A = \{t_\alpha \mid \alpha < \kappa\}$ *of* $\mathbb{R}$, *there exists an increasing function* $f : \omega \longrightarrow \omega$ *such that*

(c) $\qquad f_k \cdot t_\alpha - [f_k \cdot t_\alpha] \longrightarrow s_\alpha$ *for every* $t_\alpha \in A$.

Proof. For the given set A, assume that $\mathcal{F} = \{f^\alpha \mid \alpha < \kappa\}$ of increasing functions $: \omega \longrightarrow \omega$ be chosen so as to satisfy the conditions (a) and (b) with κ instead of ω_1. For each $f^\alpha \in \mathcal{F}$, set $F_\alpha = \omega \setminus \mathrm{range}(f^\alpha)$. Let P be the set as follows.

$$P = \{\langle N, K\rangle \mid N \text{ a finite subset of } \omega, \ K \text{ a finite subset of } \kappa\}.$$

We define a partial ordering $\langle N, K\rangle \geq \langle N', K'\rangle$ in P provided that $N \subseteq N'$, $K \subseteq K'$ and $(N' \setminus N) \cap \bigcup_{\alpha\in K} F_\alpha = \phi$.

$(P, \geq)$ satisfies c c c. For, any uncountable subset Q of P necessarily has a couple of $\langle N, K\rangle$ and $\langle N, K'\rangle$, since the set of the first members N of elements of Q is countable. It is evident that $\langle N, K\rangle \geq \langle N, K \cup K'\rangle$ and $\langle N, K'\rangle \geq \langle N, K \cup K'\rangle$.

For each $n \in \omega$ and $\alpha \in \kappa$, define $D_n = \{\langle N, K\rangle \in P \mid |N| > n\}$ and $D_\alpha = \{\langle N, K\rangle \in P \mid \alpha \in K\}$ respectively, and set $\mathcal{D} = \{D_n \mid n\in\omega\} \cup \{D_\alpha \mid \alpha\in\kappa\}$. Each member of $\mathcal{D}$ is dense in P. To see that D_n be dense, let $\langle N, K\rangle$ be any element of P. By the assumption (a) on f^αs, $\omega \setminus \bigcup_{\alpha\in K} F_\alpha$ has infinitely many elements, and therefore supplying elements of it to N, if necessary, we can take an N' which satisfies $|N'| > n$. Evidently, $\langle N', K\rangle \in D_n$ and $\langle N, K\rangle \geq \langle N', K\rangle$, since $(N' \setminus N) \cap \bigcup_{\alpha\in K} F_\alpha = \phi$.

Since $|\mathcal{D}| = \kappa < 2^{\aleph_0}$, by Martin's axiom, there is a compatible subset Q of P which meets every member of $\mathcal{D}$. Let $F = \bigcup\{N \mid \langle N, K\rangle \in Q\}$. Now, we claim that

(i) F is an infinite subset of ω

and

(ii) for all $\alpha \in \kappa$, $F \cap F_\alpha$ is finite.

(i) is evidently true, because for any $n \in \omega$, there exists an $\langle N, A\rangle$ in Q such that $|N| > n$. To show (ii), let $\langle N, K\rangle$

be any member of Q and $\langle N', K' \rangle$ be a member of $Q \cap D_\alpha$. Since Q is compatible, there is $\langle N'', K'' \rangle \in P$ such that $\langle N, K \rangle \geq \langle N'', K'' \rangle$ and $\langle N', K' \rangle \geq \langle N'', K'' \rangle$. By the latter relations and by the fact $\alpha \in K'$, we have $N \subseteq N''$ and $(N'' \setminus N') \cap F_\alpha = \phi$, from which it follows $(N \setminus N') \cap F_\alpha = \phi$. Therefore, $F \cap F_\alpha = (\bigcup\{N \mid \langle N, K \rangle \in Q\}) \cap F_\alpha \subseteq N'$, that is, $F \cap F_\alpha$ is finite.

By (i), F defines a function f which enumerates in order < its elements, that is, $f(k) = n_k$ and $F = \{n_0 < n_1 < \cdots < n_k < \cdots\}$. $\{f_k\}$ is a subsequence of $\{f_k^\alpha\}$ except for its finite terms for every $\alpha < \kappa$ by (ii), and thus (c) holds.

□

<u>THEOREM</u> 1 (MA + ¬CH). <u>For any set</u> $A \subset \mathbb{R}$, <u>if</u> $|A| < 2^{\aleph_0}$, <u>then there exists an increasing sequence</u> $\{a_k\}$ <u>such that</u>

$$\lim_{k\to\infty} e^{2\pi i a_k t} = 1 \quad \underline{\text{for every}} \quad t \in A.$$

<u>Proof</u>. Let $|A| = \kappa < 2^{\aleph_0}$ and $A = \{t_\alpha \mid \alpha < \kappa\}$. For the set A, assume that for any sequence $\{a_k\}$, the relation $\langle A, \{a_k\} \rangle \in \mathcal{T}$ implies $a_k \longrightarrow 0$. Applying Lemma 1, take a generic function f satisfying (c) and put $m_k = f_{k+1} - f_k$. Then we have

$$\begin{aligned} e^{2\pi i m_k t_\alpha} &= \exp[2\pi i(f_{k+1} - f_k)t_\alpha] \\ &= \exp(2\pi i f_{k+1} t_\alpha) \cdot \exp(-2\pi i f_k t_\alpha) \\ &\longrightarrow e^{2\pi i s_\alpha} \cdot e^{-2\pi i s_\alpha} = 1 \quad \text{for every } t_\alpha \in A \end{aligned}$$

which implies $m_k \longrightarrow 0$ by the assumption. This contradicts the definition of f.

□

As an immediate corollary to this theorem, together with Gödel's result (cf., e.g., [3]), we have the following:

It is independent of ZFC *whether the statement* (P) *holds or not for the set of constructible reals.*

Later (in §4), we will give further results on independence of the statement (P) from ZFC.

§3. Measure and category of set A

To consider the problem in connection with Lebesgue measurability and the Baire property, the following lemma is useful.

LEMMA 2. *Let* $\langle A, \{a_k\}\rangle \in \mathcal{C}$ *and* $|A| > \aleph_0$. *If a subsequence of* $\{a_k\}$ *converges to* a, *then* $a = 0$. *If* $\{a_k\}$ *does not converge to* 0, *then there exists an increasing sequence* $\{n_k\}$ *of integers such that* $\langle A, \{n_k\}\rangle \in \mathcal{C}$.

Proof. If a subsequence $\{a_{k_j}\}$ converges to a, then the condition (C) implies $e^{2\pi i a t} = 1$ for every $t \in A$. Therefore we have $at \in \mathbb{Z}$ for each $t \in A$, which implies $a = 0$ since $|A| > |\mathbb{Z}|$. Next, suppose $\{a_k\}$ does not converge to 0, then $\{a_k\}$ is unbounded by the first part of Lemma. Suppose a subsequence $\{a_{k_j}\}$ diverges to ∞. Considering its subsequence, if necessary, we may assume $\{a_{k_j}\}$ is increasing with $m_j = [a_{k_{j+1}}] - [a_{k_j}] \geq 1$ $(j=1,2,\cdots)$. On the other hand, since the sequence of decimal parts $\varepsilon_j = a_{k_j} - [a_{k_j}]$ $(j=1,2,\cdots)$ has a convergent subsequence, for brevity, we assume $\{\varepsilon_j\}$ itself converges to a limit ε in $[0, 1]$. Then, the relation

$$e^{2\pi i m_j t} = e^{2\pi i a_{k_{j+1}} t}\, e^{-2\pi i \varepsilon_{j+1} t} \cdot e^{-2\pi i a_{k_j} t}\, e^{2\pi i \varepsilon_j t}$$

implies $\lim_j e^{2\pi i m_j t} = e^{-2\pi i \varepsilon t} \cdot e^{2\pi i \varepsilon t} = 1$ for each t of A, that is, we have $\langle A, \{m_j\}\rangle \in \mathcal{C}$. Since $m_j \geq 1$, $\{m_j\}$ is unbounded on account of the first part of Lemma, so that we can pick an increasing subsequence $\{n_k\}$ of $\{m_j\}$ with $\langle A, \{n_k\}\rangle \in \mathcal{C}$. If $a_{k_j} \longrightarrow -\infty$, it is enough to consider $\{-a_{k_j}\}$ to conclude the proof.

□

<u>THEOREM</u> 2. <u>Let</u> $\langle A, \{a_k\}\rangle \in \mathcal{C}$. <u>If</u>

1) A <u>is a measurable set of positive measure</u>

<u>or</u>

2) A <u>is non-measurable</u>,

<u>then the sequence</u> $\{a_k\}$ <u>converges to</u> 0.

<u>Proof</u>. 1) Denote by μ the Lebesgue measure. Since $\mu(A) > 0$, we have $\mu(A \cap [-c, c]) > 0$ for sufficiently big $c > 0$. Assume $\{a_k\}$ does not converge to 0 and by Lemma 2 choose an increasing sequence $\{n_k\}$ of integers such that $\langle A, \{n_k\}\rangle \in \mathcal{C}$. Then, applying Riemann-Lebesgue's theorem to the characteristic function $\chi_{A\cap[-c,c]}$, we have a contradiction;

$$\mu(A \cap [-c, c]) = \int_{A\cap[-c,c]} \lim_k e^{2\pi i n_k t} dt = \lim_k \int_{A\cap[-c,c]} e^{2\pi i n_k t} dt$$

$$= \lim_k \int_{-\infty}^{\infty} \chi_{A\cap[-c,c]}(t) e^{2\pi i n_k t} dt = 0.$$

2) Suppose $a_k \not\to 0$ and choose $\{n_k\}$ as above. Consider a Borel set defined by

$$G = \{t \in \mathbb{R} \mid \lim_k e^{2\pi i n_k t} = 1\}.$$

Then, the obvious relation $\langle G, \{n_k\} \in \mathcal{C}$ implies $\mu(G) = 0$ by the first part 1). Therefore A ($\subseteq G$) is also a null set, which contradicts the non-measurability of A.

□

By a Steinhaus' theorem [11], we can show 1) in the same manner as the proof of Theorem 2', 1).

Similarly, we have

<u>THEOREM</u> 2'. Let $\langle A, \{a_k\}\rangle \in \mathcal{C}$. If

1) A *has the Baire property and is of the second category*

or

2) A *does not have the Baire property*,

then the sequence $\{a_k\}$ *converges to* 0.

<u>Proof</u>. 1) Under the assumptions of A, it is known that the algebraic sum $A - A$ contains an open neighbourhood U of 0 (Bourbaki [2, §5, Exer.27]). Therefore, noting $\langle U, \{a_k\}\rangle \in \mathcal{C}$, we see $a_k \longrightarrow 0$.

2) Suppose $a_k \not\rightarrow 0$. Then, as was shown in the proof of Theorem 2, there exists a Borel set G having properties $A \subseteq G$ and $\langle G, \{n_k\}\rangle \in \mathcal{C}$ with an increasing sequence $\{n_k\}$. Since Borel set has the Baire property, the first part 1) implies $a_k \longrightarrow 0$ if G is of the second category. Thus G should be of the first category under the assumption $a_k \not\rightarrow 0$. But this is impossible, since $A \subseteq G$ implies A is of the first category, so A has the Baire property.

□

Combining Theorem 2 and 2', we deduce that there exists a null (or meager) set A such that $\langle A, \{a_k\}\rangle \in \mathcal{C}$ implies $a_k \to 0$. Furthermore, we have instances of A on which (P) holds, even if A is nowhere dense null set. For example, so is Cantor's perfect set C, as was mentioned in Introduction. On the contrary, as to perfect set P contained in the set $C_\nu = \{t \mid t = \sum_{k=1}^{\infty} \frac{\tau_k}{3^{n_k}},\ \tau_k = 0, 2\}$ where $n_{k+1} - n_k \to \infty$, we see $\langle P, \{3^{n_k}\}\rangle \in \mathcal{C}$ by Proposition 1.

§4. Independence of the problem

For brevity, let us denote by P(A) the statement (P) on a set $A \subseteq \mathbb{R}$, namely, by P(A) we mean the following proposition:

<u>For any sequence</u> $\{a_k\}$, <u>if</u> $\langle A, \{a_k\}\rangle \in \mathcal{C}$, <u>then</u> $a_k \to 0$.

Now, we consider the set $\mathbb{R} \cap L$ of all constructible reals. It is well known by Solovay (cf., e.g. Jech [4, p.568]) that there exists a model of ZFC where $\aleph_1 < 2^{\aleph_0}$ and $\mathbb{R} \cap L$ is not Lebesgue measurable. By Theorem 2.2), $P(\mathbb{R} \cap L)$ holds in such a model. On the other hand, by Theorem 1, we see under the assumption of Martin's axiom MA and $\aleph_1 < 2^{\aleph_0}$ that $P(\mathbb{R} \cap L)$ does not hold, since $|\mathbb{R} \cap L| \leq \aleph_1$ in any model of ZFC. Therefore, $\neg P(\mathbb{R} \cap L)$ is consistent with ZFC, because $(MA + \neg CH)$ is consistent with ZFC.

Consequently, we have the following:

<u>There is a proper subgroup</u> G <u>of</u> $\mathbb{R}$ <u>such that the proposition</u> P(G) <u>is independent of</u> ZFC.

Concerning this fact, $\mathbb{R} \cap L$ might not be so interesting example in the sense that the axiom of constructibility $V = L$ implies $\mathbb{R} \cap L = \mathbb{R}$. We remark that the last fact together with Theorem 1 gives us another simple proof of independence of the problem without using Theorem 2 (or Theorem 2') and the above mentioned Solovay model (or similar models by Solovay, Vopěnka and Hrbáček [12], or Shinoda [9] with respect to the Baire property).

Furthermore, we shall show a class of subsets of $\mathbb{R}$ on which the statement (P) is independent from ZFC.

A set $A \subset \mathbb{R}$ has strong measure 0 (or property C, cf., e.g. Sierpiński [10], Kuratowski [5]), if for any sequence $\{a_n\}$ of positive reals, there is a sequence $\{I_n\}$ of intervals with length $I_n \leq a_n$ such that $A \subseteq \bigcup_{n \in \omega} I_n$. The concept of strong measure 0 was introduced by Borel [1, p.123], who called it to have <une mesure asymptotique inférieure à toute série donnée à l'advance> and conjectured "any strong measure 0 set is countable". By Laver [6], there is a generic model of ZFC in which Borel's conjecture holds. In such a model, $P(A)$ does not hold for any set A having strong measure 0.

By Lusin [7], under the continuum hypothesis $2^{\aleph_0} = \aleph_1$, there exists a subset E of $\mathbb{R}$ such that $|E| = 2^{\aleph_0}$ and $|E \cap F| < 2^{\aleph_0}$ for each nowhere dense set F. It is also known by Martin-Solovay [8], that there exists a Lusin set under the assumption of Martin's axiom and $\aleph_1 < 2^{\aleph_0}$. Now, we assume $2^{\aleph_0} = \aleph_1$ (or $MA + \neg CH$). Then, it is well-known by [10] (resp., by Strong Baire Category Theorem, see [8, p.177]) that a Lusin set E has strong measure 0. Since any Lusin set is of second category, $P(E)$ is true by Theorem 2'. Therefore by Gödel (resp., by [8]), it is consistent with ZFC that there is an uncountable set A of strong measure 0 such that P(A) is true.

Thus, we have the following conclusion:

<u>For the class</u> $\mathbb{C}$ <u>of all strong measure</u> 0 <u>sets, the proposition</u> "$\forall A \in \mathbb{C} \neg P(A)$" <u>is independent of</u> ZFC.

References

[1] E. Borel, Sur la classification des ensembles de mesure null, Bull. Soc. Math. France, vol.47(1919), 97-125.

[2] N. Bourbaki, Topologie générale, in Éléments de Mathématique, Hermann, Paris, 1958, Chap.IX, 2nd ed.

[3] K. Gödel, The consistency of the axiom choice and of the generalized continuum hypothesis, Ann. Math. Studies 3, 1940.

[4] T.J. Jech, Set theory, Academic Press, New York, San Francisco and London, 1978.

[5] C. Kuratowski, Topologie I, Warszawa 1952 (Édition Troisième, Corrigée).

[6] R. Laver, On the consistency of Borel's conjecture, Acta Math., vol.137(1976), 151-169.

[7] N. Lusin, Sur un problème de M. Baire, C.R. Acad. Sci. Paris, vol.158(1914), 1258-1261.

[8] D.A. Martin and R.M. Solovay, Internal Cohen extensions, Ann. of Math. Logic, vol.2, no.2(1970), 143-178.

[9] J. Shinoda, A note on Silver's extension, Comm. Math. Univ. Sancti Pauli, Tom.22(1973), 109-111.

[10] W. Sierpiński, Sur un ensemble non dénombrable, dont toute image continue est de mesure nulle, Fund. Math. Tom.11(1928), 301-304.

[11] H. Steinhaus, Sur les distances des points dans les ensembles de mesure positive, Fund. Math., Tom.1(1920), 93-104.

[12] P. Vopěnka and K. Hrbáček, The consistency of some theorems on real numbers, Bull. Acad. Polon. Sci., Tom.15(1967), 107-111.

Department of Mathematics
College of General Education
Nagoya University
Chikusa-ku, Nagoya 464
Japan

INTUITIONISTIC THEORIES AND TOPOSES

Tadahiro Uesu*)

The notion of theory in this paper is an extension of Lawvare's notion of algebraic theory [L] so that a theory has not only finite products but also exponentials for appropriate objects. An intuitionistic theory is defined as a theory which has truth value, conjunction and equality as morphisms.

The purpose of this paper is to explain the correspondence between toposes and intuitionistic theories. Other authors gave presentations of the correspondence between toposes and theories through formal systems of higher-order intuitionistic logic ([F],[B & J]). We, however, rather do it directly.

This paper is arranged as follows: In the first section, a usage of variables in categories is provided. In the following sections, we shall use variables according to the usage. In Section 2, we introduce the notion of theory and explain the relation between theories and languages. In Section 3, we define intuitionistic theories and deal with the correspondence between toposes and intuitionistic theories. In the final section, we give a system of higher-order intuitionistic logical calculus, and show that the completeness theorem to toposes. Other authors proposed systems of higher-order intuitionistic logical calculus which are complete to toposes also ([F],[B & J]). It is the difference from them that our system has more types, and so our system has Comprehension Axiom in a general form.

We use the following category-theoretic notation: $f:A \longrightarrow B$ denotes a

---

*) The author is partially supported by Grant-in-Aid for Co-operative Research, Project No. 434007.

morphism from A to B as usual. We regard a morphism $f:A\longrightarrow B$ as an ordered triple (f,A,B), and so we admit that there is a pair of distinct morphisms whose first components are the same. When there is no danger of confusion we identify f with $f:A\longrightarrow B$. For each pair of morphisms $f:A\longrightarrow B$ and $g:B\longrightarrow C$, we write the composition $g\circ f:A\longrightarrow C$ or simply $gf:A\longrightarrow C$.

We use the term "power" rather than the term "exponential".

We assume that for each category a representative of products of a finite family of objects is given if they exist, and a representative of power for a pair of objects is given if they exist. We write $A_1\times\cdots\times A_n$ for the representative of finite products of $A_1,\ldots,A_n$, and denote the projections by $\pi_1:A_1\times\cdots\times A_n\longrightarrow A_1,\ldots,\pi_n:A_1\times\cdots\times A_n\longrightarrow A_n$. We also write A^B for the representative of powers with base A and exponent B, and denote the evaluation morphism by $ev:B\times A^B\longrightarrow A$.

This paper is rewritten from original one according to an advice Dr. Gordon Monro.

The author is grateful to Dr. Gordon Monro for his valuable advice. The author also thanks Mrs. Keiko Momoshima and Miss Chikako Tanaka for their typing of the paper.

1. Usage of variables in categories

Variables are useful symbols in mathematics. We use them to abstract elements which have some property, as x in $\{x|P(x)\}$. We use them to define a new function from given functions, as x and y in $f(x,y)=(x+y)\times x$. We also use them to mean any elements in the domain concerned. As to category, it is seemed that variables are given no place to be active. The reason may be that in category theory we deal with arrows rather than elements. The notion of element, however, is inessential to main function of variables. Actually, it is possible to use variables in category as the traditional way. In topos theory, several authors realized the advantage of the use of variables ([M], [C], [O], [F], [J], etc.). They used languages with variables to make statements about the objects and morphisms of a given topos. In this section we provide usage of variables in categories, without using languages, so that it suits the traditional usage.

Let C be a category which has finite products, V a class of variables and $^\#$ an assignment of variables to objects of C.

For each variable x the object $x^\#$ is called the domain of x.

First, we intend to use notation as $f(x_1,\ldots,x_m)$ for each morphism $f:x_1^\#\times\cdots\times x_m^\# \longrightarrow A$. For that, we adjoin a new object $\overline{\{x_1,\ldots,x_m\}}$ to C for each finite set $\{x_1,\ldots,x_m\}$ of variables so that $\overline{\{x_1,\ldots,x_m\}}$ and $x_1^\#\times\cdots\times x_m^\#$ may be isomorphic. We denote the projections of $\overline{\{x_1,\ldots,x_m\}}$ by $x_1:\overline{\{x_1,\ldots,x_m\}}\longrightarrow x_1^\#$, $\ldots$, $x_m:\overline{\{x_1,\ldots,x_m\}}\longrightarrow x_m^\#$, and the isomophism by

$$(x_1,\ldots,x_m):\overline{\{x_1,\ldots,x_m\}}\longrightarrow x_1^\#\times\cdots\times x_m^\#.$$

So $f(x_1,\ldots,x_m)$ is the composition of the isomophism $(x_1,\ldots,x_m)$ and the morphism f. The inverse of $(x_1,\ldots,x_m)$ is denoted by $(x_1,\ldots,x_m)^{\leftarrow}$.

In the category Sets of sets and functions, the new object $\overline{\{x_1,\ldots,x_m\}}$ is realized by the direct product $\prod_{x \in \{x_1,\dots,x_m\}} x^{\#}$ with index $\{x_1,\dots,x_m\}$. Then

$$(x_1,\dots,x_m) : \begin{pmatrix} x_1 & \dots & x_m \\ a_1 & & a_m \end{pmatrix} \longmapsto (a_1,\dots,a_m)$$

and

$$(x_1,\dots,x_m)\!\leftarrow : (a_1,\dots,a_m) \longmapsto \begin{pmatrix} x_1 & \dots & x_m \\ a_1 & & a_m \end{pmatrix},$$

where $\begin{pmatrix} x_1 & \dots & x_m \\ a_1 & & a_m \end{pmatrix}$ means the element α in $\overline{\{x_1,\dots,x_m\}}$ such that

$$\alpha : x_i \longmapsto a_i \ , \text{ where } i=1,\dots,m.$$

The composition $y\!\leftarrow\! f(x_1,\dots,x_m)$ of $(x_1,\dots,x_m)$, f and $y\!\leftarrow$ is the mapping for which the following diagram commutes:

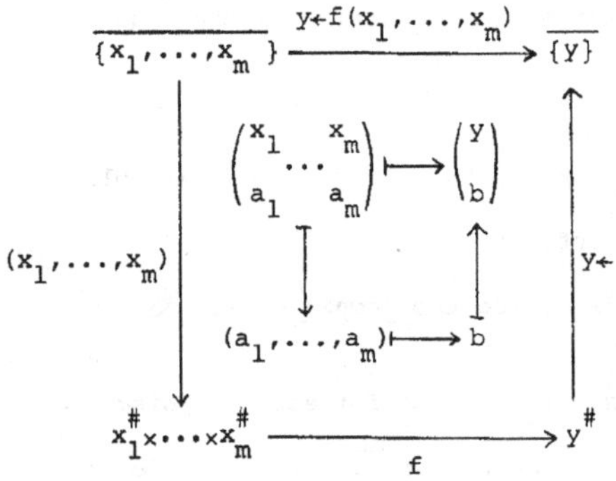

Let $\overline{C}$ be the category obtained from the category C by adjoining new objects $\overline{X}$, where X are finite subsets of V , as above. $\overline{C}$ is called the <u>system of variables</u> over C with respect to V and $^{\#}$, and denoted by C(V, $^{\#}$). For each finite subset X of V, the object $\overline{X}$ is called the <u>adjunct object</u> for X. A morphism whose domain is an adjunct object and whose codomain is an object of C is called a <u>v→morphism</u> over C. A variable x is said to <u>be contained</u> in a v→morphism $f:\overline{X}\longrightarrow A$ if $x \in X$.

We use f,g,h,...as syntactical variables which vary through morphisms in C, and $\mathfrak{f},\mathfrak{g},\mathfrak{h},\ldots$as syntactical variables which vary through v→morphisms.

1.1. Proposition. (1) For each v→morphism $\mathfrak{f}:\overline{\{x_1,\ldots,x_m\}}\longrightarrow A$, there is a unique morphism $f:x_1^{\#}\times\cdots\times x_m^{\#}\longrightarrow A$ such that

$$f(x_1,\ldots,x_m)=\mathfrak{f}.$$

(2) For each morphism $g:x_{i_1}^{\#}\times\cdots\times x_{i_m}^{\#}\longrightarrow A$, there is a unique morphism $f:x_1^{\#}\times\cdots\times x_m^{\#}\longrightarrow A$ such that

$$f(x_1,\ldots,x_m)=g(x_{i_1},\ldots,x_{i_m}),$$

where $(i_1,\ldots,i_m)$ is a permutation of $(1,\ldots,m)$.

We often use notation as $f(x,y)=g(x)$ where y is a dummy variable. It, however, is meaningless in $\overline{C}$, since $f(x,y)$ and $g(x)$ are always distinct morphisms. We intend to use such notation in $\overline{C}$. For that, we introduce the following notation:

1.2. Notation. (1) For each pair (X,Y) of finite sets of variables with $X\subseteq Y$, $X:\overline{Y}\longrightarrow\overline{X}$ denotes the morphism for which the following diagram commutes:

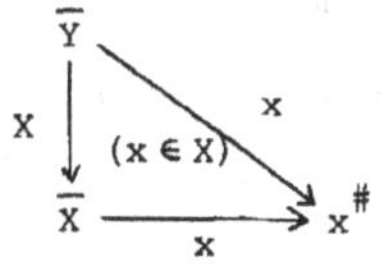

(2) For each pair $(\mathfrak{f}:\overline{X}\longrightarrow A,\ \mathfrak{g}:\overline{Y}\longrightarrow A)$ of v→morphisms, $\mathfrak{f}:\overline{X}\longrightarrow A\doteq\mathfrak{g}:\overline{Y}\longrightarrow A$ or simply $\mathfrak{f}\doteq\mathfrak{g}$ denotes that the following diagram commutes:

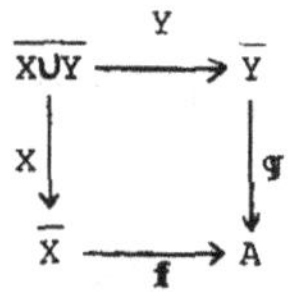

Then, $f(x,y)\doteq g(x)$ is meaningful.

Note that $\doteq$ is not necessarily a transitive relation. In Sets, if $x^{\#}$ is the empty set, then $\mathfrak{f}\doteq x$ for all v→morphisms $\mathfrak{f}$.

We also permit notation of substitution as in

$$f(x)=g(h(x),k(x)).$$

1.3. <u>Notation</u>. For n v⇸morphisms $\mathfrak{g}_1:\overline{Y_1}\to A_1,\ldots,\mathfrak{g}_n:\overline{Y_n}\to A_n$,

$$(\mathfrak{g}_1,\ldots,\mathfrak{g}_n):\overline{Y_1\cup\cdots\cup Y_n}\to A_1\times\cdots\times A_n$$

denotes the v⇸morphism for which the following diagram commutes:

$$\begin{array}{ccc}
\overline{Y_1\cup\cdots\cup Y_n} & \xrightarrow{\;Y_i\;} & \overline{Y_i} \\
{\scriptstyle(\mathfrak{g}_1,\ldots,\mathfrak{g}_n)}\downarrow & (i=1,\ldots,n) & \downarrow{\scriptstyle\mathfrak{g}_i} \\
A_1\times\cdots\times A_n & \xrightarrow[\;\pi_i\;]{} & A_i
\end{array}$$

When the domain is not stated, $(\mathfrak{g}_1,\ldots,\mathfrak{g}_n)$ is ambiguous. For example, while the domain of (x,y) is $\overline{\{x,y,z\}}$ if x and y mean the projections of $\overline{\{x,y,z\}}$, it is $\overline{\{x,y\}}$ if (x,y) means the isomorphism from $\overline{\{x,y\}}$ to $x^{\#}\times y^{\#}$. So, whenever the domain of $(\mathfrak{g}_1,\ldots,\mathfrak{g}_n)$ is not stated, we think that the domain is the adjunct object for the set of variables which are explicitly seen in $(\mathfrak{g}_1,\ldots,\mathfrak{g}_n)$. For example, the domain of (x,y) is $\overline{\{x,y\}}$, and the domain of $g(h(x),x)$ is $\overline{\{x\}}$.

Immediately we have the following proposition.

1.4. <u>Proposition</u>. If $\mathfrak{g}_1:\overline{Y_1}\to A_1\doteq\mathfrak{g}_1':\overline{Y_1'}\to A_1,\ldots,\mathfrak{g}_n:\overline{Y_n}\to A_n\doteq\mathfrak{g}_n':\overline{Y_n'}\to A_n$, then

$$(\mathfrak{g}_1,\ldots,\mathfrak{g}_n):\overline{Y_1\cup\cdots\cup Y_n}\to A_1\times\cdots\times A_n$$

$$\doteq(\mathfrak{g}_1',\ldots,\mathfrak{g}_n'):\overline{Y_1'\cup\cdots\cup Y_n'}\to A_1\times\cdots\times A_n.$$

Moreover, if $Y_1\cup\cdots\cup Y_n=Y_1'\cup\cdots\cup Y_n'$, then $(\mathfrak{g}_1,\ldots,\mathfrak{g}_n)=(\mathfrak{g}_1',\ldots,\mathfrak{g}_n')$.

1.5. <u>Notation</u>. For n v⇸morphisms $\mathfrak{g}_1:\overline{Y_1}\to y_1^{\#},\ldots,\mathfrak{g}_n:\overline{Y_n}\to y_n^{\#}$, the figure

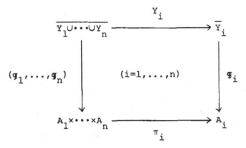

$$\begin{pmatrix} y_1 & \cdots & y_n \\ \mathfrak{g}_1 & & \mathfrak{g}_n \end{pmatrix}$$

is called a <u>substitution operator</u>. For each v→morphism $\mathfrak{f}$ of the form $f(x_1,\ldots,x_m,y_{i_1},\ldots,y_{i_k}):\{\overline{x_1,\ldots,x_m,y_{i_1},\ldots,y_{i_k}}\}\longrightarrow A$, where $x_1,\ldots,x_m,y_{i_1},\ldots,y_{i_k}$ are mutually distinct and $1\leq i_1,\ldots,i_k\leq n$, $\mathfrak{f}\begin{pmatrix} y_1 & \cdots & y_n \\ \mathfrak{g}_1 & & \mathfrak{g}_n \end{pmatrix}$ means the morphism

$$f(x_1,\ldots,x_m,\ \mathfrak{g}_{i_1},\ldots,\mathfrak{g}_{i_k}):\{\overline{x_1,\ldots,x_m\}\cup Y_{i_1}\cup\cdots\cup Y_{i_k}}\longrightarrow A.$$

Immediately we obtain the following proposition.

1.6. <u>Proposition</u>. (1) If $\mathfrak{g}_1\dot{=}\mathfrak{g}_1',\ldots,\mathfrak{g}_n\dot{=}\mathfrak{g}_n'$, and $\mathfrak{f}\dot{=}\mathfrak{f}'$, then

$$\mathfrak{f}\begin{pmatrix} y_1 & \cdots & y_n \\ \mathfrak{g}_1 & & \mathfrak{g}_n \end{pmatrix}\dot{=}\mathfrak{f}'\begin{pmatrix} y_1 & \cdots & y_n \\ \mathfrak{g}_1' & & \mathfrak{g}_n' \end{pmatrix}.$$

(2) If $\{y_1,\ldots,y_n\}-\{x_1,\ldots,x_m\}=\{y_1,\ldots,y_k\}$, then

$$\mathfrak{f}\begin{pmatrix} x_1 & \cdots & x_m \\ \mathfrak{g}_1 & & \mathfrak{g}_m \end{pmatrix}\begin{pmatrix} y_1 & \cdots & y_n \\ \mathfrak{h}_1 & & \mathfrak{h}_n \end{pmatrix}=\mathfrak{f}\begin{pmatrix} x_1 & \cdots & x_m & y_1 & \cdots & y_k \\ \mathfrak{g}_1\begin{pmatrix} y_1 & \cdots & y_n \\ \mathfrak{h}_1 & & \mathfrak{h}_n \end{pmatrix} & & \mathfrak{g}_m\begin{pmatrix} y_1 & \cdots & y_n \\ \mathfrak{h}_1 & & \mathfrak{h}_n \end{pmatrix} & \mathfrak{h}_1 & & \mathfrak{h}_k \end{pmatrix}.$$

When we define a function in terms of given functions and variables, λ-operator is convenient. We also use λ-operator in $\overline{C}$ as follows.

1.7. <u>Notation</u>. For each v→morphism $\mathfrak{f}:\overline{X}\to A$ and for each set $\{y_1,\ldots,y_n\}$ of variables with $X\subseteq\{y_1,\ldots,y_n\}$

$$\lambda y_1\cdots y_n.\mathfrak{f}:y_1^{\#}\times\cdots\times y_n^{\#}\longrightarrow A$$

denotes the composition of $(y_1,\ldots,y_n)^{\leftarrow}$, $X:\{\overline{y_1,\ldots,y_n}\}\longrightarrow\overline{X}$, and $\mathfrak{f}$. For example, the morphism $f(x_1,\ldots,x_m)$ in Proposition 1.1 (2) is denoted by $\lambda x_1\cdots x_m.g(x_{i_1},\ldots,x_{i_m})$, and the diagonal $\Delta:A\longrightarrow A\times A$ is denoted by $\lambda x.(x,x)$, where $x^{\#}=A$.

Now we intend to express the transpose $\hat{f}:C\longrightarrow A^B$ of $f:B\times C\longrightarrow A$ in terms of variables.

1.8. Notation. For mutually distinct variables $y_1,\ldots,y_n$, and an object A such that the power $A^{y_1^\# \times \cdots \times y_n^\#}$ exists, and for each v→morphism $\mathfrak{f}:\overline{X}\longrightarrow A$,

$$y_1\cdots y_n\mathfrak{f}:\overline{X-\{y_1,\ldots,y_n\}}\longrightarrow A^{y_1^\# \times \cdots \times y_n^\#}$$

denotes the morphism for which the following diagram commutes:

$$\begin{array}{ccc}
\overline{\{y_1,\ldots,y_n\}\cup X} & \xrightarrow{\quad X \quad} & \overline{X} \\
\Big\downarrow{\scriptstyle (y_1,\ldots,y_n,y_1\cdots y_n\mathfrak{f})} & & \Big\downarrow{\scriptstyle \mathfrak{f}} \\
y_1^\# \times \cdots \times y_n^\# \times A^{y_1^\# \times \cdots \times y_n^\#} & \xrightarrow[\quad ev \quad]{} & A
\end{array}$$

The morphism $y_1\cdots y_n\mathfrak{f}$ is called the abstract from $\mathfrak{f}$ with respect to $y_1,\ldots,y_n$.

Then $\lambda x.yf(y,x)$ is the transpose of $f:B\times C\longrightarrow A$, where $x^\#=C$ and $y^\#=B$.

Note that $y_1\cdots y_n\mathfrak{f}$ is distinct from $\lambda y_1\cdots y_n.\mathfrak{f}$, since while $y_1\cdots y_n\mathfrak{f}$ is a v→morphism, $\lambda y_1\cdots y_n.\mathfrak{f}$ is a morphism in C.

The following proposition can be easily proved.

1.9. Proposition. Let $\mathfrak{f}:\overline{X}\longrightarrow A$ and $\mathfrak{f}':\overline{X'}\longrightarrow A$ be v→morphisms, and $y_1,\ldots,y_n$ mutually distinct variables. Suppose that the power $A^{y_1^\# \times \cdots \times y_n^\#}$ exists.

(1) $y_1\cdots y_n ev(y_1,\ldots,y_n,x)=x$, where $x^\#=A^{y_1^\# \times \cdots \times y_n^\#}$.

(2) $ev(\mathfrak{g}_1,\ldots,\mathfrak{g}_n,y_1\cdots y_n\mathfrak{f})\doteq\mathfrak{f}\begin{pmatrix} y_1 & \cdots & y_n \\ \mathfrak{g}_1 & \cdots & \mathfrak{g}_n \end{pmatrix}$.

(3) $y_1\cdots y_n\mathfrak{f}=z_1\cdots z_n(\mathfrak{f}\begin{pmatrix} y_1 & \cdots & y_n \\ z_1 & \cdots & z_n \end{pmatrix})$, where $y_1^\#=z_1^\#,\ldots,y_n^\#=z_n^\#$, and none of $z_1,\ldots,z_n$ is contained in $y_1\cdots y_n\mathfrak{f}$.

(4) If $\{x_1,\ldots,x_m\}\cap\{y_1,\ldots,y_n\}=\phi$ and none of $y_1,\ldots,y_n$ is contained in any $\mathfrak{g}_1,\ldots,\mathfrak{g}_m$, then

$$(y_1\cdots y_n\mathfrak{f})\begin{pmatrix}x_1 & & x_m\\ & \cdots & \\ \mathfrak{g}_1 & & \mathfrak{g}_m\end{pmatrix}=y_1\cdots y_n(\mathfrak{f}\begin{pmatrix}x_1 & & x_m\\ & \cdots & \\ \mathfrak{g}_1 & & \mathfrak{g}_m\end{pmatrix}).$$

(5) If $\mathfrak{f}\doteq\mathfrak{f}'$, then $y_1\cdots y_n\mathfrak{f}\doteq y_1\cdots y_n\mathfrak{f}'$.

(6) If $y_1\cdots y_n\mathfrak{f}\doteq y_1\cdots y_n\mathfrak{f}'$, then $\mathfrak{g}\doteq\mathfrak{g}'$, where $\mathfrak{g}$ is the composition of the projection $X:\overline{X\cup\{y_1,\ldots,y_n\}}\longrightarrow\overline{X}$ and $\mathfrak{f}:\overline{X}\longrightarrow A$, and $\mathfrak{g}'$ is the composition of the projection $X':\overline{X'\cup\{y_1,\ldots,y_n\}}\longrightarrow\overline{X'}$ and $\mathfrak{f}':\overline{X'}\longrightarrow A$.

In a system of variables, $\mathfrak{f}\begin{pmatrix}y_1 & & y_n\\ & \cdots & \\ \mathfrak{g}_1 & & \mathfrak{g}_n\end{pmatrix}$ is uniquely determined by $y_1,\ldots,y_n,\mathfrak{g}_1,\ldots,\mathfrak{g}_n$ and $\mathfrak{f}$, and $y_1\cdots y_n\mathfrak{f}$ is uniquely determind by $y_1,\ldots,y_n$ and $\mathfrak{f}$ if $y_1\cdots y_n\mathfrak{f}$ exists, so the following theorem is obtained.

1.10. <u>Theorem</u>. Let $F:C_1\longrightarrow C_2$ be a functor which preserves the representatives of finite products and powers, and let $C_1(V_1,\ ^{\#})$ and $C_2(V_2,\ ^{\#\#})$ be systems of variables such that $V_1\subseteq V_2$ and $F(x^{\#})=x^{\#\#}$ for each variable x in V_1. Extend F to the system of variables $C_1(V_1,\ ^{\#})$ so that $F(\overline{X})=\overline{X}$ for each finite subset X of V_1, where $\overline{X}$ in the left is the adjunct object for X in $C_1(V_1,\ ^{\#})$, and $\overline{X}$ in the right is the adjunct object for X in $C_2(V_2,\ ^{\#\#})$. Then

(1) $$F(\mathfrak{f}\begin{pmatrix}y_1 & & y_n\\ & \cdots & \\ \mathfrak{g}_1 & & \mathfrak{g}_n\end{pmatrix})=F(\mathfrak{f})\begin{pmatrix}y_1 & & y_n\\ & \cdots & \\ F(\mathfrak{g}_1) & & F(\mathfrak{g}_n)\end{pmatrix},$$

and

(2) $$F(y_1\cdots y_n\mathfrak{f})=y_1\cdots y_nF(\mathfrak{f}),$$

provided that $y_1\cdots y_n\mathfrak{f}$ exists.

2. Theories and languages

In this section, we introduce the notion of theory which is an extension of Lawvare's notion of algelraic theory [L], and explain the relation between theory and language.

For each class Σ, Σ^* denotes the class of finite, possibly empty, strings of elements in Σ, and $\Sigma\wr\Sigma^+$ denotes the class of figures $\sigma^{\sigma_1\ldots\sigma_m}$, where $\sigma,\sigma_1,\ldots,\sigma_m \in \Sigma$ and $m=1,2,\ldots$

For a class Σ and a subclass Ψ of $\Sigma\wr\Sigma^+$, a category is a <u>theory</u> over (Σ,Ψ) if it satisfies the following conditions:

(1) The class of objects is the class $(\Sigma\cup\Psi)^*$.

(2) For each n-tuple $(t_1,\ldots,t_n)$ of elements in $\Sigma\cup\Psi$, the string $t_1\ldots t_n$ is a product of $t_1,\ldots,t_n$ with projections $\pi_1,\ldots,\pi_n$, where $n=0,1,\cdots$.

(3) Each element in Ψ of the form $\sigma^{\sigma_1\ldots\sigma_m}$ is a power with base σ, exponent $\sigma_1\cdots\sigma_m$ and evaluation morphism ev.

When Σ is a singleton and Ψ is the empty set, a theory over (Σ,Ψ) is a Lawvare's algebraic theory.

A functor F from a theory T over (Σ,Ψ) to a category C is a <u>model</u> of T in C, if $F(t_1\cdots t_n)$ is a product of $F(t_1),\ldots,F(t_n)$ with projections $F(\pi_1),\ldots,F(\pi_n)$ in C for each string $t_1\cdots t_n$ in $(\Sigma\cup\Psi)^*$, and $F(\sigma^{\sigma_1\ldots\sigma_m})$ is a power with base $F(\sigma)$, exponent $F(\sigma_1\cdots\sigma_m)$ and evaluation morphism F(ev) in C for each $\sigma^{\sigma_1\cdots\sigma_m}$ in Ψ.

For a class Σ and a subclass Ψ of $\Sigma\wr\Sigma^+$, a <u>language</u> over (Σ,Ψ) consists of the following:

(1) A sequence $\langle V_t \mid t\in\Sigma\cup\Psi\rangle$ of mutually disjoint infinite sets.

(2) A sequence $\langle O_{(\bar{s}\to t)} \mid \bar{s}\in(\Sigma\cup\Psi)^*, t\in\Sigma\cup\Psi\rangle$ of mutually disjoint sets which are also disjoint from any V_t, where $t\in\Sigma\cup\Psi$.

The elements of Σ are called sorts. The elements of $\Sigma \cup \Psi$ are called types. The elements of V_t are called variables of type t, where $t \in \Sigma \cup \Psi$. If t is a type of the form $\sigma^{\sigma_1 \cdots \sigma_m}$, where $\sigma, \sigma_1, \ldots, \sigma_m \in \Sigma$, then each variable of type t is called an m-ary variable. The elements of $O_{(\bar{s} \to t)}$ are called $(\bar{s}\to t)$-operators, where $\bar{s} \in (\Sigma \cup \Psi)^*$ and $t \in \Sigma \cup \Psi$. If the length of $\bar{s}$ is n, then a $(\bar{s}\to t)$-operator is called an n-ary operator.

In the following discussion we fix a language L over (Σ, Ψ).

Designators of L are finite strings of variables and operators defined inductively as follows:

(1) The empty string ε is a designator of type ε.

(2) Each variable of type t is a designator of type t.

(3) If e is a designator of type σ, and $x_1, \ldots, x_m$ are mutually distinct variables of types $\sigma_1, \ldots, \sigma_m$ respectively, then $x_1 \cdots x_m e$ is a designator of type $\sigma^{\sigma_1 \cdots \sigma_m}$ and also called an m-ary abstract, where $\sigma^{\sigma_1 \cdots \sigma_m} \in \Psi$.

(4) If $e_1, \ldots, e_n$ are designators of types $t_1, \ldots, t_n$ respectively, and o is a $(t_1 \cdots t_n \to t)$-operator, then $oe_1 \cdots e_n$ is a designator of type t.

(5) If $e_1, \ldots, e_m$ are designators of types $\sigma_1, \ldots, \sigma_m$ respectively, and u is a variable of type $\sigma^{\sigma_1 \cdots \sigma_m}$, then $ue_1 \cdots e_m$ is a designator of type σ.

For example, let $\Sigma = \{0,1,2,\ldots,\omega\}$, $\Psi = \{\omega^\sigma \mid \sigma \in \Sigma\}$, $\in^i$ a $(i\,i{+}1 \to \omega)$-operator, where $i=0,1,2,\ldots$, $\wedge$ a $(\omega\omega\to\omega)$-operator, $\vee$ a $(\omega\omega\to\omega)$-operator, $\supset$ a $(\omega\omega\to\omega)$-operator, $\neg$ a $(\omega\to\omega)$-operator, $\forall^\sigma$ a $(\omega^\sigma\to\omega)$-operator and $\exists^\sigma$ a $(\omega^\sigma\to\omega)$ operator, where $\sigma \in \Sigma$. Then the designators of type ω are formulas of simple type theory.

An abstract e_0 is a subabstract of a designator e if there exists a pair (ξ, η) of strings such that $\xi e_0 \eta$ is the designator e and $\xi u \eta$ is a designator for some variable u of the same type as e_0.

An occurrence of the variable x in a designator is said to be a bound occurrence if it occurs in a subabstract of the form $x_1 \cdots x_m e$ with $x \in \{x_1, \ldots, x_m\}$. Otherwise it is said to be a free occurrence.

A variable y is free for a variable x in a designator e if none of free occurrences of x in e occurs in any subabstract of the form $y_1 \cdots y_n e_o$ such that $y \in \{y_1, \ldots, y_n\}$.

A designator e' is a variant of a designator e if e' can be obtained from e by a sequence of replacements of the following types:

(1) Replace a subabstract $x_1 \cdots x_m e_o$ by an abstract $y_1 \cdots y_m e'_o$, where none of $y_1, \ldots, y_m$ occurs free in e_o, each of $y_1, \ldots, y_m$ is free for all $x_1, \ldots, x_m$ in e_o, and e'_o is the result of replacing all free occurrences of $x_1, \ldots, x_m$ in e_o by $y_1, \ldots, y_m$ respectively.

(2) Replace a subabstract of the form $x_1 \cdots x_m u x_1 \cdots x_m$ by u, where u is an m-ary variable, and vice versa.

If e' is a variant of e, we say that e and e' are homologous. The term "variant" is due to [Sh] and the term "homologous" is due to [T].

From now on we identify homologous designators.

For each n-tuple $(x_1, \ldots, x_n)$ of mutually distrinct variables and for each n-tuple $(e_1, \ldots, e_n)$ of designators, each e_i of which has the same type as x_i, the figure

$$\begin{pmatrix} x_1 & \cdots & x_n \\ e_1 & \cdots & e_n \end{pmatrix}$$

is a substitution for $\{x_1, \ldots, x_n\}$, where n=0,1,... . When n=0, the substitution $\left(\,\right)$ is called the empty substitution. For each permutation $(i_1, \ldots, i_n)$ of $(1, \ldots, n)$, we identify $\begin{pmatrix} x_1 & \cdots & x_n \\ e_1 & \cdots & e_n \end{pmatrix}$ with $\begin{pmatrix} x_{i_1} & \cdots & x_{i_n} \\ e_{i_1} & \cdots & e_{i_n} \end{pmatrix}$.

Let e be a designator and θ a substitution of the form $\begin{pmatrix} z_1 & \cdots & z_n \\ e_1 & \cdots & e_n \end{pmatrix}$. We may assume that each variable occurring free in e_i is free for z_i in e, where $i=1,\ldots,n$, and that none of $z_1,\ldots,z_n$ occurs bound in e, since we identify homologous designators. Then we let $e\theta$ be the designator which is defined as follows:

(1) If the types of $z_1,\ldots,z_n$ belong to Σ, then $e\theta$ is the designator which results from replacing all free occurrences of $z_1,\ldots,z_n$ in e by $e_1,\ldots,e_n$ respectively.

(2) If at least one of the types of $z_1,\ldots,z_n$ does not belong to Σ, then $e\theta$ is defined inductively as follows:

(a) If none of $z_1,\ldots,z_{n'}$ occurs free in e, then $e\theta$ is $e\begin{pmatrix} z_{n'+1} & \cdots & z_n \\ e_{n+1} & \cdots & e_n \end{pmatrix}$, where $1 \leq n' \leq n$.

(b) If e is z_i, then e is e_i, where $i=1,\ldots,n$.

(c) If e is an m-ary abstract of the form $x_1 \cdots x_m e^*$, then $e\theta$ is $x_1 \cdots x_m (e^*\theta)$.

(d) If e is of the form $oe^*_1 \cdots e^*_m$, where o is an m-ary operator, then e is $o(e^*_1\theta)\cdots(e^*_m\theta)$.

(e) If e is of the form $ue^*_1 \cdots e^*_m$, where u is an m-ary variable with $u \notin \{z_1,\ldots,z_n\}$, then $e\theta$ is $u(e^*_1\theta)\cdots(e^*_m\theta)$.

(f) If e is of the form $z_i e^*_1 \cdots e^*_m$ and e_i is a variable, then $e\theta$ is $e_i(e^*_1\theta)\cdots(e^*_m\theta)$, where $i=1,\ldots,n$.

(g) If e is of the form $z_i e^*_1 \cdots e^*_m$ and e_i is of the form $y_1 \cdots y_m e^{\#}$, then $e\theta$ is $e^{\#}\begin{pmatrix} y_1 & \cdots & y_m \\ e^*_1\theta & \cdots & e^*_m\theta \end{pmatrix}$, where $i=1,\ldots,n$.

Note that $e\left(\right)$ is e itself for every designator e and for the empty substitution $\left(\right)$.

2.1. <u>Proposition</u>. For each pair of substitutions $\begin{pmatrix} x_1 & \cdots & x_m \\ d_1 & \cdots & d_m \end{pmatrix}$ and

$\begin{pmatrix} y_1 & & y_n \\ e_1 & \cdots & e_n \end{pmatrix}$, and for each designator e, if $\{y_1,\ldots,y_n\}-\{x_1,\ldots,x_m\}=\{y_1,\ldots,y_k\}$, then

$$e\begin{pmatrix} x_1 & & x_m \\ d_1 & \cdots & d_m \end{pmatrix}\begin{pmatrix} y_1 & & y_n \\ e_1 & \cdots & e_n \end{pmatrix}$$

is

$$e\begin{pmatrix} x_1 & & x_m & y_1 & & y_k \\ d_1\begin{pmatrix} y_1 & & y_n \\ e_1 & \cdots & e_n \end{pmatrix} & \cdots & d_m\begin{pmatrix} y_1 & & y_n \\ e_1 & \cdots & e_n \end{pmatrix} & e_1 & \cdots & e_k \end{pmatrix}.$$

<u>Proof</u>. By induction on the length of the designator e.

A λ-<u>abstract</u> is a figure of the form

$$\lambda x_1\cdots x_m.(e_1,\ldots,e_n),$$

where $x_1,\ldots,x_m$ are mutually distinct variables, $(e_1,\ldots,e_n)$ is an n-tuple of designators other than the empty ε, m=o,1,..., and n=0,1,..., and where 0-tuple of designators means the empty string, that is, the designator ε of type ε. A λ-abstract $\lambda x_1\cdots x_m.(e_1,\ldots,e_n)$ is of <u>type</u> $(s_1\cdots s_m \to t_1\cdots t_n)$ if $x_1,\ldots,x_m$ are of types $s_1,\ldots,s_m$ respectively, and $e_1,\ldots,e_n$ are of types $t_1,\ldots,t_n$ respectively. A λ-abstract $\lambda x_1\cdots x_m.(e_1,\ldots,e_n)$ is <u>closed</u> if all variables occurring free in some of $e_1,\ldots,e_n$ belong to $\{x_1,\ldots,x_m\}$.

λ-abstracts $\lambda x_1\cdots x_m.(e_1,\ldots,e_n)$ and $\lambda x'_1\cdots x'_m.(e'_1,\ldots,e'_n)$ are <u>homologous</u> if $e_i\begin{pmatrix} x_1 & & x_m \\ x'_1 & \cdots & x'_m \end{pmatrix}$ is e'_i and $e'_i\begin{pmatrix} x'_1 & & x'_m \\ x_1 & \cdots & x_m \end{pmatrix}$ is e_i, where i=1,...,n. We also identify homologous λ-abstracts.

The <u>composition</u> of a closed λ-abstract $\lambda x_1\cdots x_i.(d_1,\ldots,d_j)$ and a closed λ-abstract $\lambda y_1\cdots y_j.(e_1,\ldots,e_k)$ is the λ-abstract

$$\lambda x_1\cdots x_i.(e_1\begin{pmatrix} y_1 & & y_j \\ d_1 & \cdots & d_j \end{pmatrix},\ldots,e_k\begin{pmatrix} y_1 & & y_j \\ d_1 & \cdots & d_j \end{pmatrix}).$$

By Proposition 2.1, we have the following proposition.

2.2 <u>Proposition</u>. The class of closed λ-abstracts have the following properties.

(1) It is closed under the operation of composition.

(2) The associative law holds for the operation of composition.

(3) The λ-abstracts of the forms $\lambda x_1 \cdots x_n.(x_1,\ldots,x_n)$ are identities.

The above proposition asserts that the class of closed λ-abstracts forms a category. So we have the following definition:

The <u>category of designators</u> for the language L is the category with objects all strings in $(\Sigma \cup \Psi)^*$, and morphisms all closed λ-abstracts. By Desig_L, we denote the category of designators for the language L.

The following proposition can be easily proved.

2.3. <u>Proposition</u>. In Desig L, $\sigma^{\sigma_1 \cdots \sigma_m}$ is a power with base σ and exponent $\sigma_1 \cdots \sigma_m$, and the evaluation morphism is

$$\lambda x_1 \cdots x_m u.ux_1 \cdots x_m : \sigma_1 \cdots \sigma_m \sigma^{\sigma_1 \cdots \sigma_m} \longrightarrow \sigma,$$

where $\sigma, \sigma_1, \ldots, \sigma_m \in \Sigma$ and $\sigma^{\sigma_1 \cdots \sigma_m} \in \Psi$.

As a corollary, we get the following theorem.

2.4. <u>Theorem</u>. The category Desig_L of designators for the language L over (Σ,Ψ) is a theory over (Σ,Ψ).

In the following, we use variables according to the usage provided in the previous section.

Let $\mathrm{Desig}_L(V, {}^{\#})$ be the system of variables over Desig_L, where V is the class of variables in L and $x^{\#}$ is t for each variable x of type t. For each closed λ-abstract $\lambda x_1 \cdots x_m.e$ of type $(t_1 \cdots t_m \to t)$, by $e:\overline{\{x_1,\ldots,x_m\}} \longrightarrow t$ we denote the composition of the isomorphism $(x_1,\ldots,x_m):\overline{\{x_1,\ldots,x_m\}} \longrightarrow t_1 \cdots t_m$ and the morphism $\lambda x_1 \cdots x_m.e:t_1 \cdots t_m \longrightarrow t$ in $\mathrm{Desig}_L(V, {}^{\#})$. This notation is consistent to the usage of the λ-operator in the previous section.

We confuse a designator e of type t and the v→morphism $e:\overline{V_e}\longrightarrow t$ in $Desig_L(V, {}^{\#})$, where V_e is the set of variables occurring free in e.

Immediately we have the following proposition.

2.5. Proposition. (1) For each m-ary abstract e_o of the form $x_1\cdots x_m e$ the v→morphism e_o is identical with the abstract from the v→morphism e with respect to $x_1,\ldots,x_m$.

(2) For each designator e of type t and each substitution $\begin{pmatrix} y_1 & \cdots & y_n \\ e_1 & \cdots & e_n \end{pmatrix}$, if e_o is the designator $e\begin{pmatrix} y_1 & \cdots & y_n \\ e_1 & \cdots & e_n \end{pmatrix}$, then the v→morphism e_o is identical with the v→morphism which results from operating the substitution operator $\begin{pmatrix} y_1 & \cdots & y_n \\ e_1 & \cdots & e_n \end{pmatrix}$ on the v→morphism e, moreover, if $V_e \subseteq \{y_1,\ldots,y_n\}$, then the v→morphism

$$e\begin{pmatrix} y_1 & \cdots & y_n \\ e_1 & \cdots & e_n \end{pmatrix}:\overline{V_{e_1}\cup\cdots\cup V_{e_n}}\longrightarrow t$$

is identical with the composition of the v→morphism $(e_1,\ldots,e_n)$ and the morphism $\lambda y_1\cdots y_n.e$.

An interpretation of L to a category C is a model of the theory $Desig_L$ in C. A frame for (Σ,Ψ) is an ordered pair (C,π) where C is a category and π is a map from $(\Sigma\cup\Psi)^*$ to the class of objects of C with the following properties:

(1) For each string $t_1\cdots t_n$ in $(\Sigma\cup\Psi)^*$, $\pi(t_1,\ldots,t_n)$ is a product of $\pi(t_1),\ldots,\pi(t_n)$.

(2) For each element σ^s in Ψ, $\pi(\sigma^s)$ is a power with base $\pi(\sigma)$ and exponent $\pi(s)$.

A structure $[\cdot]$ of L over a frame (C,π) for (Σ,Ψ) is a map from the class of operators in L to the class of morphisms in C such that for each $(t_1\cdots t_n\to t)$-operator o $[o]$ is a morphism from $\pi(t_1\cdots t_n)$ to $\pi(t)$.

2.6. Theorem. Let $[\![\cdot]\!]$ be an interpretation of L, and extend $[\![\cdot]\!]$ to the system of variables $Desig_L(V, {}^{\#})$. Then

(1) for each variable x in L, $\llbracket x \rrbracket = x$,

(2) for each m-ary abstract $x_1 \cdots x_m e$, $\llbracket x_1 \cdots x_m e \rrbracket = x_1 \cdots x_m \llbracket e \rrbracket$,

(3) for each m-ary operator o, $\llbracket o e_1 \cdots e_m \rrbracket = [o](\llbracket e_1 \rrbracket, \ldots, \llbracket e_m \rrbracket)$, where $[o] = \lambda x_1 \cdots x_m . \llbracket o x_1 \cdots x_m \rrbracket$,

(4) for each m-ary variable u, $\llbracket u e_1 \cdots e_m \rrbracket = ev(\llbracket e_1 \rrbracket, \ldots, \llbracket e_m \rrbracket, u)$,

and

(5) for each designator e and each substitution $\begin{pmatrix} x_1 & \cdots & x_m \\ e_1 & \cdots & e_m \end{pmatrix}$,

$$\left\llbracket e \begin{pmatrix} x_1 & \cdots & x_m \\ e_1 & \cdots & e_m \end{pmatrix} \right\rrbracket = \llbracket e \rrbracket \begin{pmatrix} x_1 & \cdots & x_m \\ \llbracket e_1 \rrbracket & \cdots & \llbracket e_m \rrbracket \end{pmatrix}.$$

Proof. The result is an immediate consequence of Theorem 1.10(1) and Proposition 2.5.

(5) in the above theorem is a generalized form of an elementary proposition in model theory, e.g. Proposition 1.3.18[C&K]. (1)-(4) correspond to the inductive definition of interpretations, and so we have the following theorem.

2.7. Theorem. For each structure $[\cdot]$ of L over a frame (C,π), there is a unique interpretation $\llbracket \cdot \rrbracket$ of L to C such that

$\llbracket \bar{t} \rrbracket = \pi(\bar{t})$ for each string $\bar{t}$ in $(\Sigma \cup \Psi)^*$,

and

$\lambda x_1 \cdots x_m . \llbracket o x_1 \cdots x_m \rrbracket = [o]$ for each m-ary operator o in L.

3. Intuitionistic theories and toposes

In this section we expose the correspondence between intuitionistic theories and toposes.

We use variables according to the usage provided in Section 1.

A theory T over (Σ,Ψ) is Ψ-closed if there are a map $(\ ):\Psi\to\Sigma$ and isomorphisms $CD:\psi\to(\psi)$, where $\psi\in\Psi$, in T. When such map ( ) and isomorphisms CD are specified, T is said to be Ψ-closed with respect to $(\ ):\Psi\to\Sigma$ and $CD:\psi\to(\psi)$ $(\psi\in\Psi)$, the isomorphisms $CD:\psi\to(\psi)$ are called the coding morphisms, and their inverses are called the decoding morphisms and denoted by $DC:(\psi)\to\psi$.

3.1. Proposition. Each $\Sigma\wr\Sigma^{+}$-closed theory over $(\Sigma,\Sigma\wr\Sigma^{+})$ is Cartesian-closed.

Proof. Completely trivial.

3.2. Proposition. If a theory T over (Σ,Ψ) is Ψ-closed, then $ev(x_1,\dots,x_n,DC(\mathbf{f}))\doteq ev(x_1,\dots,x_m,DC(\mathbf{g}))$ implies $\mathbf{f}\doteq\mathbf{g}$, for each pair $(\mathbf{f},\mathbf{g})$ of v→morphisms such that none of $x_1,\dots,x_m$ is contained in neither $\mathbf{f}$ nor $\mathbf{g}$.

Proof. Suppose $ev(\bar{x},DC(\mathbf{f}))\doteq ev(\bar{x},DC(\mathbf{g}))$. Then, by Proposition 1.9 (5) and (1), $DC(\mathbf{f})\doteq DC(\mathbf{g})$, so $\mathbf{f}\doteq\mathbf{g}$ since DC is an isomorphism.

3.3. Proposition. Let T be a theory over (Σ,Ψ), $(\ ):\Psi\to\Sigma$ a map, $CD:\sigma^{\bar{\sigma}}\to(\sigma^{\bar{\sigma}})$ and $Ap:\bar{\sigma}(\sigma^{\bar{\sigma}})\to\sigma$, where $\sigma^{\bar{\sigma}}\in\Psi$, morphisms in T. If each pair of morphisms $CD:\sigma^{\bar{\sigma}}\to(\sigma^{\bar{\sigma}})$ and $Ap:\bar{\sigma}(\sigma^{\bar{\sigma}})\to\sigma$ satisfies the conditions

(1) $Ap(\bar{x},DC(\alpha))=ev(\bar{x},\alpha)$, where α is a variable whose domain is $\sigma^{\bar{\sigma}}$,

and

(2) $Ap(\bar{x},\mathbf{f})=Ap(\bar{x},\mathbf{g})$ implies $\mathbf{f}=\mathbf{g}$ for each pair (f,g) of v→morphisms which contain none of variables $\bar{x}$,

then $CD:\sigma^{\bar{\sigma}}\to(\sigma^{\bar{\sigma}})$ is an isomorphism, that is, T is Ψ-closed.

Proof. Set $DC = \lambda u.\bar{x}Ap(\bar{x},u)$. Then

$$DC\circ CD(\alpha) = \bar{x}Ap(\bar{x},CD(\alpha))$$
$$= \bar{x}ev(\bar{x},\alpha), \text{ by (1)},$$
$$= \alpha \quad , \text{ by Proposition 1.9(1)}.$$

On the other hand,

$$Ap(\bar{x},CD\circ DC(u)) = ev(\bar{x},DC(u)), \text{ by (1)},$$
$$= ev(\bar{x},\bar{x}Ap(\bar{x},u))$$
$$= Ap(\bar{x},u), \text{ by Proposition 1.9(2)}.$$

Therefore, by the condition (2), $CD\circ DC(u) = u$.

This completes the proof.

For example, the theory over $(\{D_\infty\},\{D_\infty^{D_\infty}\})$ whose morphisms are continuous functions is $\{D_\infty^{D_\infty}\}$-closed, where D_∞ is Scott's domain and $D_\infty^{D_\infty}$ is the function space of D_∞ defined in [Sc]. Really, there are the coding morphism $j_\infty : D_\infty^{D_\infty} \to D_\infty$ and the application morphism $\cdot : D_\infty \times D_\infty \to D_\infty$ such that

(1) $j_\infty(f)\cdot x = f(x)$,

and

(2) $m\cdot x = n\cdot x$ implies $m = n$.

Let T be a theory over $(\Sigma,\Sigma\imath\Sigma^+)$, Ω an element of Σ, $\wedge : \Omega\times\Omega \to \Omega$ and $=_t : t\times t \to \Omega$ morphisms in T, where $t \in \{1\}\cup\Sigma\cup\Sigma\imath\Sigma^+$, 1 denotes the empty sting, 1×1 denotes 1 and $t\times t$ denotes the string tt.

In the interests of readability we write

$true:1 \to \Omega$ for $=_1 : 1\times 1 \to \Omega$,

$\curlyvee : \bar{\phi} \to \Omega$ for the composition of the isomprphism $() : \bar{\phi} \to 1$ and $true:1 \to \Omega$,

$true_A$ for the composition of the morphism $A \to 1$ and $true:1 \to \Omega$,

$\mathbf{f} =_t \mathbf{g}$ for $=_t(\mathbf{f},\mathbf{g})$,

$\mathbb{P}\wedge\mathbb{Q}$ for $\wedge(\mathbb{P},\mathbb{Q})$.

We define the equality morphisms $\equiv : t_1\cdots t_m t_1 \cdots t_m \longrightarrow \Omega$ by $\lambda xy.(\pi_1(x) =_{t_1} \pi_1(y) \wedge \cdots \wedge \pi_m(x) =_{t_m} \pi_m(y))$, where $t_1,\ldots,t_m \in \Sigma \cup \Sigma 1 \Sigma^+$, $x^{\#} = y^{\#} = t_1 \cdots t_m$, and $\pi_1 : t_1 \cdots t_m \longrightarrow t_1, \ldots, \pi_m : t_1 \cdots t_m \longrightarrow t_m$ are the projections. We write $\mathbb{f} \equiv \mathbb{g}$ for $\equiv(\mathbb{f},\mathbb{g})$.

T is an <u>intuitionistic theory</u> if it has the following properties:

(a) For all v→morphisms $\mathbb{f}$ and $\mathbb{g}$, $\mathbb{f} \equiv \mathbb{g} \doteq \Upsilon$ if and only if $\mathbb{f} \doteq \mathbb{g}$.

(b) For all v→morphisms $\mathbb{P}$ and $\mathbb{Q}$ with the same domain, $\mathbb{P} \wedge \mathbb{Q} \doteq \Upsilon$ if and only if $\mathbb{P} \doteq \Upsilon$ and $\mathbb{Q} \doteq \Upsilon$.

(c) For all v→morphisms $\mathbb{P}$ and $\mathbb{Q}$ with the same domain other than $\bar{\phi}$, if for each substitution operator θ $\mathbb{P}\theta \doteq \Upsilon$ if and only if $\mathbb{Q}\theta \doteq \Upsilon$, then $\mathbb{P} = \mathbb{Q}$.

Note that the properties (a), (b), (c) are equivalent to the following properties (a'), (b'), (c') respectively:

(a') The square

$$\begin{array}{ccc} A & \longrightarrow & 1 \\ {\scriptstyle \Delta}\downarrow & & \downarrow{\scriptstyle \text{true}} \\ AA & \xrightarrow[\equiv]{} & \Omega \end{array}$$

is a pullback for each object A.

(b') The square

$$\begin{array}{ccc} 1 & \longrightarrow & 1 \\ {\scriptstyle (\text{true},\text{true})}\downarrow & & \downarrow{\scriptstyle \text{true}} \\ \Omega\Omega & \longrightarrow & \Omega \end{array}$$

is a pullback.

(c') For all morphisms $P : A \longrightarrow \Omega$ and $Q : A \longrightarrow \Omega$, if, for each morphism $f : B \longrightarrow A$, $Pf = \text{true}_B$ if and only if $Qf = \text{true}_B$, then $P = Q$.

3.4. <u>Proposition</u>. If T is an intuitionistic theory, then the following hold.

(1) $((\mathbb{f} \equiv \mathbb{g}) \wedge \mathbb{P}\binom{x}{\mathbb{f}}) \wedge \mathbb{P}\binom{x}{\mathbb{g}} = (\mathbb{f} \equiv \mathbb{g}) \wedge \mathbb{P}\binom{x}{\mathbb{f}}$.

(2) $\mathbb{P}\wedge(\mathbb{f} \equiv \mathbb{g}) \doteq \mathbb{P}$ implies $\mathbb{P}\wedge(x\mathbb{f} \equiv x\mathbb{g}) \doteq \mathbb{P}$, where x is not contained in $\mathbb{P}$.

(3) $\mathbb{P}\wedge\Upsilon = \mathbb{P}$.

(4) $(\mathbb{P}\wedge\mathbb{Q})\wedge\mathbb{R} = (\mathbb{Q}\wedge\mathbb{R})\wedge\mathbb{P}$.

(5) $\mathbb{P}\wedge(\mathbb{Q} \equiv \mathbb{R}) = \mathbb{P}\wedge(\mathbb{P}\wedge\mathbb{Q} \equiv \mathbb{P}\wedge\mathbb{R})$.

Proof. (1) Let θ be a substitution operator $\begin{pmatrix} y_1 & & y_n \\ \mathbb{k}_1 & \cdots & \mathbb{k}_n \end{pmatrix}$. We may assume that none of $y_1,\ldots,y_n$ is x and that $\mathbb{f}$ and $\mathbb{g}$ have the same domain other than $\bar{\phi}$. Suppose $((\mathbb{f} \equiv \mathbb{g})\wedge\mathbb{P}\binom{x}{\mathbb{f}})\theta \doteq \Upsilon$. Then $\mathbb{f}\theta = \mathbb{g}\theta$ by the properties (a) and (b). Therefore, by Proposition 1.6, $\mathbb{P}\binom{x}{\mathbb{f}}\theta = \mathbb{P}\binom{x}{\mathbb{g}}\theta$. On the other hand $\mathbb{P}\binom{x}{\mathbb{f}}\theta \doteq \Upsilon$. Hence $\mathbb{P}\binom{x}{\mathbb{g}}\theta \doteq \Upsilon$. Thus $(((\mathbb{f} \equiv \mathbb{g})\wedge\mathbb{P}\binom{x}{\mathbb{f}})\wedge\mathbb{P}\binom{x}{\mathbb{g}})\theta \doteq \Upsilon$, and so, by the property (c), (1) holds.

(2) We may assume that $x\mathbb{f}\equiv x\mathbb{g}$ and $\mathbb{P}$ have the same domain other than $\bar{\phi}$. Suppose that $\mathbb{P}\wedge(\mathbb{f} \equiv \mathbb{g}) \doteq \mathbb{P}$ and $\mathbb{P}$ does not contain x. Let θ be a substitution operator $\begin{pmatrix} y_1 & & y_n \\ \mathbb{k}_1 & \cdots & \mathbb{k}_n \end{pmatrix}$ such that none of $y_1,\ldots,y_n$ is x and none of $\mathbb{k}_k,\ldots,\mathbb{k}_n$ contains x, and suppose $\mathbb{P}\theta \doteq \Upsilon$. Then $\mathbb{f}\theta \doteq \mathbb{g}\theta$ by the properties (a) and (b), and so $x(\mathbb{f}\theta) \doteq x(\mathbb{g}\theta)$. Since none of $y_1,\ldots,y_n$ is x and none of $\mathbb{k}_1,\ldots,\mathbb{k}_n$ contains x $(x\mathbb{f})\theta \doteq (x\mathbb{g})\theta$. Therefore, by the properties (b) and (c), (2) holds.

(3), (4) and (5) are trivial.

T is a Heyting theory if it satisfies the property (a) in the definition of intuitionistic theory and the properties (1)-(5) in the above proposition. So an intuitionistic theory is a Heyting theory.

Immediately we have the following proposition.

3.5. Proposition. If T is a Heyting theory, then the following hold.

(1) $\mathbb{P}\wedge\mathbb{Q} = \mathbb{Q}\wedge\mathbb{P}$.

(2) $(\mathbb{P}\wedge\mathbb{Q})\wedge\mathbb{R} = \mathbb{P}\wedge(\mathbb{Q}\wedge\mathbb{R})$.

(3) The propetty (b) in the definition of intuitionistic theory.

For each n+1-tuple $(\mathbb{P}_1,\ldots,\mathbb{P}_n,\mathbb{Q})$ of v→morphisms whose codomains are Ω,

$$\mathbb{P}_1,\ldots,\mathbb{P}_n \longrightarrow \mathbb{Q}$$

denotes

$$\mathbb{P}_1\wedge\cdots\wedge\mathbb{P}_n \doteq \mathbb{P}_1\wedge\cdots\wedge\mathbb{P}_n\wedge\mathbb{Q}.$$

If $\mathbb{Q} \doteq \Upsilon$, then we say that $\mathbb{Q}$ is valid, and write $\models \mathbb{Q}$.

$\Gamma,\Gamma',\Gamma_1,\ldots$ denote finite, possibly empty, strings of v→morphisms, whose codomains are Ω, punctuated with,.

Immediately we have the following proposition.

3.6. Proposition. If T is a Heyting theory, then the following hold:

(1) $\models \mathbf{f} \equiv \mathbf{f}$.

(2) $\mathbf{f} \equiv \mathbf{g}, \mathbb{P}\binom{x}{\mathbf{f}} \longrightarrow \mathbb{P}\binom{x}{\mathbf{g}}$.

(3) $\mathbf{f} \equiv \mathbf{g} \longrightarrow \mathbf{h}\binom{x}{\mathbf{f}} \equiv \mathbf{h}\binom{x}{\mathbf{g}}$.

(4) $\mathbb{P} \longrightarrow \mathbb{P}$.

(5) If $\Gamma,\mathbb{P} \longrightarrow \mathbb{Q}$ and $\Gamma,\mathbb{Q} \longrightarrow \mathbb{P}$, then $\Gamma \longrightarrow \mathbb{P} \equiv \mathbb{Q}$.

(6) If $\Gamma \longrightarrow \mathbf{f} \equiv \mathbf{g}$, then $\Gamma \longrightarrow x\mathbf{f} \equiv x\mathbf{g}$, where x is not contained in v→morphisms in Γ.

The figure

$$\frac{\Gamma_1 \longrightarrow \mathbb{Q}_1 \qquad \cdots \qquad \Gamma_n \longrightarrow \mathbb{Q}_n}{\Gamma \longrightarrow \mathbb{Q}}$$

denotes that if $\Gamma_1 \longrightarrow \mathbb{Q}_1,\ldots,$ and $\Gamma_n \longrightarrow \mathbb{Q}_n$, then $\Gamma \longrightarrow \mathbb{Q}$. Such figure is called an inference figure in T.

Immediately we have the following proposition.

3.7. Proposition. If T is a Heyting theory, then the following are inference figures in T:

(Sub.) $$\frac{\mathbb{P}_1,\ldots,\mathbb{P}_n \longrightarrow \mathbb{Q}}{\mathbb{P}_1\theta,\ldots,\mathbb{P}_n\theta \longrightarrow \mathbb{Q}\theta} \quad ,$$

where θ is an arbitrary substitution operator.

(Str.) $$\frac{\Gamma \longrightarrow \mathbb{Q}}{\Gamma' \longrightarrow \mathbb{Q}'} \quad ,$$

where $\mathbb{Q} \doteq \mathbb{Q}'$, for each v→morphism $\mathbb{P}$ in Γ there is a v→morphism $\mathbb{P}'$ in Γ' such that $\mathbb{P} \doteq \mathbb{P}'$, and each variable in Γ or $\mathbb{Q}$ is contained in Γ' or $\mathbb{Q}'$.

(Cut) $$\frac{\Gamma \longrightarrow \mathbb{Q} \qquad \mathbb{Q}, \Gamma \longrightarrow \mathbb{R}}{\Gamma \longrightarrow \mathbb{R}} \quad ,$$

where each variable in $\mathbb{Q}$ is contained in Γ or $\mathbb{R}$.

($\rightarrow\wedge$) $$\frac{\Gamma \longrightarrow \mathbb{P} \qquad \Gamma \longrightarrow \mathbb{Q}}{\Gamma \longrightarrow \mathbb{P} \wedge \mathbb{Q}} \quad .$$

($\wedge\rightarrow$) $$\frac{\mathbb{P}, \Gamma \longrightarrow \mathbb{R}}{\mathbb{P} \wedge \mathbb{Q}, \Gamma \longrightarrow \mathbb{R}} \quad , \quad \frac{\mathbb{Q}, \Gamma \longrightarrow \mathbb{R}}{\mathbb{P} \wedge \mathbb{Q}, \Gamma \longrightarrow \mathbb{R}} \quad .$$

($=_\Omega$) $$\frac{\mathbb{P}, \Gamma \longrightarrow \mathbb{Q} \qquad \mathbb{Q}, \Gamma \longrightarrow \mathbb{P}}{\Gamma \longrightarrow \mathbb{P} \equiv \mathbb{Q}} \quad .$$

(Abs.) $$\frac{\Gamma \longrightarrow \mathbf{f} \equiv \mathbf{g}}{\Gamma \longrightarrow x\mathbf{f} \equiv x\mathbf{g}} \quad ,$$

where x is not contained in any v→morphisms in Γ.

We introduce the logical operators in T as follows:

$\forall_A : \Omega^A \longrightarrow \Omega$ is defined by $\forall_A = \lambda\alpha.(\alpha \equiv x\Upsilon)$, where $A \in \Sigma^+$ and Σ^+ means the class $\Sigma^*-\{1\}$.

$\supset : \Omega\times\Omega \longrightarrow \Omega$ is defined by $\supset = \lambda uv.(u \wedge v \equiv u)$.

$\exists_A : \Omega^A \longrightarrow \Omega$ is defined by $\exists_A = \lambda\alpha.\forall_\Omega w(\forall_A x(ev(x,\alpha) \supset w) \supset w)$, where $A \in \Sigma^+$.

$\vee : \Omega\times\Omega \longrightarrow \Omega$ is defined by $\vee = \lambda uv.\forall_\Omega w((u \supset w) \wedge (v \supset w) \supset w)$.

$false : 1 \longrightarrow \Omega$ is defined by $false = \lambda.\forall_\Omega ww$.

$\neg : \Omega \longrightarrow \Omega$ is defined by $\neg = \lambda u.(u \supset false())$.

We write $\curlywedge : \bar{\phi} \longrightarrow \Omega$ for the composition of $() : \bar{\phi} \longrightarrow 1$ and $false : 1 \longrightarrow \Omega$.

We omit the subscript A from $\forall_A$ and $\exists_A$ when it is immaterial or clear from the context.

3.8. <u>Proposition.</u> If T is a Heyting theory, then the following hold:

(1) $\forall_A:\Omega^A \longrightarrow \Omega$ is the unique morphism for which the following inference figures hold:

$$(\rightarrow\forall) \qquad \frac{\Gamma \longrightarrow \mathbb{Q}}{\Gamma \longrightarrow \forall_A x\mathbb{Q}} \quad ,$$

where x is not contained in any v→morphism in Γ,

$$(\forall\rightarrow) \qquad \frac{\mathbb{P}\binom{x}{\mathbf{f}}, \Gamma \longrightarrow \mathbb{Q}}{\forall_A x\mathbb{P},\ \Gamma \longrightarrow \mathbb{Q}} \quad ,$$

where each variable in $\mathbf{f}$ is contained in some v→morphism in $\forall_A x\mathbb{P}, \Gamma \longrightarrow \mathbb{Q}$.

(2) $\supset:\Omega\times\Omega \longrightarrow \Omega$ is the unique morphism for which the following inference figures hold:

$$(\rightarrow\supset) \qquad \frac{\mathbb{P}, \Gamma \longrightarrow \mathbb{Q}}{\Gamma \longrightarrow \mathbb{P}\supset\mathbb{Q}} \quad ,$$

$$(\supset\rightarrow) \qquad \frac{\Gamma \longrightarrow \mathbb{P} \qquad \mathbb{Q}, \Gamma \longrightarrow \mathbb{R}}{\mathbb{P}\supset\mathbb{Q}, \Gamma \longrightarrow \mathbb{R}} \quad .$$

(3) $\exists_A:\Omega^A \longrightarrow \Omega$ is the unique morphism for which the following inference figures hold:

$$(\rightarrow\exists) \qquad \frac{\Gamma \longrightarrow \mathbb{Q}\binom{x}{\mathbf{f}}}{\Gamma \longrightarrow \exists_A x\mathbb{Q}} \quad ,$$

where each variable in $\mathbf{f}$ is contained in some v→morphism in $\Gamma \longrightarrow \exists_A x\mathbb{Q}$,

$$(\exists\rightarrow) \qquad \frac{\mathbb{P}, \Gamma \longrightarrow \mathbb{Q}}{\exists_A x\mathbb{P}, \Gamma \longrightarrow \mathbb{Q}} \quad ,$$

where x is not contained in any v→morphism in $\Gamma \longrightarrow \mathbb{Q}$.

(4) $\vee:\Omega\times\Omega \longrightarrow \Omega$ is the unique morphism for which the following inference figures hold:

$$(\rightarrow\vee) \qquad \frac{\Gamma \longrightarrow \mathbb{Q}}{\Gamma \longrightarrow \mathbb{Q}\vee\mathbb{R}} \ , \qquad \frac{\Gamma \longrightarrow \mathbb{R}}{\Gamma \longrightarrow \mathbb{Q}\vee\mathbb{R}} \ ,$$

$(\vee\rightarrow)$ $$\frac{\mathbb{P},\Gamma \longrightarrow \mathbb{R} \qquad \mathbb{Q},\Gamma \longrightarrow \mathbb{R}}{\mathbb{P}\vee\mathbb{Q},\Gamma \longrightarrow \mathbb{R}} \quad .$$

(5) false:$1 \longrightarrow \Omega$ is the unique morphism for which the following inference figure holds:

$(\curlywedge)$ $$\frac{\Gamma \longrightarrow \curlywedge}{\Gamma \longrightarrow \mathbb{Q}} \quad ,$$

where $\mathbb{Q}$ is an arbitrary v→morphism with codomain Ω.

(6) $\neg:\Omega \longrightarrow \Omega$ is the unique morphism for which the following inference figures hold:

$(\rightarrow\neg)$ $$\frac{\mathbb{P},\Gamma \longrightarrow \curlywedge}{\Gamma \longrightarrow \neg\mathbb{P}} \quad ,$$

$(\neg\rightarrow)$ $$\frac{\Gamma \longrightarrow \mathbb{P}}{\neg\mathbb{P},\Gamma \longrightarrow \curlywedge} \quad .$$

Proof. (1) $\mathbb{Q},\mathbb{Q} \longrightarrow \curlyvee$ and $\curlyvee,\mathbb{Q} \longrightarrow \mathbb{Q}$. Then, by $(=_\Omega)$ in Proposition 3.7, $\mathbb{Q} \longrightarrow \mathbb{Q} \equiv \curlyvee$. Now suppose $\Gamma \longrightarrow \mathbb{Q}$. Then, by (Cut), $\Gamma \longrightarrow \mathbb{Q} \equiv \curlyvee$. Hence, by (Abs.), $\Gamma \longrightarrow x\mathbb{Q} \equiv x\curlyvee$, that is $\Gamma \longrightarrow \forall_A x\mathbb{Q}$, provided that x is not contained in any v→morphism in Γ. Therefore the inference figure $(\rightarrow\forall)$ holds.

By Proposition 3.6 (2), $x\mathbb{P} \equiv x\curlyvee, ev(\mathbb{f},x\curlyvee) \longrightarrow ev(\mathbb{f},x\mathbb{P})$, and so $\forall_A x\mathbb{P} \longrightarrow ev(\mathbb{f},x\mathbb{P})$. Suppose $\mathbb{P}\binom{x}{\mathbb{f}},\Gamma \longrightarrow \mathbb{Q}$. Then, by (Str.), $ev(\mathbb{f},x\mathbb{P}),\Gamma \longrightarrow \mathbb{Q}$, since $ev(\mathbb{f},x\mathbb{P}) \doteq \mathbb{P}\binom{x}{\mathbb{f}}$. Therefore, by (Cut), $\forall_A x\mathbb{P},\Gamma \longrightarrow \mathbb{Q}$, provided that each variable in $\mathbb{f}$ is contained in $\forall_A x\mathbb{P},\Gamma$ or $\mathbb{Q}$. Hence the inference figure $(\forall\rightarrow)$ holds.

Suppose that $\forall'_A:\Omega^A \longrightarrow \Omega$ is a morphism for which the inference figures obtained from $(\rightarrow\forall)$ and $(\forall\rightarrow)$ by replacing $\forall_A$ by $\forall'_A$ hold. Then, since $ev(x,\alpha) \longrightarrow ev(x,\alpha)$, $\forall'_A xev(x,\alpha) \longrightarrow ev(x,\alpha)$. Hence, by $(\rightarrow\forall)$, $\forall'_A xev(x,\alpha) \longrightarrow \forall_A xev(x,\alpha)$. Similarly, $\forall_A xev(x,\alpha) \longrightarrow \forall'_A xev(x,\alpha)$. Therefore $\forall_A xev(x,\alpha) = \forall'_A xev(x,\alpha)$. On the other hand $\alpha = xev(x,\alpha)$, hence $\forall_A = \forall'_A$. Thus $\forall_A$ is uniquely determined by the inference figures $(\rightarrow\forall)$ and $(\forall\rightarrow)$.

(2)-(6) are similar to (1).

3.9. Proposition. (1) If $f(a) \equiv f(b) \longrightarrow a \equiv b$, then f is a monomorphism, provided that T is a Heyting theory.

(2) $f(a) \equiv f(b) \longrightarrow a \equiv b$ if and only if f is a monomorphism, provided that T is an intuitionistic theory.

Proof. (1) is clear.

(2) Suppose that $f: a^{\#} \longrightarrow \Omega$ is a monomorphism in T. Then, for each pair of morphisms $g: c^{\#} \longrightarrow a^{\#}$ and $h: d^{\#} \longrightarrow a^{\#}$, $f(g(c)) \doteq f(h(d))$ implies $g(c) \doteq h(d)$. Therefore, by the property (c) in the definition of intuitionistic theory, $f(a) \equiv f(b) \longrightarrow a \equiv b$.

Let $!: \Omega^B \longrightarrow \Omega^B$ is the morphism defined by

$$! = \lambda\beta.b\forall x(ev(x,\beta) \equiv (b \equiv x)).$$

Then $\exists ! b\mathbf{P} = \exists b\forall x(\mathbf{P} \equiv (b \equiv x))$.

3.10. Proposition. Suppose that T is an intuitionistic theory. Consider the square

$$\begin{array}{ccc} B & \longrightarrow & 1 \\ h\downarrow & (i) & \downarrow \text{true} \\ A & \underset{P}{\longrightarrow} & \Omega \end{array} ,$$

where $B \in \Sigma^+$ and A is an arbitrary object in T.

(1) If the square (i) is a pullback in T, then $P = \lambda a.\exists b(h(b) \equiv a)$.

(2) The following assertions are equivalent:

(a) For each morphism $Q: C \times B \longrightarrow \Omega$ such that $\models \exists ! bQ(c,b)$, there is a unique morphism $g: C \longrightarrow B$ such that $\models Q(c, g(c))$.

(b) For each monomorphism $h: B \longrightarrow A$, there is a unique morphism $P: A \longrightarrow \Omega$ for which the square (i) is a pullback.

Proof. (1) Suppose that the square (i) is a pullback. Then $P(a) \longrightarrow \exists b(h(b) \equiv a)$, since $\models P(g(c))$ implies $\models h(k(c)) \equiv g(c)$ for some morphism k. On the other hand, $\exists b(h(b) \equiv a) \longrightarrow P(a)$, since $\models P(h(b))$.

Therefore $P = \lambda a.\exists b(h(b) \equiv a)$.

(2) (a)$\Rightarrow$(b) Suppose that $h:B \longrightarrow A$ is a monomorphism and let $P = \lambda a.\exists b(h(b) \equiv a)$. Suppose $\models P(k(c))$. Then, by Proposition 3.9(2), $\models \exists! b(h(b) \equiv k(c))$. Therefore, by the assumption, there is a morphism g such that $\models hg(c) \equiv k(c)$, that is, $hg = k$. Such g is unique, since h is a monomorphism. Therefore the square (i) is a pullback, and, by (1), such P is unique.

(b)$\Rightarrow$(a) Let $\{\cdot\}:B \longrightarrow \Omega^B$ be the morphism defined by $\{\cdot\} = \lambda x.(b \equiv x)$. By the assumption (b), there is a unique morphism $P:\Omega^B \longrightarrow \Omega$ for which the square (i) is a pullback, where $A = \Omega^B$. By (1), $P = \lambda \beta.\exists b(\{b\} \equiv \beta)$. Since $\{b\} \equiv \beta = \forall x(ev(x,\beta) \equiv (b \equiv x))$, $P = \exists!$.

Now suppose $\models \exists! bQ(c,b)$. Then by the diagram

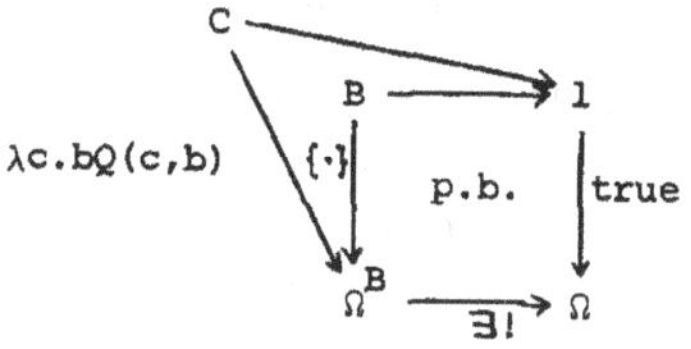

,

there is a morphism $g:C \longrightarrow B$ such that $\{g(c)\} = bQ(c,b)$. $\{g(c)\} = bQ(c,b)$ implies $(g(c) \equiv b) = Q(c,b)$, and so $\models Q(c,g(c))$.
Thus there is a unique morphism g such that $\models Q(c,g(c))$.

This completes the proof.

3.11. <u>Proposition</u>. Suppose that T is a Heyting theory. If for each morphism $P:A \longrightarrow \Omega$, there is a morphism $h:B \longrightarrow A$ such that $P = \lambda a.\exists b(h(b) \equiv a)$, then T is an intuitionistic theory.

<u>Proof</u>. By Proposition 3.5, it is sufficient to show that the property (c) in the definition of intuitionistic theory holds. Let $P = \lambda a.\exists b(h(b) \equiv a)$ and $Q = \lambda a.\exists c(k(c) \equiv a)$. Suppose that for each morphism $g:x^{\#} \longrightarrow A$ $\models P(g(x))$

if and only if $\models Q(g(x))$. Then, since $\models P(h(b))$, $\models Q(h(b))$, and so $\models \exists c(k(c) \equiv h(b))$. On the other hand, $h(b) \equiv a$, $\exists c(k(c) \equiv h(b)) \longrightarrow \exists c(k(c) \equiv a)$, hence $\exists b(h(b) \equiv a) \longrightarrow \exists c(h(c) \equiv a)$. Similarly $\exists c(k(c) \equiv a) \longrightarrow \exists b(h(b) \equiv a)$. Therefore $P = Q$. Thus the property (c) holds.

T is a <u>higher-order Heyting</u> (or <u>intuitionistic</u>) <u>theory</u> if T is a Heyting (or intuitionistic) theory, and there are a mapping $(\ \ ) : \Sigma \urcorner \Sigma^{+} \longrightarrow \Sigma$ and morphisms $Ap : \overline{\sigma}(\overline{\sigma^{\sigma}}) \longrightarrow \sigma$, where $\overline{\sigma^{\sigma}} \in \Sigma \urcorner \Sigma^{+}$, with the following property:

[Axiom of Comprehension]

$$\forall \overline{x} \exists ! y\, ev(\overline{x}, y, \alpha) \longrightarrow \exists ! u \forall \overline{x}\, ev(\overline{x}, Ap(\overline{x}, u), \alpha).$$

Such mapping $(\ \ ) : \Sigma \urcorner \Sigma^{+} \longrightarrow \Sigma$ is called the <u>type-mapping</u>, and the morphisms $Ap : \overline{\sigma}(\overline{\sigma^{\sigma}}) \longrightarrow \sigma$ are called the <u>application</u> morphisms.

Note that Axiom of Comprehension is equivalent to the following schema:

$$\forall \overline{x} \exists ! y \mathbb{P} \longrightarrow \exists ! u \forall \overline{x} (\mathbb{P}\begin{pmatrix} y \\ - \\ Ap(\overline{x}, u) \end{pmatrix}) ,$$

where u does not occur free in $\mathbb{P}$.

3.12. <u>Theorem</u>. Suppose that T is a higher-order Heyting theory.

(1) If for each morphism Q such that $\models \exists ! y Q(x,y)$ there is a morphism g such that $\models Q(x, g(x))$, then T is Cartesian-closed.

(2) Moreover, if for each morphism $P : A \longrightarrow \Omega$ there is a monomorphism $h : B \longrightarrow A$ such that $P = \lambda a. \exists b(h(b) \equiv a)$, then T is a topos.

<u>Proof</u>. (1) Let ( ) be the type-mapping of T and $Ap : \overline{\sigma}(\overline{\sigma^{\sigma}}) \longrightarrow \sigma$, where $\overline{\sigma^{\sigma}} \in \Sigma \urcorner \Sigma^{+}$, be the application morphisms of T. Then, by Axiom of Comprehension, $\models \exists ! u \forall \overline{x}(Ap(\overline{x}, u) \equiv ev(\overline{x}, \alpha))$. Therefore, by the assumption of (1), there are morphisms $CD : \sigma^{\sigma} \longrightarrow (\overline{\sigma^{\sigma}})$, where $\overline{\sigma^{\sigma}} \in \Sigma \urcorner \Sigma^{+}$, such that $\models \forall \overline{x}(Ap(\overline{x}, CD(\alpha)) \equiv ev(\overline{x}, \alpha))$. Moreover, by Axiom of Comprehension, $\forall \overline{x}(Ap(\overline{x}, u) \equiv Ap(\overline{x}, v)) \longrightarrow u \equiv v$. Thus, by Proposition 3.3, T is Cartesian-closed.

(2) By Proposition 3.11 and the assumption of (2), T is an intuitionistic

theory. Therefore, by Proposition 3.10 and the assumption of (1), for each monomorphism $h:B \longrightarrow A$ with $B \in \Sigma^+$ there is a unique morphism $P:A \longrightarrow \Omega$ for which the following square (i) is a pullback:

$$\begin{array}{ccc} B & \longrightarrow & 1 \\ h \downarrow & \text{(i)} & \downarrow \text{true} \\ A & \xrightarrow[P]{} & \Omega \end{array} .$$

Moreover, by the assumption of (2), for each morphism $P:A \longrightarrow \Omega$, there is a monomorphism $h:B \longrightarrow A$ for which the square (i) is a pullback. Thus: $1 \longrightarrow \Omega$ is a subobject classifier and so T is a topos.

3.13. Theorem. Suppose that $\mathbb{E}$ is a topos and $\tilde{\mathbb{E}}$ is a category with the following properties:

(1) $\mathbb{E}$ is a full subcategory of $\tilde{\mathbb{E}}$, and equivalent to $\tilde{\mathbb{E}}$.

(2) The class of objects of $\tilde{\mathbb{E}}$ is the class $(E \cup ElE^+)^*$, where E is the class of objects of $\mathbb{E}$.

(3) For all objects $A_1,\ldots,A_n$ in $\tilde{\mathbb{E}}$, the string $A_1\cdots A_n$ is a product of $A_1,\ldots,A_n$ in $\tilde{\mathbb{E}}$.

(4) For all objects $A,A_1,\ldots,A_n$ in $\mathbb{E}$, $A^{A_1\cdots A_n}$ is a power with base A and exponent $A_1\cdots A_n$ in $\tilde{\mathbb{E}}$.

Then $\tilde{\mathbb{E}}$ is a higher-order intuitionistic theory over (E,ElE^+). Such $\tilde{\mathbb{E}}$ exists uniquely up to isomorphism.

Proof. Since $\tilde{\mathbb{E}}$ is a topos, it is easily seen that $\tilde{\mathbb{E}}$ is an intuitionistic theory.

Let $(\ ):ElE^+ \longrightarrow E$ be the map such that $(A^{A_1\cdots A_n})$ is a power with base A and exponent $A_1\times\cdots\times A_n$ in $\mathbb{E}$ for each $A^{A_1\cdots A_n}$ in ElE^+, and let $Ap:A_1\cdots A_n(A^{A_1\cdots A_n}) \longrightarrow A$ be the composition of the isomorphism: $A_1\cdots A_n(A^{A_1\cdots A_n}) \longrightarrow A_1\times\cdots\times A_n\times A^{A_1\times\cdots\times A_n}$ in $\tilde{\mathbb{E}}$ and the evaluation morphism: $A_1\times\cdots\times A_n\times A^{A_1\times\cdots\times A_n} \longrightarrow A$ in $\mathbb{E}$.

Then Axiom of Comprehension holds in $\hat{\mathbb{E}}$, since true:$1 \longrightarrow \Omega$ is a subobject classifier, and by Proposition 3.10(2).

This completes the proof.

When T is a higher-order Heyting theory with application morphisms Ap, we write

$$(\exists x_1 \cdots x_m)\mathbf{P} \quad \text{for the v}\rightarrow\text{morphism} \quad \exists z_1 \cdots z_m \mathbf{P}\begin{pmatrix} x_1 & \cdots & x_m \\ z_1^* & & z_m^* \end{pmatrix},$$

and

$$(\exists ! x_1 \cdots x_m)\mathbf{P} \quad \text{for the v}\rightarrow\text{morphism}$$

$$\exists y_1 \cdots y_m \forall z_1 \cdots z_m (\mathbf{P}\begin{pmatrix} x_1 & \cdots & x_m \\ z_1^* & & z_m^* \end{pmatrix} \equiv (y_1 \equiv z_1 \wedge \cdots \wedge y_m \equiv z_m)),$$

where $x_1,\ldots,x_m,y_1,\ldots,y_m,z_1,\ldots,z_m$ are mutually distinct variables, none of $y_1,\ldots,y_m,z_1,\ldots,z_m$ is contained in $\mathbf{P}$, and each z_i^* denotes z_i itself if the domain of x_i is in Σ, the abstract of the form $\bar{u}Ap(\bar{u},z_i)$ if the domain of x_i is in $\Sigma \, \Sigma^+$.

The following theorem is a version of Theorem 5.3 in [F] to higher-order Heyting theory.

3.14. <u>Theorem</u>. Suppose that T is a higher-order Heyting theory, and let $\bar{T}$ be the category whose objects are morphisms with codomain Ω in T and whose morphisms from P to Q are equivalence classes $\bar{G}$ of morphisms G in T such that

$$P(\bar{x}) \longrightarrow (\exists ! \bar{y})(G(\bar{x},\bar{y}) \wedge Q(\bar{y})),$$

where G and G' are equivalent if and only if

$$P(\bar{x}),Q(\bar{y}) \longrightarrow G(\bar{x},\bar{y}) \equiv G'(\bar{x},\bar{y}).$$

The composition $\bar{H}\circ\bar{G}:P \longrightarrow R$ of $\bar{G}:P \longrightarrow Q$ and $\bar{H}:Q \longrightarrow R$ is given by the morphism

$$\lambda\bar{x}\bar{z}.(\exists\bar{y})(G(\bar{x},\bar{y}) \wedge Q(\bar{y}) \wedge H(\bar{y},\bar{z})).$$

Then $\bar{T}$ is a topos.

When T is a higher-order Heyting theory, a model J of T in a topos $\mathbb{E}$ is <u>logical</u> if $J(\mathrm{true}):J(1)\longrightarrow J(\Omega)$ is a subobject classifier, $J(\wedge):J(\Omega)\times J(\Omega)\longrightarrow J(\Omega)$ is the conjunction in $\mathbb{E}$, $J(=_t):J(t)\times J(t)\longrightarrow J(\Omega)$ are equality morphisms in $\mathbb{E}$, and $J(\mathrm{Ap}):J(\overline{\sigma}(\sigma^{\overline{\sigma}}))\longrightarrow J(\sigma)$ are evaluation morphisms in $\mathbb{E}$.

The following theorem corresponds to Theorem 8.9 in [F].

3.15. <u>Theorem</u>. Let T be a higher-order Heyting theory, $\overline{T}$ the topos defined in Proposition 3.14, and $I:T\longrightarrow\overline{T}$ the functor such that for an object A of T I(A) is the object true_A in $\overline{T}$, and for a morphism f in T I(f) is the morphism in $\overline{T}$ given by $\lambda xy.f(x)\equiv y$. Then I is a logical model of T, and for each logical model J of T in a topos $\mathbb{E}$ there is a logical functor $K:\overline{T}\longrightarrow\mathbb{E}$ unique up to isomorphism such that $K\circ I = J$.

For two higher-order Heyting theories T_i over $(\Sigma_i,\Sigma_i 1\Sigma_i^+)$ with truth value object Ω_i, conjunction $\wedge_i$, equality morphisms $=_{it}$, type mapping $(\ )_i$ and application morphisms Ap_i, where i=1,2, a functor $F:T_1\longrightarrow T_2$ is an <u>H-functor</u> if

(1) $F(\sigma)\in\Sigma_2$ for each σ in Σ_1,

(2) $F(t_1\cdots t_n) = F(t_1)\cdots F(t_n)$ for each $t_1\cdots t_n$ in $(\Sigma_1\cup\Sigma_1 1\Sigma_1^+)^*$,

(3) $F(\sigma^{\overline{\sigma}}) = F(\sigma)^{F(\overline{\sigma})}$ for each $\sigma^{\overline{\sigma}}$ in $\Sigma_1 1\Sigma_1^+$,

(4) $F(\Omega_1) = \Omega_2$,

(5) $F(\wedge_1:\Omega_1\times\Omega_1\longrightarrow\Omega_1) = \wedge_2:\Omega_2\times\Omega_2\longrightarrow\Omega_2$,

(6) $F(=_t:t\times t\longrightarrow\Omega_1) = =_{2F(t)}:F(t)\times F(t)\longrightarrow\Omega_2$ for each t in $\{1\}\cup\Sigma_1\cup\Sigma_1 1\Sigma_1^+$,

(7) $F((t)_1) = (F(t))_2$ for each t in $\Sigma_1 1\Sigma_1^+$,

and

(8) $F(\mathrm{Ap}_1:\overline{\sigma}(\sigma^{\overline{\sigma}})_1\longrightarrow\sigma)=\mathrm{Ap}_2:F(\overline{\sigma}(\sigma^{\overline{\sigma}})_1)\longrightarrow F(\sigma)$ for each $\sigma^{\overline{\sigma}}$ in $\Sigma_1 1\Sigma_1^+$.

Let HH be the category whose objects are all small higher-order Heyting theories, and whose morphisms from T_1 to T_2 are equivalence classes $|F|_H$ of H-functors $F:T_1 \to T_2$ under isomorphism.

Let $\mathbb{TT}$ be the category whose objects are all small toposes, and whose morphisms from $\mathbb{E}_1$ to $\mathbb{E}_2$ are equivalence classes $|L|_T$ of logical functors $L:\mathbb{E}_1 \to \mathbb{E}_2$ under isomorphism.

3.16. Theorem. Extend the operator $\sim$ defined in Theorem 3.13 to the functor $\sim:\mathbb{TT} \to \mathbb{HH}$, and the operator $^{-}$ defined in Theorem 3.14 to the functor $^{-}:\mathbb{HH} \to \mathbb{TT}$. Then $^{-}$ is the left adjoint of $\sim$.

Proof. The assertion follows immediately from the definition and Theorem 3.15.

4. Higher-order intuitionistic logical calculus

Let Σ be a class, Ω an element in Σ, and ( ) a map from $\Sigma\iota\Sigma^+$ to Σ. Let L_H be a language over $(\Sigma,\Sigma\iota\Sigma^+)$ with $(\bar{\sigma}(\sigma^{\bar{\sigma}})\to\sigma)$-operators $Ap^{\bar{\sigma}\sigma}$, where $\sigma^{\bar{\sigma}}\in\Sigma\iota\Sigma^+$, $(\Omega\Omega\to\Omega)$-operator $\wedge$, and $(tt\to\Omega)$-operators $=^t$, where $t\in\{1\}\cup\Sigma\cup\Sigma\iota\Sigma^+$ and 1 is the empty string.

A variable in L_H is said to be bindable if its type belongs to Σ. A formula is a designator of type Ω. A sequent is a figure of the form $\Gamma\longrightarrow\mathfrak{A}$, where $\mathfrak{A}$ is a formula and Γ is a finite, possibly empty, string of formulas punctuated with,.

In the interests of readability we introduce abbreviations for designators of L_H as follows:

$d(d_1,\ldots,d_n)$ for $Ap^{\sigma_1\cdots\sigma_n\sigma}d_1\cdots d_n d$,

$\mathfrak{A}\wedge\mathfrak{B}$ for $\wedge\mathfrak{A}\mathfrak{B}$,

$d =^t d'$ or $d \equiv d'$ for $=^t dd'$,

T for $=^1$,

$\mathfrak{A}\supset\mathfrak{B}$ for $\mathfrak{A}\wedge\mathfrak{B}\equiv\mathfrak{A}$,

$\forall\bar{x}\mathfrak{A}$ for $\bar{x}\mathfrak{A}\equiv\bar{x}T$,

$\exists x\mathfrak{A}$ for $\forall w((\forall x\mathfrak{A}\supset w)\supset w)$,

where w does not occur free in $\mathfrak{A}$,

$!x\mathfrak{A}$ for $y\forall x(\mathfrak{A}\equiv(x\equiv y))$,

where y does not occur free in $\mathfrak{A}$.

We now describe a Gentzen-type system for higher-order intuitionistic logical calculus.

As initial sequents we take all sequents of the forms:

(Equality) $\longrightarrow d\equiv d.$

$$d_1\equiv d_2,\mathfrak{A}\binom{x}{d_1}\longrightarrow\mathfrak{A}\binom{x}{d_2}.$$

(Comprehension) $\forall\bar{x}\exists! y\mathfrak{A} \longrightarrow \exists! u\forall\bar{x}\mathfrak{A}\binom{y}{u(\bar{x})}$,

where u does not occur free in $\mathfrak{A}$.

Inference figures are the following:

(Cut) $$\frac{\Gamma \longrightarrow \mathfrak{A} \quad \mathfrak{A},\Gamma \longrightarrow \mathfrak{B}}{\Gamma \longrightarrow \mathfrak{B}},$$

where each bindable variable occurring free in $\mathfrak{A}$ occurs free in some formula in $\Gamma \longrightarrow \mathfrak{B}$.

(+) $\dfrac{\Gamma \longrightarrow \mathfrak{B}}{\mathfrak{A},\Gamma \longrightarrow \mathfrak{B}}$. (-) $\dfrac{\mathfrak{A},\mathfrak{A},\Gamma \longrightarrow \mathfrak{B}}{\mathfrak{A},\Gamma \longrightarrow \mathfrak{B}}$. (X) $\dfrac{\Gamma,\mathfrak{A},\mathfrak{B},\Pi \longrightarrow \mathfrak{C}}{\Gamma,\mathfrak{B},\mathfrak{A},\Pi \longrightarrow \mathfrak{C}}$.

($=^{\Omega}$) $$\frac{\mathfrak{A},\Gamma \longrightarrow \mathfrak{B} \quad \mathfrak{B},\Gamma \longrightarrow \mathfrak{A}}{\Gamma \longrightarrow \mathfrak{A}\equiv\mathfrak{B}}.$$

($\wedge$) $\dfrac{\Gamma \longrightarrow \mathfrak{A} \quad \Gamma \longrightarrow \mathfrak{B}}{\Gamma \longrightarrow \mathfrak{A}\wedge\mathfrak{B}}$. $\dfrac{\mathfrak{A},\Gamma \longrightarrow \mathfrak{C}}{\mathfrak{A}\wedge\mathfrak{B},\Gamma \longrightarrow \mathfrak{C}}$. $\dfrac{\mathfrak{B},\Gamma \longrightarrow \mathfrak{C}}{\mathfrak{A}\wedge\mathfrak{B},\Gamma \longrightarrow \mathfrak{C}}$.

(Abstraction) $$\frac{\Gamma \longrightarrow d_1 =^t d_2}{\Gamma \longrightarrow x_1\cdots x_n d_1 \equiv x_1\cdots x_n d_2},$$

where $t\in\Sigma$ and none of $x_1,\ldots,x_n$ occurs free in any formula in Γ.

Note that we identify homologous designators.

4.1. <u>Theorem</u>. Let T be a higher-order Heyting theory, and A a class of closed formulas of L_H. Let $\|\cdot\|$ be the interpretation of L_H to T determined by a structure $[\cdot]$ such that $[Ap^{\bar{\sigma}\sigma}]$ is the application mrrphism in T for each $\bar{\sigma}^{\sigma}\in\Sigma\urcorner\Sigma^{+}$, $[\wedge]$ is the conjunction in T, $[=^{t}]$ is the equality morphism in T for each $t\in\{1\}\cup\Sigma\cup\Sigma\urcorner\Sigma^{+}$, and $\|\mathfrak{A}\|$ is valid for each formula $\mathfrak{A}$ in A. If $\mathfrak{B}$ is a formula of L_H and provable from A, than $\|\mathfrak{B}\|$ is valid.

<u>Proof</u>. The assertion follows immediately from Theorem 2.6 and the discussion in the previous section.

For a class A of closed formulas of L_H, let $\sim_A$ be the relation on the class of morphisms in the category $Desig_{L_H}$, that is, the class of λ-abstracts of L_H, defined by

$$\lambda x_1 \cdots x_m.(d_1,\ldots,d_n) \underset{A}{\sim} \lambda x_1 \cdots x_m.(d_1',\ldots,d_n')$$

if and only if $d_1 \equiv d_1' \wedge \cdots \wedge d_n \equiv d_n'$ is provable from A.

Then $\sim_A$ is a congruence relation. Now we let $\|\lambda\bar{x}.\bar{d}\|\sim_A$ be the equivalence class to which the morphism $\lambda\bar{x}.\bar{d}$ in Desig_{L_H} belongs under the equivalence relation $\sim_A$, and let T(A) be the quatient theory of Desig_{L_H} by $\sim_A$. Then T(A) is a higher-order Heyting theory and $\|\cdot\|\sim_A$ is regarded as an interpretation of L_H to T(A). Moreover we obtain the completeness theorem to higher-order Heyting theories:

4.2. <u>Theorem</u>. For each formula $\mathfrak{A}$ of L_H, $\mathfrak{A}$ is provable from A if and only if $\|\mathfrak{A}\|\sim_A$ is valid in the higher-order Heyting theory T(A).

By virtue of Theorem 3.15 and the above theorem, we have the completeness theorem to toposes:

4.3. <u>Theorem</u>. For each class A of closed formulas of L_H, there is an interpretation $\|\cdot\|$ of L_H to a topos such that, for each formula $\mathfrak{A}$ in L_H, $\mathfrak{A}$ is provable from A if and only if $\|\mathfrak{A}\|$ is valid in the topos.

References

[B & J] A. Boileau and A. Joyal, La Logique des Topos, J. Symbolic Logic, 46 (1981), pp. 6-16.

[C & K] C.C. Chang and H.J. Keisler, Model Theory, Studies in Logics and the Foundations of Mathematics, 73 (1973) (North-Holland, Amsterdam).

[C] M. Coste, Langage interne d'um Topos, Seminaire Bénabou, Université Paris-Nord (1972).

[F] M.P. Fourman, The Logic of Topoi, in: Handbook of Mathematical Logic, edited by J. Barwise, Studies in Logic and the Foundations of Mathematics, 90 (1977) (North-Holland, Amsterdam) pp. 1053-1090.

[J] P.T. Johnstone, Topos Theory, (1977) (Academic Press).

[L] F.W. Lawvere, Functional Semantics of Algebraic Theories, Proc. Nat. Acad. Sci. USA, 50 (1963), pp. 869-872.

[M] W. Mitchell, Boolean topoi and the theory of sets, J. Pure and Applied Algebra 2 (1972), pp. 261-274.

[O] G. Osius, Logical and Set Theoretical Tools in Elementary Topoi, in:Model Theory and Topoi, Springer Lecture Notes in Math., 445 (1975), pp. 297-346.

[Sc] D.S. Scott, Continuous Lattices, in: Toposes, Algebraic Geometry and Logic, Springer Lecture Notes in Math., 274 (1971).

[Sh] J.R. Shoenfield, Mathematical Logic, (1967) (Addison-Wesley).

[T] G. Takeuti, On a Generalized Logic Calculus, Japan J. Math., 23 (1953), pp. 39-96.

The Hahn - Banach theorem and a restricted inductive definition

Mariko Yasugi

Introduction

It is well-known that constructive version of analysis was initiated by Bishop in his book [3], and since then many interesting results in his line have been published. Among them is Bridges' book [5], which also contains an extensive list of relevant references. Foundational investigations of this trend have been prosperous too: typical of them are seen in [7] and [8]. Mostly they are based on intuitionistic logic, which is natural considering the nature of constructivism.

On the other hand Takeuti in his book [14] has defined a conservative extension of Peano arithmetic, in which he has developed an elementary theory of real and complex analysis. There the basic logic is classical; it is a system of finite type with the "arithmetical" comprehension, hence the cut elimination theorem holds. Real numbers are defined as Dedekind cuts of rationals, hence mathematics is developed in a natural or "classical" manner, without resorting to principles or constructions which are specifically designed for constructive purposes. Only, the whole universe is restricted to the "arithmetically definable" world. The crucial points in this practice are to avoid abstract existential statements and to select an appropriate version for a mathematical concept among classically equivalent definitions.

In the last section of [14] Takeuti comments on the limitations of elementary analysis, especially the problems which arise from the axiom of choice. We believe, however, that much of modern analysis admits a natural reconstruction in a modest extension of Takeuti's system. At any rate, all the results contributed by Bishop and his successors should be apt to such reconstruction.

It is our intention to execute such a program to see (1) what formal system is necessary for this purpose, (2) which characterization of a mathematical concept is to be taken and (3) how mathematical objects look like in our formalization. Here is our guideline.

1) The basic logic is the classical predicate calculus.

2) A mathematical theory is formalized in a system which is a modest extension of Peano arithmetic and whose proof-theoretical characterization is clear.

3) No peculiar (from a working mathematician's standpoint) terms or notions are to be introduced.

As an exemplary case, we consider here the Hahn-Banach theorem in the separable normed linear space. Bishop's construction of the Hahn-Banach extension will be formulated in a classical system of arithmetic with the axiom of a "definable" inductive definition. The method is by no means restricted to this case, and after a certain while of practice in this line we shall be able to single out typical routines. In order to indicate (1) - (3) above, we shall give account of how to connect the usual mathematical reasoning to "definable" ones in some detail, while omitting straightforward arguments in proofs. We avoid existential statements in order to show concrete forms of various objects.

Since we do not specify the space concerned, no nature of constructivity is assumed for it. Thus, the soundness of our construction is only relative to a given space. To save complication, we deal with the case where the scalar field is that of reals. For the complex case, see [3] or [5].

§1 consists of definitions of systems and axioms. In §2 the procedure to establish relative soundness is given. §3 is devoted to our major objective, a formalization of the Hahn-Banach theorem in our system. §4 outlines the consistency proof of our system.

§1 Systems and axioms

Definition 1.1. Type. 1) σ and τ are respectively atomic types of the first sort and the second sort.

2) If $\tau_0, \tau_1, \ldots, \tau_n$ are types, then $[\tau_1, \ldots, \tau_n]$ is a predicate type and $(\tau_1, \ldots, \tau_n) \to \tau_0$ is a function type.

If neither τ nor $\to$ is involved in the definition of a type, then that type is said to be of the first sort.

Definition 1.2. Language. 1) There should be countably many free variables and bound variables for types σ, τ and $[\sigma]$ respectively.

In practice, however, we do not distinguish between free and bound variables. k, ℓ, m, n and p, q, r, s, t will be used for variables of type σ, x, y, z, u, v, w for those of type τ and a, b, c, d, e for those of type $[\sigma]$. Those letters may be also used to denote terms of respective type.

2) Logical symbols are those of predicate calculus.

3) Symbols of arithmetic are the primitive symbols of FA in [14] (cf. Definitions 1.2 and 2.1 in Chapter 1 of Part II there); $0, 1, +, \cdot, =, <, N, \leq$. We shall also use abbreviated notations when necessary.

4) Symbols of a separable normed linear space with certain conditions (which are to be specified) are those listed below.

S, V, mult, add, ζ, eq, norm, f, g, ξ, λ, α, Z, h, K, j.

Types of those symbols are respectively as follows.

$[\tau]$, $[\tau]$, $([\sigma],\tau) \to \tau$, $(\tau,\tau) \to \tau$, τ, $[\tau,\tau]$, $[\tau,\sigma]$, $(\sigma) \to \tau$, $(\tau,\sigma)\to\sigma$, τ, $[\tau,\sigma]$, $[\sigma]$, $(\sigma) \to \tau$, $(\sigma)\to\sigma$, $[\sigma,\tau]$, $[\tau,\sigma]$.

5) Auxiliary symbols. (,), { , }, min.

Definition 1.3. Definability, terms, formulas and abstracts are defined as follows.

1) Free variables and constants of type σ and τ are terms.

2) If X is a symbol of type $[\tau_1,\ldots,\tau_n]$ and $\phi_1,\ldots,\phi_n$ are terms of type $\tau_1,\ldots,\tau_n$ respectively, then $X(\phi_1,\ldots,\phi_n)$ is a formula. Notice that $n = 1$ or 2 and each τ_i is either σ or τ.

3) Formation of a formula from assumed ones is as usual.

4) A formula is said to be definable if the only quantifiers it may contain are of type σ.

5) Let $F(m)$ be a definable formula where m is of type σ. Then $\min(m,F(m))$ is a (definable) term of type σ.

6) An expression of the form $\{t\}F(t)$ is called an (a definable) abstract (of type $[\sigma]$) if $F(t)$ is a definable formula and t is of type σ.

7) If X is a symbol of type $(\tau_1,\ldots,\tau_n) \to \tau_0$ and $\phi_1,\ldots,\phi_n$ are (definable) terms or abstracts of type $\tau_1,\ldots,\tau_n$ respectively, then $X(\phi_1,\ldots,\phi_n)$ is a (definable) term of type τ_0.

Notice that terms and abstracts involve quantifiers of type σ alone, hence they are definable. Note also that an abstract occurs in a formula or a term only within the scope of a symbol of function type.

Sequents are defined as usual.

Definition 1.4. Logical system L. The logical system L we are to employ is the predicate calculus of our language augmented by the comprehension rule of type $[\sigma]$ applied to (definable) abstracts. We present a few of inference rules below. See Definition 1.4 in Chapter 1, Part II of [14] for detail.

$\forall$ right of type $[\sigma]$ $$\frac{\Gamma \to \Delta,\ F(b)}{\Gamma \to \Delta,\ \forall a F(a)}\ ,$$

where b is an eigenvariable of type $[\sigma]$. (Namely, b is a free variable of type $[\sigma]$ which does not occur in Γ, Δ, $F(a)$. a is a bound variable of type $[\sigma]$.)

$\forall$ left of type $[\sigma]$ $$\frac{F(J),\ \Gamma \to \Delta}{\forall a F(a),\ \Gamma \to \Delta}\ ,$$

where J is any (definable) abstract (of type $[\sigma]$). Here J is used either as a formal expression or a meta-expression as the case may be.

$\forall$ left of type τ $$\frac{F(v),\ \Gamma \to \Delta}{\forall x F(x),\ \Gamma \to \Delta}\ ,$$

where v is any (definable) term of type τ.

$\forall$ right of type τ $$\frac{\Gamma \to \Delta,\ F(y)}{\Gamma \to \Delta,\ \forall x F(x)}\ ,$$

where y is an eigenvariable of type τ.

L is a slight modification of LS in [14].

Definition 1.5. We shall define three sets of axioms in our language ; A-C.

1) A will stand for the set of axioms of arithmetic, namely of the theory of natural numbers and rational numbers, including the axiom of mathematical induction. The primitive symbols in 3) of Definition 1.2 are thus assigned the standard meaning. See Definition 2.2 in Chapter 1 of Part II, [14]. We add the least number principle for convenience sake:

$$\forall\phi(\exists m\phi(m) \supset (\phi(\min(m,\phi(m))) \wedge \forall n(\phi(n) \supset \min(m,\phi(m)) \leq n))),$$

where ϕ is of type $[\sigma]$.

Definable arithmetic with the least number principle added is proof-theoretically equivalent with the system without it.

Since k, ℓ, m, n will be used for natural numbers, p, q, r, s, t for rationals and a, b, c, d, e for reals, we shall omit such restrictive expression as $N(k) \supset \ldots$ throughout.

2) B will stand for the set of axioms of a separable normed linear space on the real field with certain conditions, the detail of which is shown subsequently. Here B-0 to B-5 express the axioms of a separable normed linear space, and B-6 to B-10 express those concerning a continuous functional λ defined on a subspace V.

B-0. $\forall xS(x)$. (S is the space concerned.)

B-1. Equality relation on the elements of S.

B-2. $S(\zeta);S(\xi)$.

Subsequently, R(a) stands for "a is a real" (cf. Definition 3.1 in Chapter 1, Part II of [14]). Notice that R is definable.

B-3. Axioms on mult and add. For brevity as well as to conform to general mathematical practice, we write ax for mult(a,x), x+y for add(x,y) and x=y for eq(x,y). It is quite unlikely that those be confused with numerical operations and relations.

$$\forall a\forall x(R(a) \wedge S(x) \supset \exists!y(S(y) \wedge ax = y)),$$

where $\exists!y$ represents the uniqueness of y with regards to eq on S.

We shall omit restrictive R and S as above; thus the axiom above will stand as

$$\forall a\forall x\exists!y(ax = y),$$

or it can be abbreviated to

$$\forall a\forall xS(ax).$$

$$\forall x\forall y\exists!z(x+y = z)$$

or

$$\forall x\forall yS(x+y).$$

Other axioms in this category are put in one.

$\forall a\forall b\forall x\forall y$:

$$(a+b)x = ax+bx$$

$$a(x+y) = ax+ay$$

$$a(bx) = (ab)x$$

$$1x = x$$

$$x+y = y+x$$

$$x+\zeta = x$$

$$\exists!z(x+z = \zeta).$$

One can recognize that ζ stands for the zero-vector of the space S. We shall abbreviate the unique z in the last axiom by -x.

B-4. Axioms on the norm.

$$\forall xR(0,+;\{t\}norm(x,t)),$$

where R(0,+;a) represents that a is a non-negative real.
{t}norm(x,t) will be abbreviated to norm(x).

$$\forall a \forall x(\mathrm{norm}(ax) = \mathrm{abs}(a)\ \mathrm{norm}(x)),$$

where = is the relation on reals and abs(a) is the absolute value of a.

$$\forall x \forall y(\mathrm{norm}(x+y) \leq \mathrm{norm}(x)+\mathrm{norm}(y))$$

$$\forall x(\mathrm{norm}(x) = 0 \supset x = \zeta).$$

B-5. Axioms on f, g: f is a countable dense subset of S with regards to g.

$$\forall x \forall r(r > 0 \supset N(g(x,r)))$$

$$\forall x \forall r \forall s(r < s \supset g(x,r) < g(x,s))$$

$$\forall n(N(n) \supset S(f(n)))$$

$$\forall r(r > 0 \supset \forall x(\mathrm{norm}(x-f(g(x,r))) < \hat{r}),$$

where $\hat{r}$ denotes the real number corresponding to rational r. In order to save notational complication, we shall write r for $\hat{r}$ in a situation like this.

B-6. Axioms on V, a linear subspace of S.

$$\forall x(V(x) \supset S(x))$$

$$V(\zeta),\ V(\xi)$$

$$\forall x \forall y(V(x) \wedge V(y) \supset V(x+y))$$

$$\forall a \forall x(V(x) \supset V(ax)).$$

B-7. Axioms on λ, a linear functional defined on V.

$$\forall x(V(x) \supset R(\{t\}\lambda(x,t))).$$

We write $\lambda(x)$ for $\{t\}\lambda(x,t)$.

$$\lambda(\xi) = 1$$

$$\forall x \forall y(V(x) \wedge V(y) \supset \lambda(x+y) = \lambda(x)+\lambda(y))$$

$$\forall a \forall x(V(x) \supset \lambda(ax) = a\lambda(x)).$$

B-8. Axioms on α, Z, h.

$$R(+;\alpha),$$

which means that α is a positive real.

$$\forall n(S(Z(n)) \wedge \mathrm{norm}(Z(n)) = 1)$$

$$\forall r(r > 0 \supset N(h(r)))$$

$$\forall r(r > 0 \supset \alpha - r < \lambda(Z(h(r))) \leq \alpha).$$

B-9. Continuity of λ.

$$\forall x(V(x) \supset \mathrm{abs}(\lambda(x)) \leq \alpha\mathrm{norm}(x))$$

$$\forall t(\alpha(t) \equiv \exists r(\exists x(V(x) \wedge \mathrm{norm}(x) = 1$$

$$\wedge\ r < \mathrm{abs}(\lambda(x))) \wedge t < r).$$

The latter formula represents the fact that λ has norm α; thus we may express this by norm(λ) = α.

B-10. Axioms on j.

$$\forall x R(0,+;\{t\}j(x,t)).$$

We shall write j(x) for {t}j(x,t).

$$\forall x\{\forall t(j(x,t) \equiv \forall y(\lambda(y) = 0 \supset t < \mathrm{norm}(x-y)))\}.$$

This means that j(x) represents the distance from x to the null-space of λ, NS(λ). Thus we may express it by

$$d(x,NS(\lambda)) = j(x).$$

d(x,NS(λ)) is not a formal object, but the entire expression represents the fact as above.

Notice that in none of the forgoing axioms was the symbol K involved.

3) C shall stand for the axiom of a definable inductive definition.

Let I(n,x,X) be a definable expression in X (which is to be specified) which does not involve K and which will become a formula if X is appropriately substituted for by an object of type $[\tau,\sigma]$. Then the axiom of C is this:

$$\forall n\forall x(K(n,x) \supset I(n,x,\{y,m\}(m < n \wedge K(m,y))))$$

$$\wedge \forall n\forall x(I(n,x,\{y,m\}(m < n \wedge K(m,y))) \supset K(n,x)).$$

$\{y,m\}(m < n \wedge K(m,y))$ will be abbreviated to $K[n]$. It is not a formal expression, but it makes sense when substituted for X. The definition of substitution should be clear. Although I is to be specified later in §3, such a specification is irrelevant in proof-theoretical treatment as long as I is definable in X.

Let us remark that quantifiers of type $[\sigma]$ occur at the heads of axioms in the universal form.

Definition 1.6. Theory T of a "definable", separable normed linear space with certain conditions. A sequent $\Gamma \rightarrow \Delta$ of our language is said to be a theorem of T if

$$A, B, C, \Gamma \rightarrow \Delta$$

is provable in L.

Definition 1.7. Definable instantiation. Let G be an axiom in A or B. The only higher type quantifiers in G are those of type $[\sigma]$. Should G in fact contain quantifiers of type $[\sigma]$, then G must be of the form

$$\forall a_1 \ldots \forall a_n F(a_1, \ldots, a_n),$$

where $a_1, \ldots, a_n$ are of type $[\sigma]$ and $F(a_1, \ldots, a_n)$ is free of such quantifiers ($n = 1$ or 2). (See Definition 1.5.) Let $J_1, \ldots, J_n$ be any (definable) abstracts (of type $[\sigma]$). Then

$$F': \quad \forall \psi_1 \ldots \forall \psi_m F(J_1', \ldots, J_n')$$

will be called a definable instantiation of G, where J_i' is obtained from J_i by replacing all the free variables of types σ and τ occurring in J_i by appropriate bound variables, $i = 1,2$ and $\psi_1, \ldots, \psi_m$ are those bound variables. (Free variables of type $[\sigma]$

are left alone.)

Let A' and B' stand for some definable instantiations of A and B respectively (cf. Definition 1.5).

Definition 1.8. Let D be a formula without type [σ] quantifiers, and let D* be a formula obtained from D by substituting a closed definable abstract for each free variable of type [σ]. Then D* will be called a definable interpretation of D.

Let A" and B" denote A'* and B'* respectively.

Definition 1.9. Logical system M. M is that subsystem of L which does not involve variables of type [σ].

Definition 1.10. System P. P is the system M augmented by the following rule of inference and initial sequents.

1) Rule of inference: mathematical induction applied to the formulas of M.

2) Initial sequents: formulas in A" (for any instantiations and interpretations) except interpretations of MI, where MI stands for the axiom of mathematical induction in [14] (cf. Definitions 1.5, 1.7 and 1.8 above).

3) Initial sequents: formulas in C.

§2. Relative soundness

Proposition 2.1. The cut elimination theorem holds in L (cf. Definition 1.4).

Proof. The proof goes similarly to that of Theorem 1.2 in Chapter 1, Part II of [14]. Type τ quantifiers are counted in the number of "higher type quantifiers" in a formula. Notice that definability condition requires that no type τ quantifiers (let alone type [σ] ones) be involved. See [13] for detail.

Theorem 1. Let H be a formula which expresses an elementary theorem of real numbers, an elementary theorem of a separable normed linear space with a continuous linear functional or the Hahn-Banach theorem. Then

$$A, B, C \to H$$

is provable in L, hence without cuts; in other words, H is a theorem of T (cf. Definition 1.6).

Proof. For the case where H is a theorem on reals, see [14]. The proof for the second case is a routine work. The proof when H is the Hahn-Banach theorem, which is the main task of this article, will be carried out in the next section.

Proposition 2.2. Let G be any formula without type [σ] quantifiers, and suppose

$$A, B, C \to G$$

is provable in L, where G may be empty. Then there exist some definable instantiations of A and B, say A' and B' respectively such that

$$A', B', C \to G$$

is provable in L without cuts and without type [σ] quantifiers.

Proof. Due to the specific forms of the formulas involved and the cut elimination (Proposition 2.1), there are no type [σ] ∀ right inferences. If there is a type [σ] ∀ left, cross it out and then quantify over free variables of type σ and τ by ∀

(cf. Theorem 1.3 in Chapter 1, Part II of [14]).

Proposition 2.3. If for a G as in Proposition 2.2,

$$A', B', C \to G$$

is provable in L (hence without cuts), where G may be empty, then for any definable interpretation G* of G,

$$A'', B'', C \to G^*$$

is provable in M without cuts, hence without type [σ] variables at all (cf. Definitions 1.7 and 1.8).

Proposition 2.4. If {A",B",C} is consistent with M, then {A,B,C} is consistent with L.

Proof. An immediate consequence of Propositions 2.2 and 2.3.

Proposition 2.5. Let $\Gamma \to \Delta$ be a sequent of the language of M. Then $\Gamma \to \Delta$ is P-provable if and only if $A'', C, \Gamma \to \Delta$ is provable in M (cf. Definition 1.10 for P).

Proposition 2.6. If B" is consistent with P, then {A",B",C} is consistent with M.

Proof. By Proposition 2.5, where $\Gamma \to \Delta$ is $B'' \to$.

Theorem 2. P is consistent.

The proof of this theorem will be outlined in the last section.

Theorem 3. (Relative soundness of T) T is sound relative to definable interpretations of B (cf. Definition 1.6).

Proof. Suppose B" is consistent (with P). Then by virtue of Proposition 2.6 {A",B",C} is consistent with M, hence Proposition 2.4 in turn yields that {A,B,C} is consistent with L. But T is a consequence of {A,B,C} in L, hence the soundness of T.

Conclusion. Let H be as in Theorem 1, whose particular case is the Hahn-Banach theorem. Then H is sound relative to definable interpretations of the given space.

Proof. By virtue of Theorem 1, this is a special case of Theorem 3.

§3. The Hahn-Banach construction

In this section we shall carry out the proof of Theorem 1 in some detail, where H is the Hahn-Banach theorem.

We shall first present the theorem as stated by Bishop in [3] (in our formalism).

The Hahn-Banach theorem.

We can construct definable formulas $\mu(p,x,t)$ and $\beta(p,t)$ such that

$H: \forall p > 0($"$\{t\}\mu(p,x,t)$ is a linear functional of x defined on S" $\wedge \forall x(V(x) \supset \{t\}\mu(p,x,t) = \{t\}\lambda(x,t))$ $\wedge$"$\{t\}\beta(p,t)$ is the norm of $\{t\}\mu(p,x,t)$" $\wedge\ \{t\}\beta(p,t) \leq \mathrm{norm}(\lambda)+p)$.

p stands for a rational. We shall write $\mu(p,x)$ for $\{t\}\mu(p,x,t)$, $\mu(p)$ for the same when thought of as a function of x and $\beta(p)$ for $\{t\}\beta(p,t)$.

We shall show that μ and β can be constructed in our language and that H is a theorem of T.

Mathematically we refer to the method of Bishop in the proof of Theorem 4 in Chapter 9 of [3]. See also 3.3-3.5 in [5]. We do not break up the proof into several meta-mathematical steps, since concrete construction of definable objects is important from our viewpoint. In particular, we avoid forming a quotient space.

Although in our development the importance lies in formalism, we shall present the proof procedure in a semi-informal style. One should check at each construction that an object defined is definable in our sense. The form of I in C (Definition 1.5) will be made specific through the course of construction. In the subsequent propositions the statements are claimed as theorems of T, though we do not mention it.

Reminder. The reader is reminded of the notational convention specified in Definition 1.2; in particular, p, q, r, s, t will denote rationals.

We shall use the set-theoretic notations such as $x \in V$ for $V(x)$, $x \in U$ where $U = \{x;F(x)\}$ for $F(x)$, and $U \subseteq S$ for $\forall x(U(x) \supset S(x))$.

Definition 3.1. 1) $NS(\nu) = \{x;\nu(x) = 0\}$ if ν is any map from S to reals.

2) $d(x,U) = \inf\{norm(x-y);y \in U\}$ if $U \subseteq S$. Note that $d(x,U)$ is not a formal object but is used to express a relation.

Proposition 3.1. 1) j is a semi-norm in S.

2) $j(x) = (1/\alpha)abs(\lambda(x))$. In particular $j(\xi) = 1/\alpha = 1/norm(\lambda)$.

Proof. 1) is straightforward.

2) Put $\ell = h(1/k)$. Then $norm(Z(\ell)) = 1$ and $\lambda(Z(\ell)) > \alpha-(1/k)$ (B-8), where we may assume that $\alpha-(1/k) > 0$. For any x, and for any $y \in NS(\lambda)$,

(1) $norm(x-y) \geq (1/\alpha)abs(\lambda(x))$ (B-4 and B-9).

On the other hand if we put

$$z[k] = x-\lambda(x)(1/\lambda(Z(\ell)))Z(\ell),$$

then $z[k] \in NS(\lambda)$ and

$$\begin{aligned} norm(x-z[k]) &= abs(\lambda(x))(1/\lambda(Z(\ell)))norm(Z(\ell)) \\ &< abs(\lambda(x))(1/(\alpha-(1/k))). \end{aligned}$$

$z[k]$ denotes the dependence of z on k. Thus, if we put $y = z[k]$, for an appropriate k, we obtain

(2) $\forall r > 0 \exists y \in NS(\lambda)(norm(x-y) < abs(\lambda(x))((1/\alpha)-r))$,

which together with (1) implies

$$\begin{aligned} (1/\alpha)abs(\lambda(x)) &= \inf\{norm(x-y);\ y \in NS(\lambda)\} \\ &= d(x,NS(\lambda)) = j(x) \quad \text{(B-10)}. \end{aligned}$$

Henceforth various concepts relative to j, such as "j-dense", will be frequently used. The meaning of such concepts should be clear since j is a semi-norm (cf.1) of Proposition 3.1).

Proposition 3.2. f is j-dense in S.

Proof. By B-5, for every y and every $r > 0$,

$$j(y-f(g(y,r)) = d(y-f(g(y,r)), NS(\lambda))$$
$$\leq norm(y-f(g(y,r))) < r.$$

Proposition 3.3. $j(x) = 0$ if and only if $x \in NS(\lambda)$, or $NS(j) = NS(\lambda)$.

Proof. By Proposition 3.1.

Proposition 3.4. Suppose we have constructed a μ satisfying H except the coincidence of $\mu(p)$ with λ on V which also satisfies

$$\forall p > 0(\mu(p,\xi) = 1 \wedge NS(\mu(p)) \supset NS(\lambda)).$$

Then $\mu(p)$ coincides with λ on V.

Proof. For any $v \in V$, put $a = \lambda(v)$ and $w = v-a\xi$. Then $v = a\xi+w$ and the uniqueness of such an expression follows immediately. The condition above then implies $\lambda(v) = a = \mu(p,v)$.

Therefore it suffices to construct a μ satisfying the condition in Proposition 3.4.

[Assumption] Subsequently we assume $p > 0$ where p stands for a rational.

Definition 3.2. 1) $U(\lambda) = \{z;\lambda(z) = 1\} = \{z;j(z-\xi) = 0\}$ (by Proposition 3.3).

2) $d(j,x,U) = \inf\{j(x-z); z \in U\}$ for any $U \subseteq S$.

Proposition 3.5. $d(j,x,U(\lambda)) = j(x-\xi)$.

Proposition 3.6. $U(\lambda) \pm NS(j) = U(\lambda)$, where + and - are element-wise operations.

Definition 3.3. 1) $d = d(j,\zeta,U(\lambda)) = j(\xi) = d(j,\xi,NS(\lambda)) > 0$; d is definable.

2) $K(0) = \{x;j(x-\xi) < d-(p/2)\}$.

[Assumption] We shall assume $d > p$.

Proposition 3.7. 1) $K(0) \pm NS(j) = K(0)$.

2) $d(j,x,K(0)) = \max(0,j(x-\xi)-p)$.

3) $d(j,\xi,K(0)) = p/2$.

Proposition 3.8. 1) K(0) is j-convex, viz., if $x,y \in K(0)$, then $j(tx+(1-t)y-\xi) < d-(p/2)$ for any rational t, $0 \leq t \leq 1$.

2) K(0) is j-bounded, viz., j(x) has a definable upper bound $2d-(p/2)$ if $x \in K(0)$.

3) K(0) is j-open.

4) If $j(x) = 0$, then x does not belong to K(0).

5) f is j-dense in K(0).

Proof. 3) Let w be in $K(0)$, fixed. If we put $e = (d-(p/2)) -(j(w-\xi)/2)$ (>0), then, for any x satisfying $j(x-w) < e$, $j(x-\xi) < p$ holds, hence the j-neighborhood of w of radius e lies within $K(0)$.

Definition 3.4. $c(U) = \{y; \exists s > 0\ sz \in U\}$ if $U \subseteq S$, which is definable if U is.

Proposition 3.9. 1) $d(j,x,c(K(0))) = \inf\{d(j,x,tK(0)); t > 0\}$ is definable.

2) $c(K(0))$ is j-open and j-convex.

Proof. 1) $d(j,x,tK(0)) = td(j,x,K(0))$; cf. 2) of Proposition 3.7.

Definition 3.5. $\eta = -\xi$.

Proposition 3.10. 1) η does not belong to $K(0)$ and η is distinct from ζ.

2) $d(j,\eta,c(K(0))) \geq 1/2$.

Proof. 2) $d(j,\eta,c(K(0)))$ is definable by 1) of Proposition 3.9. For any $y \in c(K(0))$, writing $1/t$ for s,

$$j(y-\eta) = j(((t+1)/(t+1))t(z/t)+(-\eta)((t+1)/(t+1)))$$
$$\geq (t+1)d(j,\zeta,K(0))$$

(j is a semi-norm), which implies the inequality.

Now, we shall construct $K(n)$, $n = 0,1,2,\ldots$, each $K(n)$ being j-bounded, j-open and j-convex subset of S, where the subsequent (i)-(vi) are satisfied.

(i) d(j,z,K(n)) and d(j,x,c(K(n))) are definable for every x.

(ii) d(j,η,c(K(n))) > (1-exp(2,-n))d(j,η,c(K(n-1))), where exp(2,-n) expresses 2^{-n}.

(iii) d(j,ζ,K(n)) > 0.

(iv) d(j,f(n),c(K(n))) < 1/n or d(j,-f(n),c(K(n))) < 1/n.

(v) K(n)±NS(j) = K(n).

(vi) f is j-dense in K(n).

The conditions (ii) to (iv) are exactly (a) to (c) of Bishop's conditions in [3].

K(0) has been defined above to satisfy the condition (cf. Propositions 3.8-3.10). Assume K(n-1) has been defined.

Definition 3.6. K(n,+) = {z;∃r > 1 (f(n)-r(f(n)-z)) ∈ K(n-1))}. K(n,-) = {z;∃r > 1 (-f(n)-r(f(n)-z)) ∈ K(n-1))}.

Proposition 3.11. 1) K(n,+) and K(n,-) are j-open.
2) d(j,x,K(n,+)) and d(j,x,K(n,-)) are definable; similarly with c(K(n,+)) and c(K(n,-)).

Proof. 1) K(n-1) is j-open (the inductive hypothesis).
2) f is j-dense in K(n-1) (by(vi)). Thus

$$d(j,y,K(n,+)) = \inf\{j(y-(tf(n)+(1-t)f(m)));\ 0 < t < 1 \text{ and } f(m) \in K(n-1)\},$$

which is definable. With K(n,-) likewise.

For c(K(n,+)) and c(K(n,-)), see 1) of Proposition 3.9.

Definition 3.7. Now we define K(n).

$$
\begin{aligned}
K(n) = K(n-1) \text{ if } \quad & d(j,f(n),c(K(n-1))) < 1/n \\
& \text{or } d(j,f(n),c(K(n-1))) < 1/n; \qquad (1) \\
= K(n,+) \text{ if not}(1) \text{ and} & \\
d(j,\eta,c(K(n,+))) > & (1-\exp(2,-n))d(j,\eta,c(K(n-1))); \qquad (2) \\
= K(n,-) \text{ if not } (1), \text{ not } (2) \text{ and} & \\
d(j,\eta,c(K(n,-))) > & (1-\exp(2,-n))d(j,\eta,c(K(n-1))). \qquad (3)
\end{aligned}
$$

Note. Due to the condition (i) and Proposition 3.11, the defining formula of K in the definition above is definable (in K(n-1)). This then determines the formula I in C of Definition 1.5.

Proposition 3.12. If case (1) holds in the definition above, the conditions hold by the inductive hypotheses. If case (1) does not hold, one may assume

$$d(j,f(n),c(K(n-1))) \geq 1/(2n)$$

and

$$d(j,-f(n),c(K(n-1))) \geq 1/(2n).$$

Proposition 3.13. 1) $K(n,+) \pm NS(j) = K(n,+)$; with $K(n,-)$ likewise.

2) $K(n,+)$ and $K(n,-)$ both satisfy (iii); $d(j,\zeta,K(n,+)) > 0$ and $d(j,\zeta,K(n,-)) > 0$.

Proof. 2) Suppose $z \in K(n,+)$, viz., $f(n)-r(f(n)-z) \in K(n-1)$, $r>1$. If we put $1-t = 1/r$, then $0 < t < 1$ and $z = tf(n)+(1-t)u$ for a $u \in K(n-1)$ (u is definable from z and r). Since

$$j(z) \geq (1-t)d(j,\zeta,K(n-1))-j(tf(n))$$

and

$$j(z) \geq td(j,-f(n),c(K(n-1))) \geq (1/(2n))t$$

(Proposition 3.12), $j(z)$ has a positive definable lower bound:

$$\min(d(j,\zeta,K(n-1))(1-(a/b)), c/e),$$

where

$$a = d(j,\zeta,K(n-1))+j(f(n)),$$

$$b = (1/(2n))+d(j,\zeta,K(n-1))+j(f(n)),$$

$$c = (1/2n)d(j,\zeta,K(n-1))$$

and

$$e = (1/(2n))+d(j,\zeta,K(n-1))+j(f(n)).$$

From this follows (iii) for K(n,+). With K(n,-) likewise.

Proposition 3.14. 1) c(K(n-1)) is j-open and j-convex.
2) K(n) is j-open since K(n,+) and K(n,-) are.

Proposition 3.15. Either K(n,+) and K(n,-) satisfies (ii).

Proof. 1^0. Suppose $u \in c(K(n,+))$. Then for some s and t such that $s > 0$ and $0 < t < 1$, and $z \in K(n-1)$ (definable from u),

$$u = (t/s)f(n)+((1-t)/s)z$$
$$= qf(n)+x,\ q > 0 \text{ and } x \in c(K(n-1)).$$

If $v \in c(K(n,-))$, then

$$v = -rf(n)+y,\ r > 0 \text{ and } y \in c(K(n-1)).$$

x and y are definable.

For q and r in 1^0,

2^0. $rj(\eta-u)+qj(\eta-v) \geq (q+r)d(j,\eta,c(K(n-1)))$.

From 2^0,

3^0. $j(\eta-u)$ or $j(\eta-v) > (1-\exp(2,-(n+1)))d(j,\eta,c(K(n-1)))$, from which follows the desired conclusion.

Proposition 3.16. (iv) holds both for K(n,+) and K(n,-).

Proof. For any t, u such that $0 < t < 1$ and $u \in c(K(n-1))$,

$$d(j,f(n),c(K(n,+))) \leq j(f(n)-(tf(n)+(1-t)u))$$
$$= (1-t)j(f(n)-u),$$

which tends to 0 as t tends to 1; thus (iv) trivially holds. With K(n,-) likewise.

Proposition 3.17. 1) K(n) is j-bounded.
2) K(n) is j-convex.

Proof. 1) If a is a j-bound for K(n-1), then K(n) has a j-bound j(f(n))+a.
2) Consider the case where K(n) = K(n,+). Suppose $x,y \in K(n,+)$. Then

$$x = rf(n)+(1-r)u,$$
$$y = sf(n)+(1-s)v,$$

$0 < r, s < 1$ and $u,v \in K(n-1)$ (cf. the proof of Proposition 3.13). Suppose $0 \leq t \leq 1$.

$$tx+(1-t)y = (tr+(1-t)s)f(n)+(1-(tr+(1-t)s))w,$$

where $w = qu+(1-q)v$, $0 \leq q \leq 1$, for

$$q = (t(1-r))/((1-r)+(1-t)(r-s)).$$

Thus, $tx+(1-t)y \in K(n,t)$.

This completes the proof that Definition 3.7 gives the required inductive definition.

Proposition 3.18. 1) $\{K(n); n = 0,1,2,\ldots\}$ is increasing.
2) $c(K(n))$ is j-open and $\{c(K(n)); n = 0,1,2,\ldots\}$ is increasing.

Definition 3.8. $K = \cup \{c(K(n)); n = 0,1,2,\ldots\}$, or $x \in K$ if and only if $\exists n(x \in c(K(n)))$. $Q = \{z; -z \in K\}$.

Proposition 3.19. 1) K is j-open, j-convex and invariant under multiplication by positive scalars.
2) $d(j,x,K)$ is definable and $d(j,\eta,K) \geq p/4 > 0$.
3) ζ does not belong to K; if $j(u) = 0$, then u is not in K; $\xi \in K$; η does not belong to K.
4) $\eta \in Q$; ζ does not belong to Q.

Proof. 2) By (i) and (ii) of the conditions on K(n).

Proposition 3.20. K and Q are disjoint and j-open.

Proposition 3.21. $K \cup Q$ is j-dense in S.

Proof. f is j-dense in S (Proposition 3.2).

$$d(j,x,c(K(n))) \begin{cases} \leq j(x-f(n))+d(j,f(n),c(K(n))), \\ \leq j(x-(-f(n)))+d(j,-f(n),c(K(n))). \end{cases}$$

(1) $d(j,f(n),c(K(n))) < 1/n$

or

(2) $d(j,-f(n),c(K(n))) < 1/n$

by (v). Suppose (1) holds. Put $m = g(x,r)$. Then $j(x-f(m)) < r$. So $d(j,x,c(K(m))) \leq r+(1/m)$. But we may assume that m tends to ∞ as r tends to 0. So

$$d(j,x,K\cup Q) \leq r+(1/m),$$

which tends to 0 as r tends to 0. Thus $d(j,x,K\cup Q) = 0$ (hence the j-distance is definable). When (2) holds, the proof goes similarly.

Definition 3.9.

$$W(x) = \{z; j(z-(cx+(1-c)\eta)) = 0, \text{ where}$$
$$c = j(z-\eta)/(j(z-x)+j(z-\eta))\}.$$

Notice that c is a definable (from x and z) real and $0 \leq c \leq 1$.

We consider W(x) only when $x \in K$; thus we shall assume this subsequently until the end of Proposition 3.27.

Proposition 3.22. 1) $x \in K \cap W(x)$ and $\eta \in Q \cap W(x)$.

2) $K \cap W(x)$ and $Q \cap W(x)$ are j-convex and mutually disjoint.

3) $W(x) \pm NS(j) = W(x)$.

4) $\exists r > 0 \forall z(j(z) < r \supset x+z \in K \wedge \eta+z \in Q)$.

5) $\forall r > 0 \forall y \in W(x) \exists u \in K \cup Q(j(y-u) < \exp(r,2))$.

6) If a u in 5) belongs to K, then $x+(1/r)(y-u) \in K$.

7) $(K \cup Q) \cap W(x)$ is j-dense in W(x).

8) $K \cap W(x)$ and $Q \cap W(x)$ are one-side open segments in W(x).

9) $d(j,y,W(x)) = \inf\{j(y-(rx+(1-r)\eta)); 0 \leq r \leq 1\}$, so $d(j,y, W(x))$ is definable.

Proof. 5) By Proposition 3.21: u can be determined from x, r and y.

Definition 3.10. Y = the complement of $K \cup Q$ in S; $z \in Y$ if and only if $\neg z \in K$ and $\neg z \in Q$.

Proposition 3.23. 1) $Y \pm NS(j) = Y$.

2) $\zeta \in Y$.

3) For any y, $y \in Y$ if and only if $-y \in Y$.

4) Y is j-closed.

5) If $y \in Y$, then $ay \in Y$ for every real a.

Definition 3.11. For any $U \subseteq S$, we define bd(U) and cl(U).

1) $bd(U) = \{z; \forall r > 0 \exists n \exists m (f(n) \in U \wedge f(m) \in U)\}$.

2) $cl(U) = \{z; \forall r > 0 \exists n (f(n) \in U \wedge j(z-f(n)) < r)\}$.

Proposition 3.24. 1) $cl(K) = K \cup bd(K)$; $cl(Q) = Q \cup bd(Q)$.

2) $S = K \cup Q \cup bd(K) \cup bd(Q)$, where $\cup$ is the direct union.

3) $\zeta \in bd(K) \cap bd(Q)$.

4) $bd(K) = bd(Q)$.

5) $Y = bd(K) \cup bd(Q) = bd(K) = bd(Q)$

$= bd(K) \cap bd(Q) = cl(K) \cap cl(Q)$.

Proof. 2) By Proposition 3.21.

3) By $\neg\zeta \in K \cup Q$, $\xi \in K$, $\eta \in Q$ (Proposition 3.19), j-convexity of K and Q and 5) of Proposition 3.23, we see that any (small, positive) multiples of ξ and η belong to K and Q respectively, hence $\zeta \in bd(K) \cap bd(Q)$.

4) Suppose $y \in bd(K)$ and claim $y \in bd(Q)$; the opposite direction can be treated similarly. In case $y = \zeta$, 3) claims this. So we may assume $\neg y = \zeta$. Define

$$U(\eta,y) = \{z; j(z-(cy+(1-c)\eta)) = 0 \text{ where } c = j(z-\eta)/(j(z-y)+j(z-\eta))\}.$$

$\eta \in Q$, Q is open and $j(z-\eta) \leq j(y-\eta)$ if $z \in U(\eta,y)$. So,

$$e = \sup\{j(z-\eta); z \in U(\eta,y) \cap Q\}$$
$$= \sup\{j(ty+(1-t)\eta-\eta); 0 \leq t \leq 1,\ ty+(1-t)\eta \in Q\}$$

is definable and exists as a real. Define $u = ey+(1-e)\eta \in bd(Q) \subseteq Y$. If we can claim $e = j(y-\eta)$, then the conclusion is derived as follows.

$e = j(y-\eta)$ implies that $j(y-u) = 0$, hence $y \in bd(Q)$ since $u \in bd(Q)$.

It is now left for us to establish that $e = j(y-\eta)$.

Suppose $e < j(y-\eta)$. Then $j(y-u) > 0$. Define $U(u,y)$ similarly to $U(\eta,y)$. $U(u,y) \cap Q$ is empty by virtue of convexity of Q. We shall first claim that $U(u,y) \cap K$ is not empty under the assumption. Suppose the contrary, i.e., $U(u,y) \cap K$ is empty. Then $U(u,y) \subseteq Y$, hence $c(U(u,y)) \subseteq Y$ (cf. Definition 3.4 and 5) of Proposition 3.23). Then there is a small j-ball contained in $c(U(u,y)) \subseteq Y$ (for example, a j-ball with center $(u+y)/2$ and radius $(1/4)j(y-u)$ (>0)), contradicting the denseness of $K \cup Q$. Thus, we can claim $U(u,y) \cap K$ is not empty. Next consider the extension of $U(u,y)$ past y:

$$Ext(u,y) = \{v; \exists t > 0((y+t(y-v)) \in U(u,y)))\}.$$

$Q \cap Ext(u,y)$ is empty due to convexity of Q. If we suppose $K \cap Ext(u,y)$ is empty, then with the same reasoning as above, we get a contradiction. So $K \cap Ext(u,y)$ is not empty. This means that elements of K in $j(y-u)$ and those in $Ext(u,y)$ are separated by y, contradicting convexity of K. Therefore our major supposition that $e < j(y-\eta)$ is erroneous.

Proposition 3.25. 1) Y is a cone.

2) Y is a j-closed and j-linear j-subspace of S.

Proof. By Propositions 3.23 and 3.24.

Proposition 3.26. 1) $W(x) \cap Y$ is non-empty.

2) There is a j-unique element in $W(x) \cap Y$.

Proof. 1) Put

$$b = b[x]$$
$$= \inf\{j(\eta-(tx+(1-t)\eta));0 \leq t \leq 1 \text{ and } tx+(1-t)\eta \in K\}.$$

b is definable (in x) and $0 < b < 1$. Put $x^* = bx+(1-b)\eta$. Then $x^* \in W(x) \cap Y$, since K and Q are j-open.

2) It suffices to show that if $z = ex+(1-e)\eta \in W(x) \cap Y$, where $e = j(z-\eta)/(j(z-x)+j(z-\eta))$, then $e = b$. This can be shown by the j-convexity of K and $Y = bd(Q)$ (cf. 5) of Proposition 3.24).

Proposition 3.27. For any x in K, x can be represented as a linear combination of η and an element in $W(x) \cap Y$, and such representation is j-unique.

Proof. With x^* and $b = b[x]$ in Proposition 3.26,

$$x = (1/b)x^*+(-(1-b)/b)\eta,$$

which we abbreviate to

$$x = ax^*+c\eta.$$

Suppose x has another representation.

$$x = a'y+c'\eta, \qquad y \in W(x)+N.$$

By Proposition 3.26, $j(x^*-y)=0$, hence $j(x^*) = j(y)$ (j is a semi-norm). Using Propositions 3.25 and 3.24, we get $c = c'$, from which follows $a = a'$. So a and c are unique and y is j-unique.

Proposition 3.28. $d(j,z,Y)$ is definable and exists for every z.

Proof. $d(j,\eta,Y) = d(j,\eta,K) > 0$. Suppose $x \in K$. Then

$$d(j,x,Y) = d(j,c\eta,Y) = abs(c)d(j,\eta,Y),$$

where $x = ax^{*}+c\eta$ (Proposition 3.27).

Suppose next $z \in Q$. Then $-z \in K$, so $d(j,z,Y) = d(j,-z,Y)$.

Proposition 3.29. 1) For any x, x is j-uniquely expressed as a linear combination of η and an element of Y: $j(x-(a\eta+y)) = 0$, where a is unique and $y \in Y$ is j-unique.

2) Y is a hyperplane with an associated vector η.

Proof. 1) If $x \in K$, this has been shown in Proposition 3.27. If $x \in Y$, then $a = 0$ and $y = x$ will do. Let x be in Q. Then $-x \in K$, hence the proposition.

2) By 1) above and $d(j,\eta,Y) > 0$ (Proposition 3.28).

Now we are in the position to construct μ. We do this in several steps.

Definition 3.12. We define linear functionals on S, θ, ϕ and ψ, successively. We write $\theta(p,x)$ for $\{t\}\theta(p,x,t)$ etc.

(1) For any x in S, let $a\eta+y$ be a representation claimed in Proposition 3.29. Recall that a is unique and definable (in $p > 0$ and x).

$$\theta(p,x) = a.$$

(2) $$\phi(p,x) = \theta(p,-x).$$

(3) $\psi(p,x) = \phi(p,x)/c$,

where

$$c = \sup\{abs(\phi(p,x)); j(x) \leq 1\}$$
$$= \sup\{abs(\phi(p,f(n))); j(f(n)) \leq 1,\ n = 0,1,2,\ldots\}.$$

c is positive and definable.

It is obvious that ψ is a definable (in p) linear functional on S.

Proposition 3.30. 1) $\psi(p,\eta) < 0$.

2) The norm of $\psi(p,x)$ with regards to j exists and equals 1.

3) $\psi(p,x)$ is positive on K, hence in particular on K(0).

4) $\psi(p,y) > \psi(p,x)+d-p$ $(d = j(\xi))$ if $j(y-\xi) = 0$ and $j(x) = 0$. In particular $\psi(p,\xi) > d-p > 0$.

Proof. 4) Before we prove this, we define char(F(a),t) and negchar(F(a),t), where F(a) is any definable formula and a is a free variable which stands for reals.

$$\mathrm{char}(F(a),t):(F(a) \wedge t < 1) \vee (\neg F(a) \wedge t < 0).$$

{t}char(F(a),t) is a definable function of a and defines the characteristic function of F(a).

$$\mathrm{negchar}(F(a),t):(F(a) \wedge t < 0) \vee (\neg F(a) \wedge t < 1).$$

Now put $q = (3/4)p$ and

$$b = (p-(3/4)q)/(d-(3/4)q).$$

Put

$$m = \min(n, j(f(n)) \leq 1 \wedge 1-b < abs(\psi(p,f(n))) \leq 1),$$

which we write $m = \min(n,D(n))$. Notice that D is definable. Note that $\exists nD(n)$, hence D(m) (cf. A of Definition 1.5). Define u_0:

$$u_0 = \mathrm{char}(\psi(p,f(m)) > 0)f(m) - \mathrm{negchar}(\psi(p,f(m)) > 0)f(m),$$

which is definable. Put next

$$v_0 = (d-(3/4)q)u_0.$$

Then $\psi(p,v_0) > d-p$ and $j(v_0) \leq d-(p/2)$. Suppose $j(y-\xi) = 0$ and $j(x) = 0$. Then $\psi(p,y-x-v_0) > 0$, hence $y-x-v_0 \in K(0)$, and hence by 3) $\psi(p,y-x-v_0) > 0$. From this follows $\psi(p,y) > \psi(p,x)+d-p$.

Definition 3.13. $\chi(p,x) = \psi(p,x)/\psi(p,\xi)$. $\chi(p,x)$ is definable (in p and x).

It is obvious that x is a linear functional on S. We write $\chi(p)$ for $\chi(p,x)$ regarded as a functional of x.

Proposition 3.31. 1) $\chi(p,\xi) = 1$.
2) The norm of $\chi(p)$ is definable and $\text{norm}(\chi(p)) \leq 1/\text{abs}(\psi(p,\xi)) \leq 1/(j(\xi)-p)$.

Proof. 2) $\text{norm}(\chi(p))$

$$= \sup\{\text{abs}(\chi(p,f(n)));\text{norm}(f(n)) \leq 1\}$$

$$= \text{norm}(\psi(p))/\text{abs}(\psi(p,\xi)) \leq j\text{-norm of } \psi(p)/\text{abs}(\psi(p,\xi)).$$

Notice that the proposition above holds independent of p except for the last inequality.

Definition 3.14. Put $q = (1/2)(p/(\text{norm}(\lambda)(\text{norm}(\lambda)+p)))$. Then define μ by

$$\mu(p,x,t) \equiv \chi(q,x,t).$$

μ is a linear functional on S.

Proposition 3.32. 1) $\mu(p,\xi) = 1$.
2) $\text{norm}(\mu(p))$, which we write $\beta(p)$, is definable and

$$\beta(p) \leq \text{norm}(\lambda)+p = \alpha+p.$$

Proof. 2) By 2) of Proposition 3.31,

$$\beta(p) \leq 1/(j(\xi)-q),$$

where $j(\xi) = d = d(j,\xi,NS(\lambda)) \geq 1/\text{norm}(\lambda)$ (by continuity of λ).

Thus

$$\beta(p) \leq 1/((1/\text{norm}(\lambda))-q)$$
$$= \text{norm}(\lambda)+\text{norm}(\lambda)\sum \exp(q\,\text{norm}(\lambda),i)$$
$$< \text{norm}(\lambda)+p,$$

where the summation is taken over $i = 1,2,\ldots$.

Proposition 3.33. $NS(\mu(p)) \supseteq NS(\lambda)$.

Conclusion. The μ constructed in Definition 3.14 satisfies the Hahn-Banach condition.

Proof. By Propositions 3.32, 3.33 and 3.4.

§4. Consistency proof

We shall outline briefly a proof of Theorem 2, §2: P is consistent. The notations in this section are quite independent of those in the preceding sections.

Definition 4.1. 1) Let $(\Lambda,<)$ be a well-ordered system. We deifne a system of notations $(\Pi,\prec)$ based on $(\Lambda,<)$.

1^0. The symbol 0 is a connected element of Π.

2^0. If $\alpha \in \Lambda$ and $\beta \in \Pi$, then (α,β) is a connected element of Π.

3^0. Suppose $\alpha_1,\ldots,\alpha_n$ are connected elements of Π, $n \geq 2$. Then $\alpha_1\#\ldots\#\alpha_n$ is a non-connected element of Π.

2) We define orders $\prec$ and $\prec'$ for Π.

(1) If β is not 0, then $0 \prec \beta$ and $0 \prec' \beta$.

(2) $\#$ is interpreted as the natural sum for both $\prec$ and $\prec'$.

(3) $(\alpha,\beta) \prec' (\gamma,\delta)$ if $\alpha \prec \gamma$, or $\alpha = \gamma$ and $\beta \prec \delta$.

(4) $(\alpha,\beta) \prec (\gamma,\delta)$ if one of the following holds.

(4.1) $(\alpha,\beta) \prec' (\gamma,\delta)$ and $\beta \prec (\gamma,\delta)$.

(4.2) $(\alpha,\beta) \preceq \delta$.

Proposition 4.1. The system $(\Pi,\prec)$ is well-ordered.

$(\Pi,\prec)$ is a simplest case of the theory of ordinal diagrams, whose accessibility has been established; see §26 of [13], for example.

The consistency proof of P is an extremely simplified version of §28 of [13]. We quote the item numbers there.

Definition 28.4. Let I be ω here. Then $| <_\infty | = \omega+1$.

Definition 28.5. The rank of an occurrence of K in a formula A, $r(K;A)$ is defined to be an element of $I_\infty = I \cup \tilde{I} \cup \{\infty\}$.

Definition 28.7. The γ-degree of a formula relative to definability is defined to be a natural number; $\gamma(A) = 0$ if A is definable.

Definition 28.14. The norm of a formula, $n(A)$, is defined to be an element of $(\omega+1,\omega)$, lexicographically ordered.

Definition 28.17. The grade of a formula, $g(A)$, is defined to be an element of the system $\Lambda = (\omega,\omega,(\omega+1,\omega))$, whose lexicographical order will be denoted by $<$. $g(A) = (\gamma(A),a,n(A))$, where a is the number of eigenvariables of type τ in A.

Proposition 28.18. If

$$N(i),K(i,v) \to I(i,v,K[i])$$

or

$$N(i),I(i,v,K[i]) \to K(i,v)$$

is an initial sequent of a proof of P, and i is a constant for which $N(i)$ is provable, then

$$g(I(i,v,K[i])) < g(K(i,v)).$$

Definition 28.19. An assignment of an element of $\Pi = \Pi(\Lambda)$ to each sequent in a proof of P.

5) The lower sequent of an implicit type τ $\forall$ left is assigned $(g(F(v))+2,\sigma)$, where $F(v)$ is the auxiliary formula and σ is the element assigned to the upper sequent.

With those preparations, the rest of the consistency proof in §28 of [13] goes through for our system. It suffices to work on the system without min.

References

1. W. Ackermann, Konstruktiver Aufbau eines Abschnitts der zweiter Cantorschen Zahlenklasse, Math. Zeit. 53 (1951), 403-413.

2. S. Banach, Theóries des Opérations Lineares, Chealsea Publ. Co., N.Y. (1955).

3. E. Bishop, Foundations of Constrúctive Analysis, McGraw-Hill Book Co., N.Y. (1967).

4. E. Bishop, Mathematics as a numerical language, Intuitionism and Proof Theory, edited by J. Myhill, A. Kino and R. E. Vesley, North-Holland Publ. Co., Amsterdam (1970), 53-71.

5. D. S. Bridges, Constructive Functional Analysis, Pitman, London (1979).

6. S. Feferman, Theories of finite type related to mathematical logic, edited by J. Barwise, Studies in Logic and the Foundations of Mathematics 90, North-Holland Publ. Co., Amsterdam (1977).

7. H. Friedman, Set theoretic foundations for constructive analysis, Ann. Math. (2) 105 (1977), 1-28.

8. N. Goodman-J. Myhill, The formalization of Bishop's constructive mathematics, Toposes, Algebraic Geometry and Logic, Lecture Notes in Mathematics 274, Springer-Verlag, Berlin (1972), 83-96.

9. H. Hahn, Über lineare Gleichungssysteme in lineare Räumen, J. reine und angew. Math. 157 (1927), 214-229.

10. G. Kreisel, Analysis of the Cantor-Bendixson theorem by means of the analytic hierarchy, Bulletin of the Polish Acad. Sci. 7 (1959), 371-391.

11. G. Kreisel, The axiom of choice and the class of hyper-arithmetic functions, Dutch Acad. A, 65 (1962), 307-319.

12. W. Rudin, Real and Complex Analysis, McGraw-Hill Book Co., N.Y. (1966).

13. G. Takeuti, Proof Theory, North-Holland Publ. Co., Amsterdam (1975).

14. G. Takeuti, Two Applications of Logic to Mathematics, Iwanami Shoten Publ. Co. and Princeton Univ. Press, Tokyo (1978).

15. M. Yasugi, Arithmetically definable analysis, Proceedings of Research Institute of Mathematical Sciences, Kyoto University 180 (1973), 39-51.

University of Tsukuba, Japan